CANCER STEM CELLS

CANCER STEM CELLS

CANCER STEM CELLS

Identification and Targets

Edited by

Sharmila Bapat

WILEY

A JOHN WILEY & SONS, INC., PUBLICATION

Library of Congress Cataloging-in-Publication Data:

Cancer stem cells / [edited by] Sharmila Bapat.
   p. ; cm.
 Includes bibliographical references and index.
 ISBN 978-0-470-12201-3 (cloth)
1. Cancer cells. 2. Stem cells. I. Bapat, Sharmila.
 [DNLM: 1. Neoplastic Stem Cells.   QZ 202 C21566 2008]
 RC269.C365 2008
 362.196'994—dc22

                                        2008035481

10 9 8 7 6 5 4 3 2 1

CONTENTS

CONTRIBUTORS

CURT BALCH, School of Medicine, Indiana University, Bloomington, Indiana

DANUTA BALICKI, Hôtel-Dieu du Centre Hospitalier de l'Université de Montréal, CRCHUM, Montréal, Québec, Canada

SHARMILA BAPAT, National Centre for Cell Science, Ganeshkhind, Pune, India

RAYMOND BEAULIEU, Hôtel-Dieu du Centre Hospitalier de l'Université de Montréal, CRCHUM, Montréal, Québec, Canada

AMPARO CANO, Departamento de Bioquímica, Facultad de Medicina, Universidad Autónoma de Madrid, Instituto de Investigaciones Biomédicas "Alberto Sols," Consejo Superior de Investigaciones Científicas, Madrid, Spain

MICHAEL CHAN, Division of Human Cancer Genetics, Comprehensive Cancer Center, Ohio State University, Columbus, Ohio; Department of Life Science, National Chung Cheng University, Min-Hsiung, Chia-Yi, Taiwan

SHUBHADA V. CHIPLUNKAR, Advanced Centre for Treatment, Research and Education in Cancer, Tata Memorial Centre, Kharghar, Navi Mumbai, India

ANNE COLLINS, YCR Cancer Research Unit, Department of Biology, Heslington, University of York, York, UK

MICHAEL DEAN, National Cancer Institute–Frederick, Frederick, Maryland

STEFANIE HAGER, YCR Cancer Research Unit, Department of Biology, Heslington, University of York, York, UK

TIM H.-M. HUANG, Division of Human Cancer Genetics, Comprehensive Cancer Center, Ohio State University, Columbus, Ohio

ANJALI KUSUMBE, National Centre for Cell Science, Ganeshkhind, Pune, India

NORMAN J. MAITLAND, YCR Cancer Research Unit, Department of Biology, Heslington, University of York, York, UK

GEMA MORENO-BUENO, Departamento de Bioquímica, Facultad de Medicina, Universidad Autónoma de Madrid, Instituto de Investigaciones Biomédicas "Alberto Sols," Consejo Superior de Investigaciones Científicas, Madrid, Spain

KENNETH NEPHEW, School of Medicine, Indiana University, Bloomington, Indiana

DAVID OLMEDA,   Biotechnology Program, Centro Nacional de Investigaciones Oncológicas, Madrid, Spain

JOSÉ PALACIOS,   Molecular Pathology Program, Centro Nacional de Investigaciones Oncológicas, Madrid, Spain; Servicio de Anatomía Patológica, Hospital Virgen del Rocío, Sevilla, Spain

ANNAPOORNI RANGARAJAN,   Department of Molecular Reproduction, Development, and Genetics, Indian Institute of Science, Bangalore, India

SURAIYA RASHEED,   Department of Pathology, Keck School of Medicine, University of Southern California, Los Angeles, California

DAVID SARRIÓ,   Breakthrough Breast Cancer Research Centre, The Institute of Cancer Research, London, UK

MEERA SAXENA,   Department of Molecular Reproduction, Development, and Genetics, Indian Institute of Science, Bangalore, India

GAIL M. SEIGEL,   Ross Eye Institute, State University of New York at Buffalo, Buffalo, New York

ZHANG SHU,   School of Medicine, Indiana University, Bloomington, Indiana; Shanghai Second Medical University, Shanghai, China

PEARLLY YAN,   Division of Human Cancer Genetics, Comprehensive Cancer Center, Ohio State University, Columbus, Ohio

PREFACE

Is cancer a *stem cell disease*? At present, we are in the midst of an exciting turn of events in which the principles of stem cell biology have been applied to cancer biology. This leads to the realization that disruption of the processes of quiescence and differentiation (viewed as two sides of the coin of stem cell self-renewal) can lead to cancer. Disease progression in neoplasia is often accompanied by increasing resistance to apoptosis and therapy, development of a capacity for tissue invasion and migration, and an apparently limitless replicative potential—all features that can be described as being characteristics of a stem cell. The marriage of convenience between stem and cancer cells is thus strongly supported by the similarities between the two cell groups, and the considerable research effort that has been initiated in the field has resulted in incremental knowledge of cancer stem cells that continues to evolve rapidly.

Cancer stem cells (CSCs), first identified in acute leukemias, have now been isolated from several human malignancies, such as breast, brain, prostate, and ovarian, and also in retinoblastomas and melanomas. The initial isolation of CSCs relied on a basic knowledge of normal stem cells in the organ, assays for their functionality, and expression of specific markers on their cell membranes that provided robust mechanisms for identification of these elusive cells within the large mass of a tumor. In the absence of such specific channels for their identification, CSCs came to be identified based on the expression of characteristic genes associated with self-renewal in stem cells, including *OCT4, NANOG*, and *NESTIN*. Yet another novel approach to the isolation of CSCs within tumors was the identification of side population cells within tumors that were subsequently shown to be enriched in tumor-initiating cells.

Although the origin of CSCs is still datatable, it has been demonstrated variably that they may be generated from tissue-specific stem cells, progenitor cells, or mature cells or through fusion of tissue-residing cells with bone marrow–derived stem cells or cytotoxic T lymphocytes. Major differences between the two cell groups include the genetic and epigenetic changes in CSCs that secure and establish transforming events and make disease progression a certainty. The identification of cancer stem cells has thus opened up a new avenue in cancer biology. It is being realized increasingly that some of the limitations of current chemotherapeutics may be overcome by initiating detailed molecular investigations of the cellular hierarchies transformed in cancer. Several research groups are already exploring the design of new therapeutic strategies that could possibly target CSCs.

In this book we have sought to be as comprehensive as possible. We have detailed some of the important cancers in which CSCs have been identified. As the first book in this field, cancer biologists, stem cell researchers, and clinicians will find it a valuable repository of information. Further, we present the various approaches in the field in a highly intelligible manner, so that readers from diverse scientific fields and working environments will find it a convenient reference source in a field that is growing by leaps and bounds. It is my sincere hope that this book will stimulate readers to explore diverse ways of understanding the mechanisms by which these seemingly elusive cells evade being targeted, and thereby, help to open new pathways in the molecular aspects of biomedical research.

SHARMILA BAPAT

1 Cancer Stem Cells: Similarities and Variations in the Theme of Normal Stem Cells

SHARMILA BAPAT, ANNE COLLINS, MICHAEL DEAN, KENNETH NEPHEW, and SURAIYA RASHEED

1.1 INTRODUCTION

Then we have Beard's "germ-cell" hypothesis, in which he holds that many of the germ-cells in the growing embryo fail to reach their proper position—the generative areas—and settle down and become quiescent in some somatic tissue of the embryo. They may at some later date become active in some way, and so give rise to a cellular proliferation that may imitate the structure in which they grow, so giving rise to new growths.

—*Encyclopedia Britannica*, 1911

Tumors were first likened to embryonic cells by the Scottish embryologist John Beard, who early in the twentieth century proposed a germ cell theory of cancer, inferred from his observations.[1] The early blastocyst during its development has two distinct cell groups: the inner cell mass, which gives rise to the embryo, and an outer circle of trophoblastic cells, which enable implantation of the embryo into the uterus and subsequent formation of placenta[2] (Fig. 1.1). The trophoblast cells are thus invasive and metastatic, which are also key characteristics of cancer cells. Beard postulated that the presence of pancreatic enzymes in the embryo and the mother actually restricts the invasion of trophoblast cells into surrounding tissues, and claimed that cancer arose from straying by such cells due to a failure of the enzymatic surveillance mechanisms. He went on to demonstrate that injections of the pancreatic proteolytic enzyme trypsin could destroy cancer cells in patients.[3]

Although the hypothesis and consequent cancer treatment generated considerable attention, its very nature was such that it could not be confirmed reproducibly and Beard's ideas remained largely unaccepted by his peers. However, his

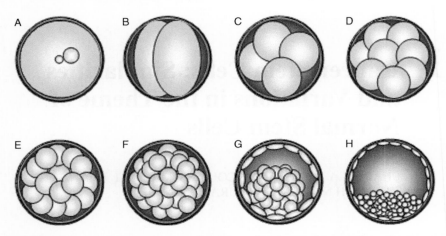

FIGURE 1.1 Developmental stages from normal early vertebrate embryo development to the blastocyst stage: (A) single-celled zygote; (B) two-celled embryo; (C) four-celled embryo; (D) early morula; (E) compacted morula; (F) late morula; (G) early blastocyst; (H) late blastocyst. (*See insert for color representation of figure.*)

hypothesis has recently regained prominence, with similar observations that relate cancer cells and disease progression with the characteristics of stem cells. The finding that tumors often express surface antigens (cancer/testis antigens) that are otherwise expressed only at the embryonic–extraembryonic stage[4] lends additional support to Beard's theory. A stem cell is characterized by its unique capacity to differentiate into multiple cellular lineages and to self-renew in an undifferentiated state[5] (Fig. 1.2). It has now been shown definitively in certain cancers that a small subset of cells exist in a tumor with similar regenerative and self-renewal mechanisms that enable tumor formation and progression.[6] This tiny subset of cells, referred to as *cancer stem cells* (CSCs), is also considered to be more chemoresistant than the bulk of tumor cells and is thus more difficult to target and eradicate.[7] Thus, CSCs need to be specifically targeted and eliminated to achieve tumor ablation, a concept that has begun to revolutionize approaches to cancer therapy and drug design. Although distinctly removed from Beard's original proposal, the theory is currently evolving, with increasing evidence in several types of cancers.

1.2 STEM CELLS IN THE LIFE OF AN ORGANISM

The unicellular zygote is recognized to be the first stem cell in a human life and is identified as being totipotent, due to its ability to generate an entire organism.[8] However, during further embryonic development, a gradient of decreased regenerative potential is produced and distributed in specific stem cell compartments in a developing embryo. This decrease in regenerative capability leads to a subsequent loss of totipotency but a concurrent retention of pluripotency by stem cells

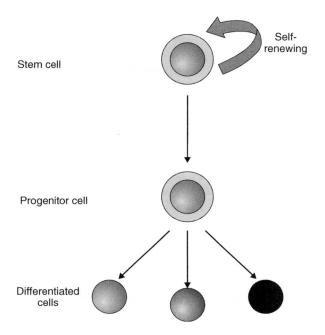

FIGURE 1.2   Basic outline of stem cell functioning.

in the embryo. A neonate is equipped with stem cells which for the most part are multipotent (i.e., committed to tissue-specific potential).[9] Some stem cells, however, retain a higher level of pluripotency even in the adult (e.g., subsets of bone marrow−derived stem cells[10] which are reported to have trans-differentiation potential).

1.2.1 Stem Cells in Early Development and Fetal Life

The early embryo generated from the fertilized egg cell exists initially more or less as a "ball of cells" called a *morula* (Fig. 1.1). The first recognized differentiation event in development occurs at the late morula stage, when the outer cell layer of the embryo adopts epithelial features and initiates the formation of a blastocyst that contains two cell types: trophectoderm and inner cell mass (ICM). The trophectoderm gives rise to the trophoblast, while the ICM undergoes a second differentiative step to form the epiblast and primitive endoderm,[11,12] thereby progressing toward the blastocyst stage. The epiblast (sometimes referred to as the primitive ectoderm) further gives rise to the embryo, while the primitive endoderm develops into the extraembryonic endoderm, which provides nutrients and developmental cues to the embryo and contributes to development of the yolk sac. The ICM cells at the blastocyst stage are no longer totipotent but are recognized as being pluripotent.[8]

The next recognized landmark during embryonic differentiation is gastrulation, when the epiblast is transformed into the three germ layers of the embryo (i.e., ectoderm, mesoderm, endoderm) and the basic body plan of the animal is established. In mouse, gastrulation is initiated around 6.5 days postcoitum, when the surface ectoderm cells undergo an epithelial–mesenchymal transition, delaminate, and migrate through the primitive streak.[13] Ingression of progenitors for the embryonic mesoderm begins around the midstreak stage with the formation of progenitors that will give rise to the lateral plate mesoderm of the upper body as well as cardiac and cranial mesoderm.[14] Within the epiblast, nascent mesodermal cells migrate anteriorly and laterally. Early in gastrulation, cells leaving the streak contribute largely to the mesoderm of extraembryonic tissues: the yolk sac, amnion, and allantois.[15] Thus, there is a temporal progression of mesoderm commitment. In addition, the site of ingression of progenitors into the primitive streak is regionalized. For example, cells fated to form extraembryonic mesoderm ingress into the posterior portion of the streak, while lateral plate and paraxial mesoderm ingress into the middle and anterior regions, respectively.[16] The primitive streak thus functions as a posterior organizing center. How different mesodermal populations are set aside to form specific lineages is not well understood.[17]

On transplantion into the proximal epiblast, the distal epiblast cells (which normally differentiate into neuroectoderm and surface ectoderm) colonize the extraembryonic and posterior mesoderm of the embryo. At lower frequencies, cells from the distal epiblast and the extraembryonic ectoderm can be respecified to primordial germ cells,[18] which normally arise from cells of the proximal–posterior epiblast. Signals from both the extraembryonic ectoderm and the visceral endoderm are required to establish normal patterning of the underlying epiblast.[19] Although it is now evident that reciprocal signaling interactions between the epiblast and extraembryonic tissues play critical roles in the induction and maintenance of embryonic patterning during gastrulation,[20] how these complex events are orchestrated is still not fully understood. The remaining events of embryogenesis, a major part of which is organogenesis, rely on the functioning of localized stem cells that together are committed to regenerate specific organs in response to the surrounding microenvironment or niche. Thus, each of these stem cell "nests" will eventually undergo a programmed sequence of events, including cell divisions, migration, and apoptosis, to generate an organ at a specific location with respect to the other embryonic tissues and organs within the embryo.

1.2.2 Stem Cells in the Adult Organism

One of the most important functions of stem cells in postembryonic development is to repair and compensate for the loss of damaged tissues in the body. This dynamic process continues throughout life, and the unspecialized stem cells survive in each organ by creating their own niches.[21,22] The effective maintenance of a population of healthy stem cells within a particular tissue involves the concurrent operation of multiple genetic and epigenetic factors, along with stringent controls within each niche that create perfect harmony among the different cells of various organ systems.[23] Adult stem cells are thus central to tissue homeostasis and are identified through three distinctive properties:

1. Self-renewal: the ability to undergo division and form new cells with a potential identical to the mother cell[24,25]

2. Differentiation: the ability to give rise to a heterogeneous population of cells arranged in a hierarchical manner; includes various tissue-specific lineages, thereby building up the requisite critical mass toward replenishing the tissue of short-lived, differentiated cells

3. Homeostasis: the ability to regulate and balance differentiation and self-renewal in the tissue or organ

This unique combination of properties imparts to stem cells a continuing role during the entire life of an organism and further ensures that almost all developed tissues in the body, from the growing fetus to the adult, harbor stem cells. The property of self-renewal has gained recognition as a defining characteristic of stem cells and involves an asymmetrical cell division into two cell types, with an equal distribution of the gene pool among the daughter cells.[26,27] One of the daughter cells returns to the stem cell niche and remains quiescent until further signals for division are received through the microenvironment. The other cell is a progenitor cell committed to a differentiation pathway[28] (Fig. 1.2) that initiates the generation of cells toward fulfilling the needs of tissue-specific regeneration. This is achieved through the production of transit-amplifying (TA) cells, which ultimately commit to terminal differentiation (Fig. 1.3). The interposition of the TA population allows requisite cell amplification while eliminating the need for frequent cycling of stem cells, which holds the danger of mutations. Early TA cells are slow-cycling and undergo a few cell divisions, some of which may be asymmetric; late TA cells undergo a state of rapid, yet restricted cell proliferation that effectively builds up the critical mass necessary for differentiation. This arrangement of tissue regenerative processes as a hierarchy thus ensures that tissue-specific stem cells retain a long-term regenerative capacity,[29] whereas progenitors characterized by a finite division capacity eventually stop dividing and differentiate along tissue-specific lineages into mature cell types, and on completing their functionality, ultimately undergo apoptosis.[30] The process also ensures a long-term genetic stablility, as a large majority of the de novo mutations arising during the processes of cell proliferation, commitment, and differentiation in the hierarchy will undergo a process of natural elimination, as all the differentiated cells will ultimately undergo apoptosis. In this way, the property of self-renewal and hierarchial arrangement of stem cell derivatives allows a long-lived to indefinite life span of genetically stable stem cells within an organ.

Stem cell divisions can also occur in a symmetrical manner, which enables the size of the stem cell pool to be increased at certain stages during development or after wounding[31]; for example, stem cells within the bone marrow (i.e., the hematopoeitic stem cells) maintain relatively rapid division rates in order to meet the demand for a turnover of the large numbers of different blood cells required

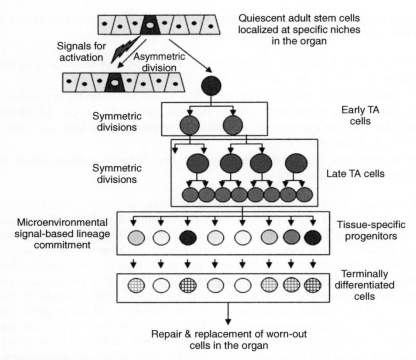

FIGURE 1.3   Stem cell hierarchies in normal tissue regeneration. (From ref. 30.) (*See insert for color representation of figure.*)

by the body. Stem cells from other tissues, such as those in the skin and colon, maintain a slow but steady or constant growth rate toward replenishing tissue; yet other stem cells, in tissues such as the brain, remain quiescent at most times and are activated only on stimulation by tissue damage or by hormonal exposure resulting in a change in the physiological state. The regulation of stem cells in an adult is thus tightly controlled, to allow for the growth replenishment of tissues and to permit repair of damaged tissues. The properties of stem cells described above also impart to them a continuing regenerative potential: the competence to display a specialized differentiation capability, the ability to give rise to the large number of cell types within the embryo and the adult organism. A decrease in this differential potential is evident as we move from the totipotent single-celled embryo to pluripotent ICM cells: late embryonic and fetal stem cells. Pluripotency is replaced by tissue-specific multipotential capability as somatic cell populations are formed, and subsequently it is retained only in those cells allocated to the germline and certain adult stem cells. This critical balance of self-renewal versus differentiation is achieved by the gene expression programs of stem cells, which regulate transcriptional activation and inactivation mechanisms to allow cells to maintain a pluripotent state but also permit differentiation into more specialized states. A disruption in such homeostatic mechanisms is believed to lead to the abnormal state of cancer.

1.3 CANCER STEM CELLS

Most cancer cells divide rapidly and can be grown indefinitely in culture as immortal cells and express a plasticity of differentiation, similar to embryonic stem (ES) cells and adult stem cells. In fact, before the advances that led to an understanding of the developmental plasticity of ES cells, embryonal carcinoma and teratocarcinoma cells (derived from germ cell tumors and known to differentiate to give rise to cells of many lineages) were used as in vitro models for studies relating to development and differentiation.[32] In 1976, Beatrice Mintz and Ralph Brinster showed independently that teratocarcinomas could also give rise to normal chimeric mice.[33,34]

The realization of an involvement of stem cells in cancer has been built up over the last several decades—almost a century ago if John Beard is indeed considered "father of cancer stem cell biology." However, the first direct evidence for the existence of CSCs came from the work of John Dick and his co-workers, who identified the presence of CSCs in acute lymphocytic leukemia through extensive cell cloning and demonstration of their self-renewing capacity—a critical property of all stem cells.[25] CSC identification in leukemia has, in fact, changed the way that many scientists view cancer[35,36] and has led to their isolation from some solid tumors. CSCs represent only about 1 % of the tumor but appear to be the only cells capable of generating a new tumor in immune compromized mouse models. Whereas stem cells themselves may be difficult to isolate from an already differentiated healthy tissue such as the breast, it has been possible to isolate stem cell–derived clonal tumor cells by engrafting tumor tissues in suitable animal models such as breast tumor–derived cells transplanted in the mammary fat pads of nonobese diabetic/severely compromised immunodeficient mice (NOD/SCID) mice.[37,38] Additional studies have presented data indicating that established cell lines also contain a minor population of cells with properties similar to those of stem cells.[39,40]

1.3.1 Activation of Stem Cells and Cancer

The modern era has seen the cloning of many genes mutated in the germline of patients in families segregating cancer as a genetic trait. These *tumor suppressor genes* are thought to be critical in preventing the formation of cancer. However, it is unclear why mutations in a gene such as *RB* or *TP53*, expressed in all the cells of a human being, should give rise to tissue-specific cancers such as retinoblastoma or breast cancer. In one report, in which tumors derived from embryonic tissue from tissues under hormonal control or renewal tissues such as skin and gut were distinguished, it is suggested that the action of tumor suppressor genes can be explained by their differential effects on stem cells derived from varied tissues.[41] Further, the oncogenic potential of different resident stem cells may be distinct since the genetic or epigenetic factors vary from person to person and even between organs in the same person. A question thus arises: How do CSCs arise in tissues and progress to give rise to a new organ (i.e., the tumor)?

Box 1-1   Possible Origins of CSCs in Tissues Leading to the Occurrence of Cancer

- Transformation of stem cells residing in a tissue, leading to altered growth and differentiation properties
- Transformation of a local pool of early progenitors that reacquire self-renewal properties
- Series of effective mutations that render committed transient-amplifying progenitor or differentiated somatic cells within a tissue immortal (de-differentiation)
- Fusion of circulating bone marrow–derived stem cells with tissue-residing cells

---

To address this issue among related issues, research over the last decade has attempted to associate cellular mechanisms with mutagenic effects within tissues leading to the emergence of CSCs (Box 1-1). The possibilities that emerge include the following:

1. *Stem cells as targets of transforming mutations.* Through the acquisition of abnormal growth and differentiation properties during its long-lived residence within a tissue, a multipotent tissue stem cell may, following a series of aberrant events, give rise to a CSC. A disruption of the stem cell niche with a shift toward growth-promoting signals rather than growth-inhibiting signals results in dominant stem cell activation rather than the *transient activation* that is required for normal tissue homeostasis. This could occur by hormonal stimulation, recurrent posttissue damage, inflammation, radiation, chemicals, infections, inactivation of tumor suppressor gene(s), and/or activation of oncogene(s). The change in the tissue microenvironment leads to a chronic activation of stem cells and results in their long-term proliferation. Such chronically dividing stem cells could become vulnerable to additional genetic events. The recognized effects of these events include autonomous growth, loss of cell cycle regulation, and resistance to apoptosis, which are all well-understood properties of cancer cells.[42] Cancer can thus be thought of as a disease resulting from the abnormal growth of stem cells resulting from their chronic activation followed by genetic insults, culminating in transformation.

2. *Progenitor cells as targets of transforming mutations.* A CSC need not be derived from a bona fide stem cell, but instead, can arise from tissue-specific

early progenitors (i.e., early TA cells following oncogenic transformation). This would retain a minimal number of changes to retain stem cell properties, as they are derived directly from stem cells; and accompanying events that support the transformation process may be identical to those for stem cell transformation.

3. *De-differentiation of committed progenitors or differentiated cells.* In yet other cases, there is evidence that committed progenitors (i.e., late TA cells, or even differentiated cells) may also reacquire the property of self-renewal to give rise to cells with stemlike properties that are pluripotent. The latter is described as a de-differentiation phenomenon that is very commonly reported in plants and also to a certain extent in animals with lower levels of tissue and organ complexities. More recently, the process has also been demonstrated in tissues derived from higher vertebrates such as mice,[43–45] wherein transfection of activated oncogenes were shown to transform murine fibroblasts into cells with stem cell–like properties, indicating that non-stem cells can be converted into CSCs.

4. *Fusion of tissue-specific stem cells with circulating bone marrow stem cells.* The finding that fusion of circulating bone marrow–derived stem cells with differentiated tissue cells can create cells with self-renewal capacity leads to yet another speculation on the origin of CSCs.[46] This is believed to involve the mobilization of bone marrow–derived cells either at the wrong time and/or their incorporation at the wrong place within other tissues, as a first step toward transformation. Moreover, several CSCs express the pluripotency and self-renewal markers expressed characteristically by hematopoietic stem cells. Whether this reflects a global commonality of expression of these markers or an outcome of the fusion process is not yet very clear.

Of all the possibilities noted above, adult stem cells giving rise to cancer is an attractive hypothesis, given that the classic multistep model of carcinogenesis requires a long-lived cell in which multiple genetic hits can occur. However, regardless of whether the cell of origin is a normal adult stem cell or progenitor or a differentiated tissue cell, and regardless of the mechanism of its emergence, CSCs are defined by their stem cell–like properties (Box 1-2). Moreover, despite the diversity in possible origins of CSCs, all the processes argue for dynamic changes within the tissue leading to transformation.

Box 1-2   Characteristics Shared by Normal and Cancer Stem Cells

- Capacity for asymmetric divisions (self-renewal), which generates a quiescent stem cell and a committed progenitor and contributes toward developing a critical mass of cells
- Regulation of self-renewal by similar signaling pathways (Wnt, Sonic Hedgehog, and Notch) and at the epigenetic level by Polycomb genes (*BMI-1* and *EZH2*)

- Expression of factors such as Oct4, Nanog, and Sox2 and also nodal and cancer testis-specific antigens (CTAs), which maintain a functional plasticity by promoting pluripotency and immortality
- Capacity to arrange a hierarchy of cellular derivatives that includes progenitors and differentiating cells
- Extended telomeres and telomerase activity that increases the cellular life span
- Expression of ABC transporters, contributing to cellular resistance against specific growth-inhibitory drugs
- Predisposition for growth factor independence through secretion of growth factors and cytokines
- Stimulation of angiogenesis through secretion of angiopoietic factors
- Expression of similar surface receptors (e.g., CXCR4, CD133, $\alpha 6$ integrin, c-kit, c-met, LIF-R) that are either identified as stem cell markers or are associated with homing and metastases

1.3.2 Isolation and Identification of Cancer Stem Cells

Based on the perpetuation of cytogenetic abnormalities during serial transplantation in culture and in animal models, it was reported in 1957 that ascitic fluid possessed possible cancer stem cells.[47] Soon after, Kleinsmith and Pierce demonstrated that transplantable teratocarcinomas were derived from a single pluripotent cell.[48] Similar to tissue stem cells, CSCs are believed to be capable of asymmetric cell division (i.e., they can give rise to one self-renewing stem cell and one daughter cell committed to differentiation). It is postulated, however, that CSCs are also capable of symmetric division[6] (to generate two daughter stem cells), as is the case with normal stem cells when they are required to proliferate rapidly, such as during inflammation or wound healing. All cancers are now believed to be composed of a mixture of self-renewing stem cells, transiently amplifying progenitors, and proliferative cells with a shorter life span that can undergo limited differentiation. Cumulatively, all these factors make the identification and isolation of the initial rare stem cell population within tumors a challenging issue in cancer biology.

Research in the field of stem cell biology has traditionally focused on the hematopoietic system, where detailed lineages have been elucidated and various surface molecules associated with the components of the hierarchy have been identified as specific markers.[49,50] Following these well-accepted hierarchies of blood-forming lineages, studies attempted to identify leukemic stem cells using known *cluster of differentiation* (CD) cell surface markers.[51,52] Subsequently, in a study of acute myeloid leukemia (AML), Lapidot et al. successfully identified putative leukemia-initiating cells using limiting dilution analysis.[25] This study demonstrated that in AML, malignant leukemic stem cells are probably derived from immature (uncommitted) bone marrow cells, which express a similar cell surface marker profile (i.e., $CD34^+/CD38^-$) and originate from a similar bone marrow niche. Similarly, in multiple myeloma, a subpopulation of $CD138^-/CD34^-$ cells were demonstrated to be clonogenic in vitro and capable of propagating tumors upon serial passaging in NOD/SCID mice as compared to $CD138^+$ cells, which were nontumorigenic.[53] More recently, in AML, a population of cells within the $CD34^+/CD38^-$ cell fraction have been identified further on the basis of its $CD96^+/CD90^-$ expression as being present only in leukemic samples and not in normal bone marrow, constituting a major leap in the specific identification of leukemic stem cells.[54]

Apart from the hematopoietic system, the presence of stem cells in tissues with a high cellular turnover, such as skin, gut, testis, and the respiratory tract, has been reported widely. More recently, the presence of stem cells in solid, slow-growing tissues and organs has gained recognition.[7,28,55] The specific expression of surface molecules (e.g., CXCR4, CD133, EPCAM, CD44; Table 1.1) by normal stem cells has provided a mechanism to separate out the rare cancer stem cells from within the tumor mass. In solid tumors, Al-Hajj et al. demonstrated a subpopulation of human breast cancer cells also capable of inducing tumors in nude mice.[35] Those putative breast CSCs were immunophenotyped as $CD44^+/CD24^{lo/-}$ and were capable of initiating tumorigenesis at a density of merely 200 cells, in contrast to unsorted cells, which generally require a density of at least 10^6 cells per injection. In brain tumors, CD133 expressing candidate CSCs have been isolated that are similar to normal (neural) stem cells. The $CD133^+$ CSCs were also capable of driving tumorigenesis in mouse models at low cell numbers (about 100 cells).[56] Since then, several reports exist of the isolation and identification of stem cells based on the expression of surface markers from other cancers, such as prostate, lung, and colon (Table 1.1).

TABLE 1.1 Surface Markers Used for Isolation and Identification of Cancer Stem Cells

Type of Cancer	Markers	Refs.
AML	$CD34^+$, $CD38^-$	57
	$CD96^+$, $CD90^-$	54
Neural	$CD133^+$	56
Mammary	$CD44^+$, $Cd24^-$	35
Prostate	$CD133^+$, β-integrin	58
Lung	$Sca1^+$, $CD45^-$, $CD31^-$, $CD34^+$	59
Hepatocellular	$CD133^+$	60
Pancreas	$CD44^+$, $CD24^+$, ESA^+	61
Colorectal	$EpCAM^{high}$, $CD44^+$	62
Colon	$CD133^+$	63,64
Liver	$CD133^+$	65

Recently, however, one commentary questioned the reliability of the commonly used methods for identifying CSCs.[66] This included concerns regarding enzyme treatments during cell preparation from tumors, isolation based on CD cell surface profiles that are identical to normal tissue stem cells, genomic instability that could result in plasticity in the "stemness" of specific tumor cells, and the routine use of nonphysiologic microenvironments to assess the tumorigenicity of engrafted cells. Despite these objections, however, the CSC hypothesis has gained wide acceptance for the study of such neoplastic progenitors.

1.3.3 De Novo Generation of a New Organ (Tumor) by Transformed Stem Cells

Analogies between normal and CSCs in terms of cellular, biochemical, and molecular events are currently being resolved at a basic level, and considerable work is in progress to describe these processes in greater detail during transformation. Nonetheless, similarities and differences in the functioning of normal and cancer stem cells at the tissue level, whereby a consolidation of the events described above contributes toward either maintenance of homeostasis or development of abnormal tumor tissue, are unclear. It is, however, understood that both normal and tumorigenic stem cells give rise to phenotypically heterogeneous populations that exhibit various degrees of proliferation and differentiation capabilities. The heterogeneity in tumor tissues may be attributed to both continuing mutagenesis and aberrant differentiation of cancer cells, the latter being a variation of normal differentiation driven by cancer stem cells. This disrupted differentiation is often accompanied by aberrant expression patterns (e.g., the expression of germ cell markers in epithelial ovarian cancer).[67]

Two general hypotheses have been put forth to describe tumor formation based on a microdissection of the events described above. The first, *stochastic cancer stem cell model*, postulates that each population of cells within a heterogeneous tumor has an equal but extremely low tumorigenic potential.[68] In such a scenario, tumor progression is a continuous process based on the positive selection of genetically unstable clones that confer a survival advantage on a tumor within its surrounding microenvironment. The stochastic model thus also accounts for the emergence of drug resistance during chemotherapy through selection of cells with genotypes that allow survival from the intended drug insult[69] (e.g., DNA damage by platinum drugs). Under this model, however, isolation of tumor progenitors would not be reproducible, as their existence would be random. A recent example is the proposed stochastic model of gastrointestinal stem cells, suggesting that all stem cells in a niche are descendents of a single common ancestral stem cell, and that intestinal crypts expand further clonally by crypt fission, forming two daughter crypts.[70] The same mechanism has been postulated to contribute to the expansion of mutated stem cells and CSC clones in the colon and in the entire gastrointestinal tract.

The fact that organ-specific stem cells derive all of the differentiated cells within a given tissue has led to the proposal of a stem cell hierarchical model for tissue development, maintenance, and repair, as discussed earlier (Fig. 1.3). Deriving from this, the *hierarchial cancer stem cell model* puts forth the suggestion that the tumorigenic potential of tumors is limited to a very small clonogenic population of cells (i.e., the CSCs).[7] The large population of cells within a tumor are descendants of these CSCs, do not have self-renewal capacity, and are organized in the form of a hierarchy. Thus, the model posits that not all cells in a tumor are equal and that the tumor-initiating cells are a rare subset with a distinct phenotype. This hierarchical model helps explain why most tumors are heterogeneous despite their clonal origin. This approach also accounts for the perplexities faced by researchers in establishing a permanent cell line from primary tumors or in recreating tumors in animal models.

1.4 SELF-RENEWAL AND DIFFERENTIATION IN CSCS

The steady-state expression of a stem cell in an adult organism is one of quiescence. On receiving specific signals from its microenvironment, it can be induced to undergo asymmetric stem cell division, which generates an identical stem cell (self-renewal) and a progenitor with decreased plasticity (the latter divides further symmetrically to generate several TA cells that are all committed to differentiation). Current knowledge indicates that evolutionary conserved mechanisms regulate this process. Beginning at the late blastocyst stage in embryogenesis and further in adult tissues, timing of self-renewal and regulation of loss of pluripotency are crucial for the orderly functioning of stem cells and the elimination of potentially teratogenic cells. A central question that remains is: How can a single cell divide to produce two progeny cells with different fates?

Many studies examining the mechanisms regulating asymmetrical divisions have focused on mitotic cleavage orientation and the uneven distribution of cell-fate-determining molecules such as Numb and ACBD3 within the two cells resulting from the division of a stem cell.[71] In neural stem cells and progenitors, both proteins interact through an essential Numb domain. ACBD3 associates with the Golgi apparatus in neurons and interphase progenitor cells but becomes cytosolic after Golgi fragmentation during mitosis, when Numb activity is needed to distinguish the two daughter cells. Accordingly, cytosolic ACBD3 can act synergistically with Numb to specify cell fates, and its continuing presence during the progenitor cell cycle inhibits neuron production. This represents a mechanism for coupling cell fate specification and cell cycle progression.[72] In addition to neuronal stem cells, Numb has also been reported to regulate cell fate in other stem cells (e.g., skeletal muscle satellite cells).[73] Similarly, PAR proteins are also suggested to control asymmetric cell division in a wide range of organisms and somatic cell types[74]; other asymmetrically segregating proteins include[75] CD53, CD133, L-selectin, Lamp-2, and CD71.

Studies of asymmetric division have relied heavily on the lessons learned in the regulation of *Drosophila* germline and somatic stem cells. Early in the cell cycle in larval neural stem cells, the two centrosomes become unequal: One organizes an aster that stays near the apical cortex for most of the cell cycle, while the other loses microtubule-organizing activity and moves extensively throughout the cell until shortly before mitosis, when it organizes the second mitotic aster near the basal cortex. Upon division, the apical centrosome remains in the stem cell while the other segregates to the differentiating daughter.[76] Almost along identical lines, developmentally programmed asymmetric behavior and inheritance of mother and daughter centrosomes underlies the stereotyped spindle orientation and asymmetric outcome of stem cell divisions in the *Drosophila* male germline.[77] More recently, in mammalian systems, reports indicate that the fate of daughter cells is determined by their relative orientation within the stem cell niche (e.g., in dividing satellite cells in skeletal muscle, on division the daughter cell attached to the basal lamina remains a stem cell, whereas the daughter that loses contact with the basal lamina up-regulates Myf5 and becomes a committed myogenic cell).[78]

Although some of the molecular events associated with asymmetrical divisions as outlined above have been described in normal stem cells, further work is required to identify the similarities and variations of those that determine events in CSC self-renewal.

1.5 CSC PLASTICITY AS REGULATED BY INTRINSIC AND EXTRINSIC STEM CELL FACTORS

In contrast to normal stem cells, it has been theorized that CSCs undergo genomic alterations that allow them to escape cell cycle regulation and achieve growth factor, anchorage independence, and resistance to apoptosis, besides contributing to dysregulation of self-renewal and expansion.[79] The acquisition of each of these characteristics is complementary to the others and requires a suitable microenvironment in which the transformed CSCs are believed to proliferate and differentiate into an entire tumor.[80] This understanding implies that the plasticity gained by CSCs is regulated by a cooperative effect of cell intrinsic (autocrine) factors, which may either involve changes in DNA sequences/copy number of genes or gene silencing through methylation or altered chromatin architecture (genetic and epigenetic effects), together with cell extrinsic (paracrine/derived from the tumor microenvironment) factors.

1.5.1 Stem Cell Intrinsic Factors: Genetic and Epigenetic Effects

A disturbed balance in gene regulation of tissue stem cells promoting self-renewal and/or aberrant differentiation is characteristic of cancer.[81] However, uncontrolled expansion of stem cells by itself may not produce fully invasive tumors.[82] Thereby, proliferating proto-oncogenic stem cells appear to require at least one additional permanent genetic mutation to drive them along a trajectory toward transformation.[83] This could be achieved either through oncogene activation or by silencing of tumor suppressor genes, which effectively supplements the perturbed shift toward self-renewal; continuing mutagenesis would further ensure clone amplification and disease progression. Toward an understanding of such changes in stem cells leading to cancer, several models have been developed based on rapidly self-renewing stem cells in human colon crypts. Thus, in familial adenomatous polyposis the initial germline mutation is in the "gatekeeper" tumor suppressor gene adenomatous polyposis coli. Second hits in this gene involve a stochastic and sequential accumulation of genetic and epigenetic defects that over time are selected for by the position of the earlier mutation. Extensive work indicates that crypts harboring mutated stem cells are clonal units that may develop a selective advantage, eventually leading to niche succession. Aberrant crypt foci are formed that contain several such clonal crypts generated through crypt fission involving the longitudinal division of each crypt into two daughter units. In such models the coexistence of defective progeny and hierarchies consequently generated marks a transition between pre-tumor and tumor progression.[84]

Recent studies argue that in a stem cell triggered into division, self-renewal requires the expression of certain "stemness" genes, whereas genes associated with differentiation must be repressed. Conversely, in the progenitor cell committed to differentiate, some of the stemness genes are switched off, whereas specific lineage-associated genes involved in differentiation begin to be expressed. The classical *stem cell self-renewal signature* is defined by the expression of three transcription factors identified primarily in ES cells (i.e., Oct4, Nanog, and Sox2), now known to be crucial for the phenomenon of pluripotency.[85] The downstream regulation mediated by these three key molecules in the cell, including their feedback/autoregulatory and feedforward loops are critical in the establishment and maintenance of pluripotency.[86–88] Other genes involved in self-renewal besides these three master regulators include *Stella, FGF4, BMP4, Stat3, UTF1*, and *Rex1*.[89,90] All or combinations of these determinants of pluripotency are almost always detected in malignant cancer tissues. Although not much is known regarding their mechanisms of regulation in CSCs, it is expected that most of their gene targets and therefore, effects of these genes would be similar to those in normal stem cells. Thus, the current understanding of the role of these factors in cancer is derived from the corresponding data in normal stem cells (mostly in embryonic stem cell systems).

Toward counteracting self-renewal and driving cell fate determination, a loss of pluripotency is required which is induced by the activation of several pathways that perturb the levels of the pluripotent regulators above and impose alternate cellular fates, such as those associated with Cdx2 and GCNF function.[8] In ES cells, the induction of GCNF expression facilitates differentiation through inhibition of the pluripotent state by repression of *OCT4* and *NANOG* regulatory genes. Other factors involved in regulating Oct4 expression are Oct4 itself, Sox2, SP1, RAR, SF1, COUP TF I and II, and LRH-1.[91–94] Oct4, Nanog, Sox2, FGF4, and Stella repression was observed upon retinoic acid-mediated differentiation.[95] However, at present the regulation of Sox2 during differentiation remains unclear.

Several members of the Polycomb group (PcG) proteins have been identified as important regulators of cell self-renewal and cell fate determination decisions. These proteins act as transcriptional repressors by regulating chromatin remodeling, especially in association with stem cell function, and are also reported to be altered in various cancers.[96] Recently, Lee et al. have identified several genes that were bound by the human Polycomb repressive complex 2 (PRC2) subunit SUZ12, display H3K27 (lysine 27 on histone 3) trimethylation, and were repressed transcriptionally in human ES cells.[97] A significant subset of these genes was occupied by Oct4, Sox2, and Nanog. Disruption of H3K27 methylation resulted in de-repression of most of these genes.[98] Further data suggest that specific genes are primed for expression in pluripotent cells and that occupancy by opposing epigenetic marks are involved in this process. That this also occurs at many *NANOG*, *OCT4*, and *SOX2* target genes suggests that regions of H3K4 (lysine 4 on histone 3) methylation, within overlapping regions of extended H3K27 trimethylation, are characteristically associated with active and inactive chromatin states, respectively, in ES cells (a detailed version is presented in Chapter 10). These bivalent modifications occupy regions that correspond to developmentally regulated genes usually silenced in ES cells and also tumor suppressor genes. H3K27 methylation deficiency also results in premature expression of repressed genes. On the other hand, increased repression due to persistent H3K27 methylation that is further supplemented with H3K9 (lysine 9 on histone 3) methylation is a feature of transformation and has been reported in several cancer cell lines[99] and in embryonal cancer.[100] The genetic and epigenetic mechanisms of gene regulation during self-renewal and differentiation, however, represent a striking commonality between normal and cancer stem cells.

1.5.2 Stem Cell Extrinsic Effects: Niche Effects and Microenvironmental Signaling

The stem cell niche is loosely defined as stem cells surrounded by other differentiated cells within a tissue at defined locations. The niche consists of heterologous cell types that harbor stem cells and influence their fate through direct contact, thereby functioning to balance the quiescence and activation of stem cells,[101] the key to homeostatic regulation of stem cells, yet supporting ongoing tissue regeneration. The niche is thus a physical anchoring site for stem cells and generates factors including certain extracellular matrix (ECM) components and signaling molecules that control stem cell number, proliferation, and fate determination.[79,102,103] Examples of stem cell niches include the hair follicle bulge compartment[104] and the crypts of Lieberkuhn, located at the base of villi within the intestinal epithelium.[105]

A majority of mutagenic agents described to confer a risk for cancer also perturb the normal stem cell niche and tissue homeostasis, besides inducing changes in the DNA of some stem cells and impairing or enhancing some of their characteristic properties, such as self-renewal or their intrinsic ability for DNA repair. These stem cells with altered DNA may remain in a state of perpetual activation by intercellular communications *in cis* (through ECM proteins) or *in trans* (through autocrine or paracrine factors) produced by niche cells. Such continuous activation further accentuates altered gene expression and protein profiles that drive these stem cells toward differentiation, cell death, and degenerative or transformation pathways. Another function of the tumor niche is the active recruitment of new endothelial and stromal cells into tumors that is essential for developing a pro-angiogenic environment that enhaces tumor survival under adverse conditions.[106]

Very recently, the stringent regulation imposed on stem cells through their niche have begun to be identified. A normal niche may evolve to become proto-oncogenic, or in fact may be the prime feature in tumorigenesis and therefore represent an oncogenic state. The concept of proto-oncogenic stem cell niches is best exemplified in lung cancer, wherein different histological types of neoplasias have been correlated with stem/progenitor populations at specific tissues in the lung[107] (e.g., the putative stem cell populations originating in the bronchiole mucosa in pulmonary neuroendocrine cells of lung-derived neuroepithelial bodies have been proposed as likely originating cells for small cell lung carcinoma[108–110]. Similarly, a direct relationship between proximal airway basal progenitors and cells associated with carcinogenesis in murine models for human squamous cell carcinoma is suggested,[111,112] while bronchoalveolar cell and central bronchiolar adenocarcinomas are believed to originate from bronchioalveolar stem cells that reside at the bronchoalveolar duct junctions in lung tissues.[113–115]

Oncogenic stem cell niches have just recently been described. The retinoblastoma (RB) gene, dentified as a tumor suppressor, has been shown to regulate HSCs extrinsically by maintaining the competence of the adult bone marrow to support their growth and normal homeostatic hematopoiesis. Loss of RB expression in the niche and myeloid cells leads to degradation of the osteoblastic niche and consequent displacement of HSCs. The latter then undergo rapid expansion and mobilization to the spleen, promoting myeloid development that ultimately culminates in myeloproliferative disease[116] (MPS). Along similar lines, mice deficient for another gene [i.e., the retinoic acid receptor γ ($RAR\gamma$)], develop MPS driven by a $RAR\gamma$-deficient microenvironment.[117] The MPS phenotype in both cases (i.e., loss of RB and $RAR\gamma$) continues through the life span of the mice and is more pronounced in older mice. Moreover, the disease cannot be propagated through successive transplantation of HSCs from MPS-affected mice to normal mice, identifying the disease to be extrinsic to tumor-derived HSCs.

Another common feature of tumor and tissue stem cells is utilization of similar signal pathways that normally control cell fate during early embryogenesis. Such regulatory signal molecules, including components of the Notch, Wnt, and Hedgehog pathways, bone morphogenetic proteins,[90] fibroblast growth factor, leukemia inhibitory factor, and transforming growth factor-β,[118–122] have been shown to play roles in controlling stem cell self-renewal and in regulating lineage fate in different systems. In numerous tumors, however, the signaling cascades initiated by these molecules have now been demonstrated to be dysregulated (e.g., in skin, liver, colorectal, and pancreatic cancers, Wnt signaling has been demonstrated to be aberrant[123,124]). In ovarian cancer, the Wnt signal transducer β-catenin is over-expressed at an advanced stage of tumor progression.[125] The Hedgehog cascade, well known as a regulator of patterning during embryonic development,[126] has been associated with breast,[127] ovarian,[128] and prostate cancers,[129] whereas Notch overstimulation has been strongly implicated in T-cell malignancies.[130] It has been demonstrated further that various lineage determination molecules within these signaling pathways exhibit a significant degree of crosstalk.[131] An important difference in the signals between normal and transformed states is that those in normal tissues are transiently expressed stem cell–activation signals, whereas in cancer, these signals dominate and lead to a state of long-term or permanent activation.[132]

The molecules and underlying machinery used by normal stem cells for homing or mobilization and CSCs for invasion and metastasis are also realized to be similar. For example, during HSC activation and mobilization, matrix metalloproteinase-9 (MMP-9) is required for proteolysis of the extracellular matrix components and converting stem cell factor from a membrane-bound form into a free form, which then promotes HSC proliferation and mobilization through a c-Kit receptor. Intriguingly, the molecules of the MMP family are considered as key players in the process of cancer cell metastasis. Additionally, cell surface receptors and the ligands required for their activation, such as SDF1 and CXCR4, are also expressed during normal stem cell homing and mobilization as well as cancer cell metastasis.

1.6 CONCLUSIONS AND FUTURE PERSPECTIVES

We have been presenting the widely accepted view that malignancy results from a complex network of interactions between altered cellular genes and numerous exogenous and endogenous factors of a tissue-derived cell. The probability of the accumulation and effects of such changes and interactions is almost certainly highest in long-lived tissue stem cells, but is equally effective in progenitors or differentiated cells that acquire an immortal phenotype. Unfortunately, it is not precisely known how CSCs arise in tissues or which of the resident tissue cells would be more susceptible to the transformation process. Thus, at a very fundamental level, we have yet to determine the extent to which stem cell biology is relevant to all types of human cancers. A detailed update in understanding of stem cell involvement in the major cancers is presented in the following chapters; however, it would be rather naive to believe that CSCs represent a universal modality for emergence and progression. Further studies employing cellular lineage tagging of putative stem and cancer progenitors in tightly regulated transgenic mouse models for specific cancer types are thus sorely needed. Once these cells and lineage relationships are better understood, one can hope to better understand the process.

Several concerns also remain regarding the strategy of CSC identification using xenotransplantation models that demonstrate the presence of a self-renewing population. Since highly purified, FACS-isolated tumor cells are used in these experiments, the model remains imperfect, as in addition to a horde of other host factors, evolution of a tumor *in situ* involves complex interactions between a stem cell poised on the brink of transformation and its microenvironment. Thus, although most findings make it highly likely that dysregulation of cell fate–determining pathways contributes to the formation of aberrant stem cells and their hierarchies in cancer, identification of crosstalk between the extracellular environment and the regulation of pluripotency at the level of transcription is seriously restricted in the current models.

It is also to be expected that the role of CSCs in migration and metastases would be much different than their role in primary tumors. Such differential regulation of migrating CSCs and those in metastases is also not yet identified. Once additional insights into the biology of candidate CSCs, as well as of those subpopulations that initiate metastases, become apparent with further characterization and validation of their gene and protein expression profiles, as well as the role of these molecules in tumor progression and clinical responses, it will become clear whether engrafting activity (the key feature of identification protocols today) is an accurate reflection of stem cell activity. However, it is certain that advances in our knowledge of the properties of stem cells have definitely culminated in a realization for specific targeting and eradication of CSCs, and that treatment of bulk disease is an insufficient panacea for cancer. The development of a new generation of treatments to target the rare CSCs is thus critical, but poses formidable challenges:

1. Ideally, a therapy should target unique CSC pathways and "turn back the clock" from a state of disease to one of normal tissue and organ homeostasis.

2. A concern in achieving the goal stated above is that normal stem and progenitor cells actually prove to be more sensitive than CSCs to the effects of current chemotherapeutic drugs. This provides a competitive advantage to CSCs and makes their positive selection quite likely, leading to the emergence of drug-resistant clones. Delineation of the effects of the new drug regimes on the evolution of CSCs is thereby imperative.

3. In cases where clinical remission is achieved, the presence of drug-resistant CSCs that have "escaped" chemotherapy would initiate a relapse. This necessitates the development of sensitive methods for detection of residual CSCs for follow-up in patients in remission. The establishment of diagnostic endpoints by which treatment success can be measured is thus required.

A culmination of the understanding of CSC biology will thus aid the development of more effective and targeted therapies to treat this astoundingly complex and devastating disease.

REFERENCES

1. Beard J. Embryological aspects and etiology of carcinoma. *Lancet.* 1902; 1:1758–1761.
2. Yamanaka Y, Ralston A, Stephenson RO, Rossant J. Cell and molecular regulation of the mouse blastocyst. *Dev Dyn.* 2006; 235:2301–2314.
3. Beard J. The action of trypsin upon living cells of the Jensen sarcoma. *Brt Med J.* 1906; 1:140–141.
4. Costa FF, Le Blanc AK, Brodin BB. Cancer/testis antigens, stem cells and cancer. *Stem Cells.* 2007; 25:707–711.
5. McCulloch EA, Till JE. Perspectives on the properties of stem cells. *Nat Med.* 2005; 11:1026–1028.
6. Wicha MS, Liu S, Dontu G. Cancer stem cells: an old idea—a paradigm shift. *Cancer Res.* 2006; 66:1883–1890.
7. Reya T, Morrison SJ, Clarke MF, Weissman IL. Stem cells, cancer, and cancer stem cells. *Nature.* 2001; 414:105–111.
8. Ralston A, Rossant J. Genetic regulation of stem cell origins in the mouse embryo. *Clin Genet.* 2005; 68:106–112.
9. Tiedemann H, Asashima M, Grunz H, Knöchel W. Pluripotent cells (stem cells) and their determination and differentiation in early vertebrate embryogenesis. *Dev Growth Differ.* 2001; 43:469–502.
10. Jiang Y, Jahagirdar BN, Reinhardt RL, et al. Pluripotency of mesenchymal stem cells derived from adult marrow. *Nature.* 2002; 418:41–49.
11. Tam PP, Loebel DA. Gene function in mouse embryogenesis: get set for gastrulation. *Nat Rev Genet.* 2007; 8:368–381.
12. Hill MA. Early human development. *Clin Obstet Gynecol.* 2007; 50:2–9.
13. Mikawa T, Poh AM, Kelly KA, Ishii Y, Reese DE. Induction and patterning of the primitive streak, an organizing center of gastrulation in the amniote. *Dev Dyn.* 2004; 229:422–432.
14. Kinder SJ, Tsang TE, Quinlan GA, Hadjantonakis AK, Nagy A, Tam PP. The orderly allocation of mesodermal cells to the extraembryonic structures and the anteroposterior axis during gastrulation of the mouse embryo. *Development.* 1999; 126:4691–4701.
15. Lu CC, Brennan J, Robertson EJ. From fertilization to gastrulation: axis formation in the mouse embryo. *Curr Opin Genet Dev.* 2001; 11:384–392.
16. Kinder SJ, Loebel DA, Tam PP. Allocation and early differentiation of cardiovascular progenitors in the mouse embryo. *Trends Cardiovasc Med.* 2001; 11:177–184.
17. Baron MH. Early patterning of the mouse embryo: implications for hematopoietic commitment and differentiation. *Exp. Hematol.* 2005; 33:1015–1020.
18. Yoshimizu T, Obinata M, Matsui Y. Stage-specific tissue and cell interactions play key roles in mouse germ cell specification. *Development.* 2001; 128:481–490.
19. Belaoussoff M, Farrington SM, Baron MH. Hematopoietic induction and respecification of A-P identity by visceral endoderm signaling in the mouse embryo. *Development.* 1998; 125:5009–5018.

20. Brennan J, Lu CC, Norris DP, Rodriguez TA, Beddington RS, Robertson EJ. Nodal signalling in the epiblast patterns the early mouse embryo. *Nature.* 2001; 411:965–969.

21. Domen J, Weissman IL. Self-renewal, differentiation or death: regulation and manipulation of hematopoietic stem cell fate. *Mol Med Today.* 1999; 5:201–208.

22. Moore KA, Lemischka IR. Stem cells and their niches. *Science.* 2006; 311:1880–1885.

23. Fuchs E, Segre JA. Stem cells: a new lease on life. *Cell.* 2000; 100:143–155.

24. McCulloch EA, Siminovitch L, Till JE. Spleen-colony formation in anemic mice of genotype WW. *Science.* 1964; 144:844–846.

25. Lapidot T, Sirard C, Vormoor J, et al. A cell initiating human acute myeloid leukaemia after transplantation into SCID mice. *Nature.* 1994; 367:645–648.

26. Blau HM, Brazelton TR, Weimann JM. The evolving concept of a stem cell: entity or function? *Cell.* 2001; 105:829–841.

27. Slack JM. Stem cells in epithelial tissues. *Science.* 2000; 287:1431–1433.

28. Raff M. Adult stem cell plasticity: fact or artifact? *Annu Rev Cell Dev Biol.* 2003; 19:1–22.

29. Braun KM, Watt FM. Epidermal label-retaining cells: background and recent applications. *J Invest Dermatol Symp Proc.* 2004; 9:196–201.

30. Bapat SA. Evolution of cancer stem cells. *Semin Cancer Biol.* 2007; 17:204–213.

31. Morrison SJ, Kimble J. Asymmetric and symmetric stem-cell divisions in development and cancer. *Nature.* 2006; 441:1068–1074.

32. Lotem J, Sachs L. Epigenetics and the plasticity of differentiation in normal and cancer stem cells. *Oncogene.* 2006; 25:7663–7672.

33. Mintz B, Illmensee K. Normal genetically mosaic mice produced from malignant teratocarcinoma cells. *Proc Natl Acad Sci U S A.* 1975; 72:3585–3589.

34. Brinster RL. Participation of teratocarcinoma cells in mouse embryo development. *Cancer Res.* 1976; 36:3412–3414.

35. Al-Hajj M, Wicha MS, Benito-Hernandez A, Morrison SJ, Clarke MF. Prospective identification of tumorigenic breast cancer cells. *Proc Natl Acad Sci U S A.* 2003; 100:3983–3988.

36. Hemmati HD, Nakano I, Lazareff JA, et al. Cancerous stem cells can arise from pediatric brain tumors. *Proc Natl Acad Sci U S A.* 2003; 100:15178–15183.

37. Al-Hajj M, Clarke MF. Self-renewal and solid tumor stem cells. *Oncogene.* 2004; 23:7274–7282.

38. Dontu G, Al-Hajj M, Abdallah WM, Clarke MF, Wicha MS. Stem cells in normal breast development and breast cancer. *Cell Prolif.* 2003; 1:59–72.

39. Hirschmann-Jax C, Foster AE, Wulf GG, et al. A distinct side population of cells with high drug efflux capacity in human tumor cells. *Proc Natl Acad Sci U S A.* 2004; 101:14228–14233.

40. Setoguchi T, Taga T, Kondo T. Cancer stem cells persist in many cancer cell lines. *Cell Cycle.* 2004; 3: 414–415.

41. Knudson AG Jr, Strong LC, Anderson DE. Heredity and cancer in man. *Prog Med Genet.* 1973; 9:113–158.

42. Beachy PA, Karhadkar SS, Berman DM. Tissue repair and stem cell renewal in carcinogenesis. *Nature.* 2004; 432:324–331.

43. Okita K, Ichisaka T, Yamanaka S. *Nature.* 2007; 448:313–317.

44. Wernig M, Meissner A, Foreman R, et al. in vitro reprogramming of fibroblasts into a pluripotent ES-cell-like state. *Nature.* 2007; 448:318–324.

45. Takahashi K, Yamanaka S. Induction of pluripotent stem cells from mouse embryonic and adult fibroblast cultures by defined factors. *Cell.* 2006; 126:663–676.

46. Houghton J, Stoicov C, Nomura S, et al. Gastric cancer originating from bone marrow–derived cells. *Science.* 2004; 306:1568–1571.

47. Makino, S. Further evidence favoring the concept of the stem cell in ascites tumors of rats. *Ann N Y Acad Sci.* 1956; 63:818–830.

48. Kleinsmith LJ, Pierce GB. Multipotentiality of single embryonal carcinoma cells. *Cancer Res.* 1964; 24:1544–1551.

49. Baum CM, Weissman IL, Tsukamoto AS, Buckle AM, Peault B. Isolation of a candidate human hematopoietic stem-cell population. *Proc Natl Acad Sci U S A.* 1992; 89:2804–2808.

50. Spangrude GJ, Aihara Y, Weissman IL, Klein J. The stem cell antigens Sca-1 and Sca-2 subdivide thymic and peripheral T lymphocytes into unique subsets. *J Immunol.* 1988; 141:3697–3707.

51. Civin CI, Gore SD. Antigenic analysis of hematopoiesis: a review. *J Hematother.* 1993; 2:137–144.

52. Spits H, Lanier LL, Phillips JH. Development of human T and natural killer cells. *Blood.* 1995; 85:2654–2670.

53. Matsui W, Huff CA, Wang Q, et al. Characterization of clonogenic multiple myeloma cells. *Blood.* 2004; 103:2332–2336.

54. Hosen N, Park CY, Tatsumi N, et al. CD96 is a leukemic stem cell–specific marker in human acute myeloid leukemia. *Proc Natl Acad Sci U S A.* 2007; 104:11008–11013.

55. Smalley M, Ashworth A. Stem cells and breast cancer: a field in transit. *Nat Rev Cancer.* 2003; 3:832–844.

56. Singh SK, Hawkins C, Clarke ID, et al. Identification of human brain tumor initiating cells. *Nature.* 2004; 432:396–401.

57. Bonnet D, Dick JE. Human acute myeloid leukemia is organized as a hierarchy that originates from a primitive hematopoietic cell. *Nat Med.* 1997; 3:730–737.

58. Collins AT, Berry PA, Hyde C, Stower MJ, Maitland NJ. Prospective identification of tumorigenic prostate cancer stem cells. *Cancer Res.* 2005; 65:10946–10951.

59. Kim CF, Jackson EL, Woolfenden AE, et al. Identification of bronchioalveolar stem cells in normal lung and lung cancer. *Cell.* 2005; 121:823–835.

60. Suetsugu A, Nagaki M, Aoki H, Motohashi T, Kunisada T, Moriwaki H. Characterization of CD133$^+$ hepatocellular carcinoma cells as cancer stem/progenitor cells. *Biochem Biophys Res Commun.* 2006; 351:820–824.

61. Li C, Heidt DG, Dalerba P, et al. Identification of pancreatic cancer stem cells. *Cancer Res.* 2007; 67:1030–1037.

62. Dalerba P, Dylla SJ, Park IK, et al. Phenotypic characterization of human colorectal cancer stem cells. *Proc Natl Acad Sci U S A.* 2007; 104:10158–10163.

63. O'Brien CA, Pollett A, Gallinger S, Dick JE. A human colon cancer cell capable of initiating tumor growth in immunodeficient mice. *Nature.* 2007; 445:106–110.

64. Ricci-Vitiani L, Lombardi DG, Pilozzi E, et al. Identification and expansion of human colon-cancer-initiating cells. *Nature.* 2007; 445:111–115.

65. Ma S, Chan KW, Hu L, et al. Identification and characterization of tumorigenic liver cancer stem/progenitor cells. *Gastroenterology.* 2007; 132:2542–2556.

66. Hill RP. Identifying cancer stem cells in solid tumors: case not proven. *Cancer Res.* 2006; 66:1891–1895.

67. Bapat SA, Mali AM, Koppikar CB, Kurrey NK. Stem and progenitor-like cells contribute to the aggressive behavior of human epithelial ovarian cancer. *Cancer Res.* 2005; 65:3025–3029.

68. Nowell PC. The clonal evolution of tumor cell populations. *Science.* 1976; 194:23–28.

69. Kruh GD. Introduction to resistance to anticancer agents. *Oncogene.* 2003; 22:7262–7264.

70. Yen TH, Wright TA. The gastrointestinal tract stem cell niche. *Stem Cell Rev.* 2006; 2:203–212.

71. Miyata T. Asymmetric cell division during brain morphogenesis. *Prog Mol Subcell Biol.* 2007; 45:121–142.

72. Zhong W. The mammalian Golgi regulates numb signaling in asymmetric cell division by releasing ACBD3 during mitosis. *Cell.* 2007; 129:163–178.

73. Shinin V, Gayraud-Morel B, Gomès D, Tajbakhsh S. Asymmetric division and cosegregation of template DNA strands in adult muscle satellite cells. *Nat Cell Biol.* 2006; 8:677–687.

74. Nikolaou S, Gasser RB. Extending from PARs in *Caenorhabditis elegans* to homologues in *Haemonchus contortus* and other parasitic nematodes. *Parasitology.* 2007; 134:461–482.

75. Beckmann J, Scheitza S, Wernet P, Fischer JC, Giebel B. Asymmetric cell division within the human hematopoietic stem and progenitor cell compartment: identification of asymmetrically segregating proteins. *Blood.* 2007; 109:5494–5501.

76. Rebollo E, Sampaio P, Januschke J, Llamazares S, Varmark H, González C. Functionally unequal centrosomes drive spindle orientation in asymmetrically dividing *Drosophila* neural stem cells. *Dev Cell.* 2007; 12:467–474.

77. Yamashita YM, Mahowald AP, Perlin JR, Fuller MT. Asymmetric inheritance of mother versus daughter centrosome in stem cell division. *Science.* 2007; 315:518–521.

78. Kuang S, Kuroda K, Le Grand F, Rudnicki MA. Asymmetric self-renewal and commitment of satellite stem cells in muscle. *Cell.* 2007; 129:999–1010.

79. Li L, Neaves WB. Normal stem cells and cancer stem cells: the niche matters. *Cancer Res.* 2006; 66:4553–4557.

80. Clarke MF, Fuller M. Stem cells and cancer: two faces of Eve. *Cell.* 2006; 124:1111–1115.

81. Caussinus E, Hirth F. Asymmetric stem cell division in development and cancer. *Prog Mol Subcell Biol.* 2007; 45:205–225.

82. Hochedlinger K, Yamada Y, Beard C, Jaenisch R. Ectopic expression of Oct-4 blocks progenitor-cell differentiation and causes dysplasia in epithelial tissues. *Cell.* 2005; 121:465–477.

83. Knudson AG. Antioncogenes and human cancer. *Proc Natl Acad Sci U S A.* 1993; 90:10914–10921.

84. Gostjeva EV, Thilly WG. Stem cell stages and the origins of colon cancer: a multidisciplinary perspective. *Stem Cell Rev.* 2005;1:243–251.

85. Campbell PA, Perez-Iratxeta C, Andrade-Navarro MA, Rudnicki MA. Oct4 targets regulatory nodes to modulate stem cell function. *PLoS ONE.* 2007; 2:e553.

86. Boyer LA, Lee TI, Cole MF, et al. Core transcriptional regulatory circuitry in human embryonic stem cells. *Cell.* 2005; 122:1–10.

87. Loh YH, Wu Q, Chew JL, et al. The Oct4 and Nanog transcription network regulates pluripotency in mouse embryonic stem cells. *Nat Genet.* 2006; 38:431–440.

88. Rodda DJ, Chew JL, Lim LH, et al. Transcriptional regulation of nanog by Oct4 and Sox2. *J Biol Chem.* 2005; 280:24731–24737.

89. Constantinescu S. Stemness, fusion and renewal of hematopoietic and embryonic stem cells. *J Cell Mol Med.* 2003; 7:103–112.

90. Ying QL, Nichols J, Chambers I, Smith A. BMP induction of Id proteins suppresses differentiation and sustains embryonic stem cell self-renewal in collaboration with Stat3. *Cell.* 2003; 115:281–292.

91. Masui S, Nakatake Y, Toyooka Y, et al. Pluripotency governed by Sox2 via regulation of Oct3/4 expression in mouse embryonic stem cells. *Nat Cell Biol.* 2007; 9:625–635.

92. Fuhrmann G, Chung AC, Jackson KJ, et al. Mouse germline restriction of Oct4 expression by germ cell nuclear factor. *Dev Cell.* 2001; 1:377–387.

93. Barnea E, Bergman Y. Synergy of SF1 and RAR in activation of Oct-3/4 promoter. *J Biol Chem.* 2000; 275:6608–6619.

94. Okumura-Nakanishi S, Saito M, Niwa H, Ishikawa F. Oct-3/4 and Sox2 regulate *Oct-3/4* gene in embryonic stem cells. *J Biol Chem.* 2005; 280:5307–5317.

95. Gu P, LeMenuet D, Chung AC, Mancini M, Wheeler DA, Cooney AJ. Orphan nuclear receptor GCNF is required for the repression of pluripotency genes during retinoic acid–induced embryonic stem cell differentiation. *Mol Cell Biol.* 2005; 25(19):8507–8519.

96. Pietersen AM, van Lohuizen M. Stem cell regulation by polycomb repressors: postponing commitment. *Curr Opin Cell Biol.* 2008; 20:201–207.

97. Lee TI, Jenner RG, Boyer LA, et al. Control of developmental regulators by Polycomb in human embryonic stem cells. *Cell.* 2006; 125:301–313.

98. Azuara V, Perry P, Sauer S, et al. Chromatin signatures of pluripotent cell lines. *Nat Cell Biol.* 2006; 8:532–538.

99. Johnson BV, Rathjen J, Rathjen PD. Transcriptional control of pluripotency: decisions in early development. *Curr Opin Genet Dev.* 2006; 16:447–454.

100. 100.Ohm JE, McGarvey KM, Yu X, et al. A stem cell–like chromatin pattern may predispose tumor suppressor genes to DNA hypermethylation and heritable silencing. *Nat Genet.* 2007; 39:237–242.

101. Ho AD, Wagner W. The beauty of asymmetry: asymmetric divisions and self-renewal in the haematopoietic system. *Curr Opin Hematol.* 2007; 14:330–336.

102. Watt FM, Hogan BL. Out of Eden: stem cells and their niches. *Science.* 2000; 287:1427–1430.

103. Scadden DT. The stem-cell niche as an entity of action. *Nature.* 2006; 441:1075–1079.

104. Ohyama, M, Terunuma A, Tock CL, et al. Characterization and isolation of stem cell–enriched human hair follicle bulge cells. *J Clin Invest.* 2006; 116:249–260.

105. Bach SP, Renehan AG, Potten CS. Stem cells: the intestinal stem cell as a paradigm. *Carcinogenesis.* 2000; 21:469–476.

106. Ganss R. Tumor stroma fosters neovascularization by recruitment of progenitor cells into the tumor bed. *J Cell Mol Med.* 2006; 10:857–865.

107. Giangreco A, Groot KR, Janes SM. Lung cancer and lung stem cells: strange bedfellows? *Am J Respir Crit Care Med.* 2007; 175:547–553.

108. Meuwissen R, Linn SC, Linnoila RI, Zevenhoven J, Mooi WJ, Berns A. Induction of small cell lung cancer by somatic inactivation of both Trp53 and Rb1 in a conditional mouse model. *Cancer Cell.* 2003; 4:181–189.

109. Minna JD, Kurie JM, Jacks T. A big step in the study of small cell lung cancer. *Cancer Cell.* 2003; 4:163–166.

110. Reynolds SD, Giangreco A, Power JH, Stripp BR. Neuroepithelial bodies of pulmonary airways serve as a reservoir of progenitor cells capable of epithelial regeneration. *Am J Pathol.* 2000; 156:269–278.

111. Hong KU, Reynolds SD, Watkins S, Fuchs E, Stripp BR. Basal cells are a multipotent progenitor capable of renewing the bronchial epithelium. *Am J Pathol.* 2004; 164:577–588.

112. Schoch KG, Lori A, Burns KA, Eldred T, Olsen JC, Randell SH. A subset of mouse tracheal epithelial basal cells generates large colonies *in vitro. Am J Physiol Lung Cell Mol Physiol.* 2004; 286:631–642.

113. Johnson L, Mercer K, Greenbaum D, et al. Somatic activation of the K-ras oncogene causes early onset lung cancer in mice. *Nature.* 2001; 410:1111–1116.

114. Stevens TP, McBride JT, Peake JL, Pinkerton KE, Stripp BR. Cell proliferation contributes to PNEC hyperplasia after acute airway injury. *Am J Physiol.* 1997; 272:486–493.

115. Giangreco A, Reynolds SD, Stripp BR. Terminal bronchioles harbor a unique airway stem cell population that localizes to the bronchoalveolar duct junction. *Am J Pathol.* 2002; 161:173–182.

116. Walkley CR, Olsen GH, Dworkin S, et al. A microenvironment-induced myeloproliferative syndrome caused by retinoic acid receptor γ deficiency. *Cell.* 2007; 129:1097–1110.

117. Walkley CR, Shea JM, Sims NA, Purton LE, Orkin SH. Rb regulates interactions between hematopoietic stem cells and their bone marrow microenvironment. *Cell.* 2007; 129:1081–1095.

118. Nichols J, Chambers I, Taga T, Smith A. Physiological rationale for responsiveness of mouse embryonic stem cells to gp130 cytokines. *Development.* 2001; 128:2333–2339.

119. Batlle E HJ, Beghtel H, van den Born MM, et al. β-Catenin and TCF mediate stem cell positioning in the intestinal epitheleum by controlling the expression of EphB/ephrinB. *Cell.* 2002; 111:251–263.

120. Boiani M, Scholer H. Regulatory networks in embryo-derived pluripotent stem cells. *Nat Rev Mol Cell Biol.* 2005; 6:872–884.

121. Chickarmane V, Troein C, Nuber UA, Sauro HM, Peterson C. Transcriptional dynamics of the embryonic stem cell switch. *PLoS Comput Biol.* 2006; 2:1080–1092.

122. Burdon T, Smith A, Savatier P. Signalling, cell cycle, and pluripotency in embryonic stem cells. *Trends Cell Biol.* 2002; 12:432–438.

123. Doucas H, Garcea G, Neal CP, Manson MM, Berry DP. Changes in the Wnt signalling pathway in gastrointestinal cancers and their prognostic significance. *Eur J Cancer.* 2005; 41:365–379.

124. Kolligs FT, Bommer G, Goke B. Wnt/beta-catenin/tcf signaling: a critical pathway in gastrointestinal tumorigenesis. *Digestion.* 2002; 66:131–144.

125. Rask K, Nilsson A, Brannstrom M, et al. Wnt-signalling pathway in ovarian epithelial tumors: increased expression of beta-catenin and GSK3beta. *Br J Cancer* 2003; 89:1298–1304.

126. Tickle, C. Patterning systems—from one end of the limb to the other. *Dev Cell.* 2003; 4:449–458.

127. Katano, M. Hedgehog signaling pathway as a therapeutic target in breast cancer. *Cancer Lett.* 2005; 227:99–104.

128. Sanchez P, Clement V, Ruiz i Altaba A. Therapeutic targeting of the Hedgehog-GLI pathway in prostate cancer. *Cancer Res.* 2005; 65:2990–2992.

129. Hopfer O, Zwahlen D, Fey MF, Aebi S. The Notch pathway in ovarian carcinomas and adenomas. *Br J Cancer.* 2005; 93:709–718.

130. Wilson A, Radtke F. Multiple functions of Notch signaling in self-renewing organs and cancer. *FEBS Lett.* 2006; 580:2860–2868.

131. Duncan AW, Rattis FM, DiMascio LN, et al. Integration of Notch and Wnt signaling in hematopoietic stem cell maintenance. *Nat Immunol.* 2005; 6:314–322.

132. Ross J, Li L. Recent advances in understanding extrinsic control of hematopoietic stem cell fate. *Curr Opin Hematol.* 2006; 13:237–242.

2 Leukemic Stem Cells

SHARMILA BAPAT

2.1 INTRODUCTION

The concept of stem cell involvement in the regeneration of adult tissue arose in the early twentieth century. However, it was not until a few decades later, in the early 1960s, that the idea was shown to be a reality. Becker and co-workers demonstrated the clonal nature of a small group of quiescent stem cells present in the bone marrow, which, on stimulation, could also give rise to various types of blood cells.[1] Along similar lines, the concept that cancer could be a stem cell disease was hypothesized for a few decades before definitive proof-of-concept experiments established unequivocally the involvement of stem cells in initiating a tumorigenic state, following the identification of similarities between hematopoietic stem cells (HSCs) and their leukemic counterparts, which propagate the disease successfully.[2]

Our current understanding of leukemic stem cells (LSCs) is thus based on studies spanning almost a century of research in hematology and leukemia. The development of clonal assays for all major hematopoietic lineages, together with the availability of multiparameter fluorescence-activated cell sorting (FACS) has enabled the prospective purification and/or enrichment of HSCs. Moreover, the cell surface expression of specific molecules of the various cellular derivatives of the classical HSC hierarchy, including stem cells, progenitors, and differentiated cells and their functional readouts [e.g., as in vitro colony-forming units (CFUs) in colony-initiating assays as well as competitive repopulating assays in animal models][3−5] further enables the reproducibility of data. These developments have now not only been applicable to leukemia research, but the techniques established for the identification of leukemia-initiating cells are considered as the "gold standards" of cancer stem cell research and have been extended to the field of solid tumors, such as brain, breast, ovary, colon, pancreas, lung, and prostate cancer. The importance of studies in LSCs is heightened further due to the realization that almost every important aspect of cancer stem cell function, such as self-renewal, cell signaling, maturation arrest, and generation of aberrant cellular hierarchies,

Cancer Stem Cells: Identification and Targets, Edited by Sharmila Bapat

identified as being dysregulated in leukemia, are also applicable to the remaining cancers. Thereby, LSC research is considered as a path-breaking model in cancer stem cell research.[6]

2.2 DYSREGULATION OF HEMATOPOIESIS IN LEUKEMIA

Development from an HSC to a mature cell of the hematolymphoid system involves dynamic loss of self-renewal capacity, proliferation ability, and lineage potentials that gradually shift from pluripotent to multipotent and finally, unipotent capacities. In leukemias, although this downshift may not be altered completely, a retention or restoration of the self-renewal and proliferative capabilities is evident. However, before attempting to identify the extent of disruption of normal hematopoiesis in leukemia, it is necessary to understand how both processes are regulated on the cellular and molecular levels.

2.2.1 Normal Hematopoietic Stem Cell Hierarchies

During a human lifetime, a huge mass of blood cells of different types must be produced. This massive cell production relies on a small fraction of cells in the bone marrow (i.e., the HSCs) and is maintained through a continuous cell turnover involving cell proliferation, differentiation, and apoptosis. Although HSCs reside predominantly in the bone marrow, they are also found in peripheral blood at very low densities, where they play a physiological role in certain situations, such as functional reengraftment of unconditioned bone marrow.[7]

A myriad of molecular and cellular factors are involved in maintaining the steady state of hematopoietic tissue. Despite this complex regulation, malignancy is usually avoided because the cell turnover is organized ingeniously as a hierarchy that is quite similar to the one outlined in Fig. 2.1. At the base of this hierarchy is a self-renewing long-term stem cell that retains HSC properties (LT-HSC) which undergoes an asymmetric division to generate another LT-HSC and a short-term stem cell (ST-HSC). The ST-HSC, in turn, retains self-renewal for a short time, although these are believed to be symmetric divisions and give rise to two ST-HSC cells.[8] ST-HSCs further generate the multipotential progenitors (MPPs), which undergo differentiation to give rise to oligolineage-restricted progenitors through functionally irreversible maturation steps[9] (Fig. 2.1).

Two types of oligolineage-restricted progenitors have been identified in the hematopoietic system: the common lymphoid progenitors (CLPs), which give rise to T- and B-lymphocytes and natural killer cells,[10] and the common myeloid progenitors (CMPs), which generate the myelomonocytic progenitors (GMPs) and the megakaryotic erythroid progenitors (MEPs). The former, on differentiation, produce monocytes, macrophages, and granulocytes. The MEPs, while preferentially generating megakaryocytes, platelets, and erythrocytes, also maintain the potential for B-cell lineage differentiation at an extremely low frequency.[11] Interestingly, both CMPs and CLPs can give rise to dendritic cells,[12,13] suggesting the existence

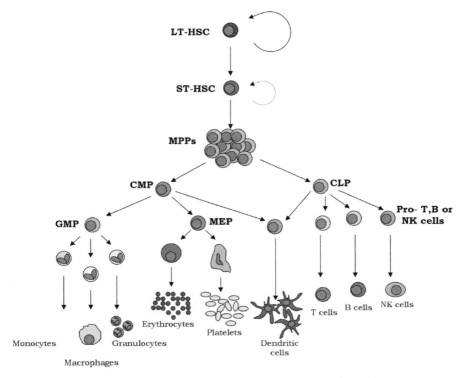

FIGURE 2.1   Normal hematopoietic cell hierarchy. (*See insert for color representation of figure.*)

of an alternative commitment pathway to the mutually exclusive developmental pathways for the myeloid and lymphoid lineages.

It is thus apparent that resurrection of the hematopoietic orchestra involves a multitude of cell progenitors that are poised at specific locations along a defined trajectory to yield a wide array of differentiated blood cell types while maintaining the HSC pool. Using specific combinations of cell surface markers, the hematopoietic progenitor cells are separable as pure populations and have been shown to be devoid of detectable self-renewal activity but are characterized by increased mitotic activity.[14]

However, under altered circumstances (e.g., those requiring rapid cell turnover, posttransplantation, during aging), variations in the cell hierarchy do occur. For instance, the LT-HSCs may undergo a symmetrical type of division to yield two LT-HSCs, leading to an overall increase in the numbers of HSCs. However, like the other cell types, this increase is transient, as the number of HSCs is also regulated by programmed cell death or apoptosis.[15] Another deviation that is derived is an exit of the HSCs from a self-renewal mode to generate two progenitor cells that are committed for differentiation; ultimately, this results in clone depletion. The latter variation is believed primarily to be the mechanism by which mutations may naturally be depleted, especially in aging populations.

2.2.2 Understanding Aberrant Hierarchies in Leukemia

Today it is realized that leukemia represents a dysregulation of the homeostatic mechanisms of the bone marrow and lymphatic systems that involves a disruption of the mechanisms regulating the organized hierarchy described above. Certain types of leukemias have thus been described as newly formed aberrant hematopoietic tissue generated from transformed cells that have either retained or reacquired the capacity for self-renewal, proliferation, and resistance to apoptosis through accumulated mutations, and can thus be described as LSCs.

The concept suggests, therefore, that leukemia follows an aberrant and poorly regulated process of organogenesis analogous to that of normal HSCs. The HSC and LSC pools both have extensive self-renewal potential and the ability to give rise to new hematopoietic tissues, with a definitive heterogeneity that correlates with a distinctive set of phenotypic characteristics and proliferative potentials, despite the fact that those are abnormal in the latter case. Some of the heterogeneity in the tumor arises as a result of continuing mutagenesis and/or epigenetic changes, but another portion is likely to arise from aberrant differentiation of the CSCs. Because most leukemias are believed to have a clonal origin, LSCs must give rise to phenotypically diverse progeny, including a few LSCs with an indefinite life span (LT-LSC); short-term progenitors with an extended proliferative potential, depending on the grade of the disease (ST-LSC); and finally, some cells that differentiate and undergo apoptosis. A distinctive difference between the normal and leukemic hierarchy is that the cellular components of the latter remain in circulation longer, with extended proliferation but decreased differentiation capabilities. The leukemic state is thus very efficient in terms of cell mass production but is compromised by the functionality of the cells produced.

The high turnover of cells during hematopoiesis that depends on several cell intrinsic and extrinsic factors, including stromal and microenvironment-related factors, is a tightly controlled process that achieves homeostasis within the body. However, cell division is an inherently error-prone process, and genetic alterations, mostly mutations, are known to arise frequently during DNA replication and in association with aging. A large majority of such mutations are assumed to be lost during consequent cycles of self-renewal or division; alternatively, the mutation may be lethal to the cell and lead to clone depletion. In direct contrast, certain mutations that import pro-survival features to the stem cell, such as a proliferative advantage, enhanced self-renewal, growth factor independence, or resistance to apoptosis, would result in dominance of the cell clone within the heterogeneous population in the bone marrow. Such a dominating clone, with further acquisition of additional transforming features, has the potential to develop into a LSC clone.

Apart from spontaneously occurring gene mutations, chromosomal alterations, including trisomies, insertions, deletions, rearrangements, inversions, and translocations, are known to occur extensively in leukemia; in fact, specific associations have been established as guidelines in disease subtyping. Chromosomal alterations may either be inherited, or induced by such external factors as alkylating drugs, ionizing radiations, and chemicals, and are realized to be an underlying feature in the emergence of a LSC, as they often mediate alterations in the structure and/or regulation of cellular oncogenes. Retroviral infections are also believed to be capable of leading to leukemogenesis via an insertional mutagenesis of oncogenes.[16] LSC identity is thus largely dictated by the nature of the oncogenic events and how these events perturb the essential processes of self-renewal, proliferation, differentiation, and survival. The genetic lesions in LSCs result in a block of differentiation (maturation arrest) that allows cells belonging to a certain clone not only to continue proliferation and to resist apoptosis, but also to accumulate in large numbers, which is not a feature associated with normal hematopoietic cellular hierarchies.

Increasing evidence now indicates that three key biological changes affect the development of leukemic clones: (1) an increased probability of differentiation defects at the level of the most primitive LSCs or early progenitors, (2) an increased turnover rate of the leukemic progenitors at all stages of differentiation, and (3) an increased ability to survive under conditions of growth factor deprivation. Such a model accounts for the long latent period for the development of certain leukemias, such as chronic myeloid leukemia (CML), as well as suggesting why stem cells may persist in large numbers but still fail to compete in contributing to the daily output of mature blood cells in patients with disease.[17]

Another requirement for HSCs and LSCs is the ability to avoid telomere shortening through the action of the telomerase complex.[18,19] Telomerase activity is also repressed during differentiation of maturation-sensitive human tumor cell lines but not in resistant human tumor cell lines.[20] Thus, under normal conditions, an inverse relationship between the degree of differentiation and telomerase activity exists and is regulated in an HSC by restricting the number of cell divisions that it can undergo in a human lifetime. The most important component of the telomerase complex is human telomerase reverse transcriptase (hTERT), which is often overexpressed in tumor cells. Alternatively, telomerase activity and hTERT mRNA expression may not be increased significantly, but variations in mRNA splicing may play a significant role in determining the overall length of telomeres and thereby the functional hTERT levels that restrict telomere shortening despite a rapid turnover of cells, and lengthen the life span of the transformed clone.[21]

2.2.3 Types of Leukemia

Leukemia thus represents a heterogeneous group of monoclonal diseases that arise from hematopoietic stem and progenitor cells in the bone marrow or other hematopoietic organs.[22] As described above, the underlying causes for clonal expansion are poorly understood but definitely involve several genetic and epigenetic changes. All these events culminate in an imbalance created by the accumulation of large numbers of cells of an individual lineage. In addition, the abnormal cells may secrete factors that perturb the homeostatic mechanisms and inhibit normal hematopoiesis, thus generating the symptoms of leukemia.

Pancytopenia is typical and results in part from the physical replacement of normal marrow elements by immature cells. Following replacement of the bone marrow, the abnormal cells spill into the circulation and infiltrate other organs.

The various types of leukemia are identified primarily by the identity of the target cell that undergoes transformation. When the cancer develops in the lymphocytes (lymphoid lineage), it is called *lymphocytic leukemia*, while cancer of the granulocytes or monocytes (myeloid lineage) is described as *myelogenous leukemia*. The rate of disease progression is an additional feature used for classification. In chronic leukemias, the immature hematopoietic progenitors (colloquially referred to as *blasts*) accumulate slowly and continue to form functional hematopoietic cells; because of this, disease progression is slow. On the other hand, in acute leukemias, the progenitors proliferate rapidly, resulting in blast accumulation, decreased turnover of mature hematopoietic cells, and rapid disease progression. Normal and leukemic severe compromised immune deficiency (SCID)-repopulating cells (SRCs) coexist in the bone marrow and peripheral blood from patients in the chronic phase, whereas a dominance of leukemic SRCs is evident in blast crisis.[23] Thus, the main classes of leukemia include (1) chronic myeloid leukemia, (2) acute myeloid leukemia, (3) chronic lymphocytic leukemia, (4) acute lymphoblastic/lymphocytic leukemia.

Chronic Myeloid Leukemia (CML)   CML is a malignant, clonal myeloproliferative disorder, affecting the myeloid, erythroid, and megakaryocytic blood elements, that accounts for 15 to 20 % of all cases of leukemia. This was the first hematological malignancy to be identified in association with a specific chromosomal abnormality[24] and involves a translocation between chromosomes 9 and 22, now recognized as the Philadelphia chromosome (Ph1). The molecular consequence of Ph1 is the fusion of the *ABL* gene (chromosome 9) to the *BCR* (breakpoint cluster region on chromosome 22) sequences, giving rise to a chimeric *BCR-ABL* gene. The product of this fusion functions as a constitutively activated tyrosine kinase and is involved in an endless list of pathways found to be altered in CML (including transforming functions). Consequently, the BCR-ABL kinase becomes a molecular target for drug design as well as monitoring therapeutic efficacies. Existing therapies for CML include the use of busulfan, hydroxyurea, interferon-α, imanitib, or allogenic stem cell transplantation.

In early chronic-phase CML, the apoptosis and differentiation pathways are often intact; however, the disease invariably progresses from a chronic phase to an accelerated acute phase, also referred to as *blast phase* or *blast crisis*. This is accompanied by a loss of capacity of the transformed cell clone for terminal differentiation and is characterized by massive blast and promyelocyte counts, leukocytosis, and splenomegaly, along with acquisition of additional cytogenetic abnormalities in addition to the Ph1 abnormality. Although initially it was believed that CML arises as a consequence of the clonal expansion of HSCs that express the *BCR-ABL* fusion gene,[25-29] it is now realized that progression to blast crisis is associated with expansion of the myeloid progenitor fraction, which consists mainly of GMPs that acquire some degree of self-renewal capacity rather than expansion of their own pool.[30–34] Moreover, although *BCR-ABL* correlates with the development of CML, additional genetic and/or epigenetic events are required for progression to blast crisis, including further *BCR-ABL* amplification and activation of β-catenin in the GMPs,[34,35] which effectively result in acquisition of resistance to apoptosis and an escape from innate and adaptive immune responses.

Acute Myeloid Leukemia (AML)   AML has been described as a clonal disorder in which transformation and uncontrolled proliferation of an abnormally differentiated, long-lived myeloid progenitor cell results in high circulating numbers of immature blood forms and replacement of normal marrow by malignant cells. A detailed analysis of the various chromosomal alterations in this disorder suggest that AML alleles may be divided broadly into two categories:

1. *Genetic alterations conferring a proliferative and/or survival advantage on hematopoietic precursors.* These involve either *gain of function* or activating mutations in *KIT, FLT3, SHP2,* or *RAS* family members; or *loss of function* or knock-out mutations, as in *NF1.*

2. *Genetic alterations impairing hematopoietic differentiation.* The best studied of such aberrations in AML include multiple translocations targeting the core binding factor (CBF), and those involving the retinoic acid receptor alpha (RARα) locus. CBF is a heterodimeric transcription factor comprising *AML1 (RUNX1)* and CBFβ subunits and is necessary for normal hematopoietic differentiation. Multiple translocations that target *CBF*, including *AML1-ETO, CBFβ-SMMHC,* and *TEL-AML1*, result in a loss of function of CBF, and consequently, an imbalance in definitive hematopoietic development. Multiple translocations involving the RARα locus with either *PML, PLZF, NPM, NuMA,* and *STAT5B:X* genes have been identified in AML.[36] Most, if not all, of these have been reported to result in a block of differentiation at the promyelocyte stage.

However, it is realized that no single type of chromosomal aberration can be considered as being pivotal toward the development of AML; extensive genotyping studies, in fact, lead to the reasoning that the pathogenesis of AML involves more than one mutation. AML may thus be concluded to represent a complex and heterogeneous disease resulting from multiple mutations, deletions, numerical aberrations, and translocations.

Chronic Lymphocytic Leukemia (CLL)   CLL represents a monoclonal expansion of lymphocytes; in a majority of cases, B-lymphocytes are involved. The neoplastic cell is a hypoproliferative, immunologically incompetent small lymphocyte. A primary involvement of the bone marrow is characteristic of the disease, with secondary release into the peripheral blood. The circulating lymphocytes further selectively infiltrate the lymph nodes, spleen, and liver. Constitutive integrin expression, activation state of the cell, and effective modification of integrin-mediated function by cytokines determine the different patterns of clinical disease observed in CLL.[37] With disease progression, symptoms of lymphadenopathy, splenomegaly, hepatomegaly, and secondary immune deficiency with hypogammaglobulinemia are frequently observed. Over the last decade, it has become apparent that there is diversity in types of CLL in patients based on the occurrence of hypermutation of their immunoglobulin genes.[38,39] CD38 expression has been described as an important prognostic marker in CLL,[40] while CD40–CD154 interactions are important within and among the malignant cells, and lead to occurrence of autoimmune syndromes in some cases of CLL.[41]

B-cell chronic lymphocytic leukemia (B-CLL) is a disease of proliferating mono-clonal B-cells that have a distinctive phenotype[42] (i.e., CD19^{hi}CD5$^+$ CD23$^+$ IgM^{lo}). The progressive rise of mature CD5$^+$ B-lymphocytes, despite the low proportion of proliferating cells, has led to the notion that B-CLL is related primarily to defective apoptosis. Since the chemokine receptor CXCR4 (impor-tant in homing of normal HSCs, progenitor cells, lymphocytes, and monocytes) is overexpressed and functionally active in B-CLL, it may contribute to the tropism of B-CLL cells for the bone marrow stroma.[43] An in-depth analysis of individual VH rearrangements in patients with mutated and germline CLL revealed that the latter group expressed a rearranged VH sequence usage pat-tern similar to that reported in ALL patients, whereas the mutated CLL group aligned with the controls. This suggests that germline B-CLL may derive from a pool that has been unable to follow or complete the normal pathway of B-cell differentiation.[44] However, CFU-GMs are not increased in peripheral blood of patients with CLL,[45] indicating perhaps that the block occurs farther down-stream in the differentiation hierarchy. Moreover, the microenvironment is also likely to play a prominent role in disease manifestation because the malignant cells are observed to accumulate progressively invivo, whereas they undergo rapid spontaneous apoptosis when cultured in vitro.[46]

Hairy cell leukemia (HCL) is an uncommon malignant disorder of small B-lymphocytes that gets its name from the presence of cytoplasmic projections in these cells. The association of hairy cell leukemia with other neoplasms, mainly non-Hodgkin's lymphomas, is well known; it has recently been suggested that association of HCL and small lymphocytic lymphoma might be much more frequent than expected.[47]

T-cell malignancies are rare and include many different types of disorders, such as T-cell prolymphocytic leukemia, T-cell large granular lymphocytic leukemia, adult T-cell leukemia/lymphoma, cutaneous T-cell lymphoma, and peripheral T-cell lymphoma.[48] The mature T- and NK-cell leukemias are even more uncommon neoplasms derived from mature or postthymic T-cells.[49]

Acute Lymphoblastic/Lymphocytic Leukemia (ALL)   ALL is a malignant clonal disorder of the bone marrow lymphopoietic precursor cells that results in a high number of circulating blasts that lack the potential for differentiation and maturation. An inhibition of the normal development of hematopoietic cell elements occurs, followed by replacement of the normal marrow by malignant cells, and the potential for leukemic infiltration of the central nervous system and abdominal organs. The clinical presentation is dominated by progressive weakness and fatigue secondary to anemia, infection secondary to leukopenia, and bleeding secondary to thrombocytopenia. When 50 % of the bone marrow is replaced, peripheral blood cytopenias are observed.

T-cell prolymphocytic leukemia (T-PLL) is a rare aggressive postthymic malignancy associated with chromosomal translocations involving the T-cell receptor gene and one of two proto-oncogenes (TCL-1 and MTCP-1) in the majority of cases.[48] In some cases of T-PLL, the leukemic cells undergo secondary transformation, resulting in different phenotypes of cutaneous T-cell lymphoma, such as mycosis fungoides, lymphomatoid papulosis, and cutaneous CD30$^+$ anaplastic large cell lymphoma.[50]

Adult T-cell leukemia–lymphoma (ATLL) is a lymphoproliferative disease of malignant T-cells, often associated with a retroviral infection by human T-cell leukemia. Unlike most leukemias, bone marrow infiltration is not always present in adult T-cell leukemia–lymphoma, although lymph nodes are widely affected.

T-cell acute lymphoblastic leukemia (T-ALL) is a form of pediatric leukemia that is reported to be caused by approximately 12 distinct chromosomal translocations that lead to aberrant expression of as many different cellular genes.[51] Recent work has shown that the majority of human acute T-cell lymphoblastic leukemias and T-ALL have gain-of-function mutations in NOTCH1, a type I transmembrane receptor that normally signals through a γ-secretase-dependent mechanism and relies on ligand-induced regulated intramembrane proteolysis.[52]

2.3 IDENTIFICATION AND ISOLATION OF CANCER-INITIATING CELLS IN LEUKEMIA

From previous sections it becomes obvious that LSCs are defined as rare, self-renewing cells that (1) keep on accumulating during late-stage disease (crisis); (2) generate a phenotypic, morphologic, and functional leukemic cell hierarchy; (3) express lineage-specific antigens; and (4) exhibit more altered microenvironmental interactions than do normal HSCs. However, specifically targeting LSCs in order to eradicate the disease starting from its source necessitates their identification as a first step. One of the earliest markers reported was cytoplasmic aldehyde dehydrogenase (ALDH) activity exclusively in hematopoietic stem and progenitor cells; this was seen to be down-regulated further during differentiation.[53] Since then, ALDH activity has also been reported in LSCs, with a superior NOD/SCID engrafting potential.[54,55]

The association of distinct molecules on the cell membrane, with specific stages in the normal cellular regenerative hierarchy, has served to provide some markers for LSCs. The first such report was the association of expression of CD34$^+$–CD38$^-$ with AML-initiating cells that had a capacity to home and repopulate the bone marrow on transplantation and generate leukemic blasts in SCID mouse models.[56] It was further shown that this cell could generate the entire leukemic population in a hierarchial manner similar to the normal hematopoietic one.[2] Soon after, cells with the same immunophenotype were also identified in ALL.[57]

A limitation of this identification scheme was that the same cell surface markers were associated with both normal and LSCs. This necessitated a more stringent definition of the LSC than the $CD34^+CD38^-$ phenotype. Further research thus led to reports of additional putative markers, including CD123,[58] CD13, and CD44, variable amounts of CD33, c-kit, and CD133,[59] and adenosine triphosphate–binding cassette (ABC) transporters.[60] The expression of CD44 has been identified specifically in association with its role as a key regulator of AML-LSCs and is reported to be involved in the interaction of LSCs with the niche toward maintaining their stem cell properties; it has further been suggested to provide a therapeutic target for the eradication of LSCs.[61]

More recently, on employing a signal sequence trap strategy to isolate cell surface molecules expressed on human AML-LSCs, CD96 has been suggested to be a promising LSC-specific marker.[62] CD96 is a cell surface antigen that plays a role in tumor immunosurveillance[63] and has now been reported to be useful in characterizing a subgroup of immature AML at the divergence of T-cell and myeloid lineage.[64] FACS analysis demonstrated a higher frequency of CD96 expression in AML samples than the normal HSC-enriched populations ($Lin^-CD34^+CD38^-CD90^+$). Transplantation of $CD96^+$ and $CD96^-$ cell fractions into irradiated newborn $Rag2^{-/-}$ $\gamma^{-/-}$ mice showed significant levels of engraftment in bone marrow with only $CD96^+$ cells. Thus, the CD96 molecule is a promising candidate and may serve as an LSC-specific therapeutic target.[62]

Yet another group has recently identified a unique subpopulation of Ph1-positive ALL cells coexpressing markers of endothelial cells (VE-cadherin, PECAM-1, and Flk-1) and committed B-lineage progenitors.[65] Stromal cell contact promoted VE-cadherin, stabilized β-catenin, and up-regulated *BCR-ABL* fusion gene expressions in these Ph^+VE-cadherin$^+$ LSCs, indicating a unique response of this fraction of cells to microenvironment-derived self-renewing and proliferative cues.

2.4 MOLECULAR REGULATION OF ABERRANT HIERARCHIES

Resolution of the developmental hierarchy established by HSCs has been supported by identification of the molecular mechanisms regulating lineage commitment within a hematopoietic system. This regulation involves a dynamic change in the expression of specific transcriptional regulators, growth factors, and their receptors, whose effects determine lineage commitment and cell maturation. Interestingly, the involvement of the same molecules is now being identified with progression and maintenance of stem cells in a leukemic state.

2.4.1 Signaling Pathways Deregulated in Leukemia

The concept of a *stem cell niche* followed observations that stem cell interactions with their specific microenvironmental elements could be central to the regulation and maintenance of stem cell self-renewal and differentiation capacities. The HSC niche is an anatomical unit located in the endosteum within the bone marrow cavity, composed of osteoblasts, osteoclasts, and stromal fibroblasts. Various studies have shown that osteoblasts are important players in providing HSCs with extrinsic cues. The intimate relationship between a stem cell (normal or leukemic) and its microenvironment is brought into view with identification of specific molecular players involved in stem cell homing, migration, and adhesion, and their crosstalk with cellular components of the niche. The molecules involved set off a cascade of signals that are relayed from the microenvironment to the transcription level within the stem cell and finally manifested as cellular effects. The recent example of normal HSCs being tumorigenic under the influence of tumor stroma serves to emphasize the importance of the niche in leukemogenesis[66,67] (detailed in Chapter 1). Moreover, several development-associated transcriptional factor families are also aberrantly implicated in the transformation process.[68] The major signaling pathways identified in normal as well as transformed processes are recognized to involve several key players of the Wnt, Notch, Sonic Hedgehog, TGF-β, RARα, and Stat5 families. Further, crosstalk between pathways has also been noted to be significant in leukemogenesis; for example, deregulation in the form of hyperactive Shh and Wnt with repressed Notch and Hox pathways involving Stat3, Gli3, β-catenin, CyclinD1, Hes1, HoxA10, and p21 are construed to act synergistically and form an important hub in CML progression.[69]

Wnt Signaling   Wnts are a large family of secreted glycoproteins involved in embryonic development, stem cell self-renewal, proliferation, differentiation, and oncogenesis. The classical Wnt signaling cascade inhibits the activity of the enzyme GSK3β, augmenting β-catenin translocation to the nucleus and transcription of the LEF/TCF target genes. β-Catenin has been shown to preserve CD34 expression and impair myelomonocytic differentiation and clonogenic proliferation of AML-CFC or CFU-L in methylcellulose.[70] Wnt signaling is active in CLL; the Wnt/β-catenin–regulated transcription factor LEF-1 and its downstream target cyclin D1 are overexpressed in CLL. The various ligands that initiate the Wnt cascade (i.e., Wnt3, Wnt5b, Wnt6, Wnt10a, Wnt14, and Wnt16, as well as the Wnt receptor Fzd3) are also highly expressed in CLL compared with normal B-cells.[71] Silencing of SFRPs by CpG island methylation is suggested to be a possible mechanism contributing to aberrant activation of Wnt signaling in CLL.[72]

Translocation products (e.g., the fusion proteins in AML) also block hematopoietic differentiation through activation of the Wnt pathway in HSCs.[73,74] AML1-ETO, PML-RARα, or PLZF-RARα induce the expression of Wnt signaling–associated genes.[74,75] Flt3 and Wnt-dependent signaling pathways synergize in some AML patients, resulting in myeloid transformation.[76] Activation of TCF/LEF signaling also decreases growth factor withdrawal–induced apoptosis of normal progenitors[77]; crosstalk between the Wnt and the adhesion-dependent signaling pathways through GSK3β is reported to govern the chemosensitivity of AML.[78] Methylation of Wnt inhibitors is associated with activation of the Wnt-signaling pathway[79] (e.g., *DKK1* methylation in the progression of AML and CBF leukemia[80] and *DKK3* methylation at an early stage of ALL[81]). ALL associated with the t(1;19) fusion protein E2A-Pbx1 targets Wnt16b, which promotes transformation, enhanced cell survival, and leukemogenesis through the canonical Wnt pathway.[82–84]

The noncanonical Wnt/Ca^{2+} pathway also suppress cyclin D1 expression and negatively regulates B-cell proliferation in a cell autonomous manner. Wnt5a hemizygous mice develop myeloid leukemias and clonal B-cell lymphomas that display loss of Wnt5a function in tumor tissues. Further, analysis of human primary leukemias reveals deletion of the WNT5A gene and/or loss of WNT5A expression in a majority of patient samples. These results suggest that Wnt5a suppresses hematopoietic malignancies.[85]

Notch Signaling   The presence of Notch receptors and ligands as well as their downstream targets at critical stages of hematopoietic development is suggestive of the involvement of this signaling pathway in cell-fate decisions during embryonic hematopoiesis, intrathymic T-cell development, B-cell development, and peripheral T-cell function.[86] This pathway is often described as being at a crossroad since it is associated with enhanced self-renewal of HSCs as well as altered lymphoid development, indicating that it may influence both stem cells and the fate of the CLPs.[87] Notch activation in leukemia is reported either through gain-of-function mutations (e.g., in FBW7[88]) or is triggered by fusion proteins [e.g., MLL-MAML2,[89] t(1;22) translocation,[90] t(7;9)[91]] that are commonly associated with T-cell lymphoblastic leukemias and lymphomas (T-ALL). The Notch pathway also acts in a combinatorial manner with developmental/oncogenic transcriptional factors in T-ALL pathogenesis; Notch1 mutations have been identified along with HOX11, HOX11L2 mutations, or SIL-TAL1 expression.[92]

Besides self-renewal, the other effects mediated by Notch signaling in leukemia include the transcriptional control of PTEN and regulation of the PI3K-AKT pathway[93]; ubiquitin-mediated proteolysis of E47 and Tal1/SCL protein[94]; interaction with c-Myc[95]; positive regulation of the mTOR pathway[96]; maintenance of high SKP2 levels to reduce p27Kip1, which leads to more rapid cell cycle progression[94]; Hes1 activation, which mediates growth arrest and apoptosis in B-cells[97]; and so on. A finely tuned interplay between Notch3 and pre-TCR pathways is known to converge on regulation of NF-κB activity, leading to differential NF-κB subunit dimerization, which regulates distinct gene clusters involved in either cell differentiation or proliferation/leukemogenesis.[98]

Clonal expansion and transformation mediated by Notch are dependent on pre-T-cell-specific (TCR) signals associated with the development of CD4$^+$CD8$^+$ T-cells in the bone marrow,[99] including PKCθ,[100] which mediates activation of NF-κB, inhibition of the E2A pathway, and cooperation with Myc, E2A-PBX, and Ikaros dysregulation.[101] Thereby, the NF-κB pathway is realized to be a potential target of future therapies of T-ALL.[102] Notch activation induced by its ligand, Delta-1, is also reported to reduce TNF-α-induced growth suppression and apoptosis through suppression of the sequential activation of caspase 8, caspase 3, and PARP induced by TNF-α.[103] Sustained GATA-2 expression by Notch1 also inhibits differentiation of hematopoietic cells[104]; yet other pathways regulated by Notch1 mediate differentiation-independent cell death.[105]

Other Pathways   TGF-β has been shown to inhibit the proliferation of quiescent hematopoietic stem cells and stimulate differentiation of late progenitors to erythroid and myeloid cells. Insensitivity to TGF-β is implicated in the pathogenesis of many myeloid and lymphoid neoplasms. Loss of extracellular TGF receptors and disruption of intracellular TGF-β signaling by oncogenes is seen in a variety of malignant and premalignant states.[106] Expression of the fusion protein Bcr-Abl leads to hyperresponsiveness of myeloid cells to TGF-β[107]; the mutation *BCR/ABL*-Y177 activates Ras-Akt in human hematopoietic progenitor cell transformation leading to CML.[108] The amplitude of Ras pathway signaling is also a determinant of myeloid cell fate; moderate Ras activation in primitive hematopoietic cells is suggested to be an early event in leukemogenesis.[109]

Of the cytokine/receptor–initiated signaling pathways activated in hematopoietic cells, the Janus kinase/signal transducer and activator of the transcription (STAT) pathway plays a major role. In normal hematopoietic stem cells, STAT5 is required for robust competitive repopulation and proliferative responses to early acting cytokines. In particular, interactions with Grb2-associated binding protein have linked STAT5 with the PI3K pathway and its downstream signaling.[110] PI3K-Akt activation is also associated with development of chemoresistance in AML blasts through a mechanism involving p53-dependent suppression of MRP1 expression.[111] Activation of STAT5 in CD34$^-$c-Kit$^+$Sca-1$^+$Lin$^-$ HSCs leads to a drastic expansion of multipotential progenitors and promotes HSC self-renewal ex vivo. In sharp contrast, STAT3 was demonstrated to be dispensable for HSC maintenance invivo, and its activation facilitated lineage commitment of HSCs invitro.[112]

2.4.2 Self-Renewal of Normal and Leukemic Stem Cells

One of the earliest features identified in leukemia is enhanced self-renewal capacity of transformed cells.[113] Activation of β-catenin in CML GMPs appears to mediate this self-renewal activity as well as the leukemic potential of LSCs.[34] The polycomb transcriptional repressor Bmi1[114] has also been identified as being critical in the self-renewal function of HSCs as well as LSCs. Expression of the homeobox transcription factors *HOXA9* and *MEIS1A* (shown to reproducibly generate myeloid leukemia in mouse models) in mouse fetal liver cells indicated that while Hoxa9 alone blocks the differentiation of nonleukemogenic myeloid cell–committed progenitors, coexpression with Meis1 is required for the production of AML-initiating progenitors, which also transcribe a group of HSC genes, including *CD34* and *FLT3* (thus defined as Meis1-related leukemic signature genes).[115] Together, these genes induce leukemia in Bmi1-deficient mice with the same kinetics and characteristics as in normal mice, implying that Bmi1 is not necessary for either leukemia initiation or progression. However, LSCs from Bmi1-deficient mice could not self-renew, resulting in far fewer leukemic cells in the blood of primary mouse recipients and an inability to produce leukemic cells in secondary recipients. This result clearly identifies self-renewal as an essential component of the development of leukemia, distinct from the potent differentiation block and proliferation induced by *HOXA9* and *MEIS1A*. Further, the Bmi1 targets p16^{Ink4a} and p19^{Arf} were occasionally altered in such deficient cells. This could to some extent compensate for the loss of Bmi1, but also hints that other targets of Bmi1 may also be involved in the process of self-renewal. More recently it has also been shown that the tumor suppressor PTEN drives the self-renewal of normal HSCs and the formation of leukemia cells through different mechanisms,[116] such as a cell-context-dependent, PTEN-mediated regulation of mTOR.[117,118] Despite the extensive ongoing research in this aspect of stem cell biology, the precise mechanisms controlling HSC and LSC self-renewal are still not precisely understood.

2.4.3 Epigenetic Effects

In addition to the cell intrinsic and extrinsic events defined so far (genetic regulation and signaling, respectively) that regulate molecular processes governing hematopoiesis, an extensive interplay between lineage-specific transcription factors, DNA methylation, and covalent histone tail modifications (e.g., acetylation, methylation, phosphorylation, SUMOylation, and ubiquitination), which constitute the *histone code*, has also been realized. Adult T-cell leukemia/lymphoma, characterized as one of the most aggressive human neoplasias, is thought to be caused by both genetic and epigenetic alterations to the host cellular genes of T-cells infected with human T-cell leukemia virus type I (HTLV-I).[119] Similarly, abnormal Wnt signaling due to both types of alterations in ALL establishes a group of patients with a significantly adverse prognosis.[120] Epigenetic factors affect chromatin structure and gene transcription, allowing the developmental potential of hematopoietic stem cells to be regulated dynamically, and are often targets for dysregulation in many hematological malignancies.

Hypo/Hypermethylation of Gene Promoters   Hypermethylation of gene promoters leading to their silencing has been reported in certain tumor suppressor and lineage specification genes, cell cycle checkpoints, and so on, and is reported to contribute effectively to blast accumulation in leukemic crisis. Some such methylated gene promoters include *p73* in ALL and Burkitt's lymphoma[121]; AKAP12 in leukemic myeloblasts[122]; *DLX3* in pediatric B-ALL patients[123]; PTPRO in CLL tumorigenesis[124]; *PCDH10*, a tumor suppressor that is variably silenced in lymphomas (e.g., Burkitt's, diffuse large B-cell, Hodgkin's, nasal natural killer/T-cell) and also some leukemia cell lines but not in normal peripheral blood mononuclear cells or lymph nodes[125]; RARβ2 in primary AML blasts[126]; and GGH promoter methylation, which results in a significantly higher accumulation of MTX polyglutamates in nonhyperdiploid B-lineage ALL.[127] Similarly, the genes *DLC-1, PCDHGB7, CYP27B1, EFNA5,* and *CCND1* were frequently hypermethylated in non-Hodgkin's lymphoma but not in benign follicular hyperplasia[128]; allelic silencing at the tumor suppressor locus 13q14.3 occurs in B-cell CLL[129]; *AhR* promoter hypermethylation is reported in primary ALL[130]; *lamin A/C* hypermethylation is associated with poor outcome in diffuse large B-cell lymphomas[131]; *TWIST2* is reported to be differentially methylated in CLL cells relative to Ig V(H) mutational status[132]; and the CD26 antigens are also silenced in ATL.[133] Interestingly, *SOCS1* gene silencing caused by methylation of CpG islands in CML is reversed to an unmethylated status in molecular remission,[134] providing a marker for predicting patient prognosis. *DAPK1* expression of the CLL allele is down-regulated by 75 % in germline cells due to increased HOXB7 binding. Additional promoter methylation results in further loss of expression, contributing to the CLL phenotype.[135]

Although hypermethylation is more frequently associated with tumorigenesis, hypomethylation leading to activation of proto-oncogenes is also reported [e.g., hypomethylation of *PRAME* (a tumor-associated antigen) plays a significant role in the progression of CML,[79] activation of HOX11 in T-cell acute lymphoblastic leukemia (regardless of the involvement of translocation)[136]]. Identification of differential combinations of up- and down-regulated genes through microarrays is also being studied increasingly in leukemias [e.g., 11 genes (*DCC, DLC-1, DDX51, KCNK2, LRP1B, NKX6-1, NOPE, PCDHGA12, RPIB9, ABCB1,* and *SLC2A14*) were differentially methylated between ALL patients and controls[137]].

Chromatin-Regulating Genes   The role of higher-order chromatin assembly is currently being realized in development-, lineage differentiation-, and tumorigenesis-associated gene expression. Deletion of *L3MBTL* is associated with myeloid malignancies; the gene represents a candidate 20q target and is also a member of the Polycomb family of proteins, which together with the trithorax proteins, bring about coordinated regulation of patterns of gene activity.[138] An increase in histone H3K4 (lysine 4 on histone 3) and H3K79 (lysine 27 on histone 3) methylation within the *Bmi1* promoter provides proof of an epigenetic mechanism for histone modifications in *SALL4*-mediated *Bmi1* gene deregulation, which is associated with enhanced self-renewal.[139] Similarly, altered regulation of DNMT3b and HMTaseI expression and the methylation-dependent binding proteins, and MBD2 and MeCP2 expressions, correlated with stages of B-CLL, suggest that these changes may represent deregulation of the epigenetic repertoire.[140]

GSK-3 inhibition abrogates NF-κB binding to its target gene promoters (XIAP, Bcl-2), in part through epigenetic modification of histones, and thereby enhances apoptosis in CLL B-cells ex vivo. This identifies GSK-3 as a potential therapeutic target in the treatment of CLL.[141] MLL1 (a histone methyltransferase) localizes with RNA polymerase II to the 5' end of most protein-coding genes where histone H3K4 trimethylation occurs. In addition to this 5' proximal binding behavior, MLL1 also occupies an extensive domain within a transcriptionally active region of the *HoxA* cluster. This dual role of MLL1 as a start site–specific global transcriptional regulator and participation in larger chromatin domains at the *Hox* genes is important in maintenance of cellular identity in hematopoiesis.[142,143]

Chromosomal Rearrangements Contributing to Epigenetic Events   Although leukemia-derived fusion proteins such as MOZ-TIF2 (rearranged in chromosome translocations associated with AML) promote self-renewal of LSCs, recent studies indicate that murine MOZ and MORF are important for proper development of hematopoietic and neurogenic progenitors, respectively, thereby highlighting the importance of epigenetic integrity in safeguarding stem cell identity.[144] Both proteins have intrinsic histone acetyltransferase activity and are components of quartet complexes with noncatalytic subunits containing the bromodomain, plant homeodomain-linked finger domain, and proline–tryptophan–tryptophan–proline containing domain, three types of structural modules characteristic of chromatin regulators. Similarly, the α-catenin tumor suppressor in HSCs is believed to provide a growth advantage that contributes to human MDS or AML with del(5q).[145] The myeloid maturation block by AML1-MTG16 is associated with *CSF1R* epigenetic down-regulation.[146] Another fusion protein (i.e., MLL-AF10) results in leukemic transformation in an hDOT1L methyltransferase activity–dependent manner, which results in up-regulation of Hoxa9, concomitant with hypermethylation of H3K79. Thus, mistargeting of hDOT1L plays an important role in MLL-AF10-mediated leukemogenesis.[147]

It is envisaged that changes in the epigenetic program are multifactorial in nature, and a larger part of the epigenetic repertoire is as yet not well understood. However, the discovery that epigenetic marks can be reversed by compounds targeting aberrant transcription factor/coactivator/corepressor interactions and histone-modifying activities provides the basis for an exciting field in which the epigenome of cancer cells may be manipulated with potential therapeutic benefits.

2.4.4 microRNA in Leukemia Development

MicroRNAs (miRNAs) are a novel class of small noncoding RNA molecules that regulate gene expression by inducing degradation or translational inhibition of target mRNAs. Increasing experimental evidence also supports the idea of aberrant miRNA expression in cancer pathogenesis.[148] Many miRNAs are reported to be differentially regulated in various hematopoietic lineages. Three miRNAs (i.e., miR-10a, miR-10b, and miR-196a-1) showed a clear correlation with *Hox* gene expression,[149] which is important in lineage demarcation. Other hematopoietic lineage-specific or lineage-enriched miRNAs include miR-181c, miR-181d, miR-191, and miR-136; miRNA clusters include miR-29c, miR-302, miR-98, miR-29a, and let-7a-1.[150] MiR-181a expression correlates strongly with AML, while miR-181a, let-7a, and miR-30d are consistently underexpressed in CLL.[151] MicroRNAs may also be either tumor suppressors (e.g., miR-15a/miR-16-1 locus, miR-142) or have oncogenic properties (e.g., miR-155 and miR-17/miR-92 clusters).[152] The miR-15a/miR-16 locus (13q14.3) regulates Bcl2 negatively at a posttranscriptional level, thereby inducing apoptosis, and is frequently deleted or down-regulated in patients with B-cell CLL,[153–156] while miR-142 is at a translocation site in aggressive B-cell leukemia. miR-let-7 also functions as a tumor suppressor and may inhibit cancers by regulating oncogenes and/or genes that control cell differentiation or apoptosis.[157]

Up-regulation of miR-15a, miR-15b, miR-16-1, let-7a-3, let-7c, let-7d, miR-223, miR-342, and miR-107, and down-regulation of miR-181b expressions in APL patients and cell lines during all-*trans* retinoic acid (ATRA) treatment have been reported.[158] Deregulation of the *TCL1* oncogene is a causal event in the pathogenesis of aggressive CLL; expression levels of miR-29 and miR-181 generally correlate inversely with Tcl1 expression in CLL samples.[159] miR155 and BIC (its host gene) have been reported to accumulate in human B-cell lymphomas, especially in diffuse large B-cell lymphomas, Hodgkin lymphomas, and certain types of Burkitt lymphomas; it leads to polyclonal expansion, favoring the capture of secondary genetic changes for full transformation.[160] miR-21 is also dramatically overexpressed in leukemia, although the corresponding genomic loci are not amplified; whereas miR-150 and miR-92 are significantly deregulated.[161]

A new role for a leukemia fusion protein linking the epigenetic silencing of a microRNA locus to the differentiation block of leukemia has recently been reported.[162] Human granulocytic differentiation is controlled by a regulatory circuitry involving miR-223, miR-107, and the transcriptional factors NFI-A and C/EBPα.[163] miR-223 is also a direct transcriptional target of AML1/ETO, the t(8;21)-generated AML fusion protein. By recruiting chromatin remodeling enzymes at an AML1-binding site on the pre-miR-223 gene, AML1/ETO induces heterochromatic silencing of miR-223. Ectopic miR-223 expression, RNAi against AML1/ETO, or demethylating treatment enhances miR-223 levels and restores cell differentiation. A BCR⁻ABL⁻c-MYC⁻miR-17-92 pathway has also been reported; consequently in three Bcr⁻Abl⁺ cell lines, expression of the polycistronic miR-17-92 cluster miRNAs was specifically down-regulated by both imatinib treatment and anti-BCR⁻ABL RNAi.[164] In unilineage monocytic culture generated by hematopoietic progenitor cells, microRNAs 17-5p, 20a, and 106a are down-regulated, whereas AML1 is up-regulated at the protein but not at the mRNA level. These miRNAs target AML1, which promotes M-CSFR transcription. In addition, AML1 binds the miRNA 17-5p-92 and 106a-92 cluster promoters and transcriptionally inhibits the expression of miRNA 17-5p-20a-106a. These studies indicate that monocytopoiesis is controlled by a circuitry involving sequentially miRNA 17-5p-20a-106a, AML1, and M-CSFR, whereby miRNA 17-5p-20a-106a functions as a master gene complex interlinked with AML1 in a mutual negative feedback loop.[165]

miRNA expression profiling of CLL and ALL revealed the five most highly expressed miRNAs to be miR-128b, miR-204, miR-218, miR-331, and miR-181b-1 in ALL, and miR-331, miR-29a, miR-195, miR-34a, and miR-29c in CLL. The miR-17-92 cluster was also found to be up-regulated in ALL, as reported previously for some types of lymphomas. The differences observed in gene expression levels were validated for miR-331 and miR-128b in ALL and CD19+ samples, suggesting the potential roles of these miRNAs in hematopoiesis and leukemogenesis.[166]

Kis2 ncRNAs (Non-coding RNAs) are the pri-miRNA of the oncogenic miR-106-363; their overexpression results in miR-106a, miR-19b-2, miR-92-2, and miR-20b accumulation and rising miR-92 and miR-19 levels. Myosin regulatory light chain−interacting protein, retinoblastoma-binding protein 1-like, and homeodomain-interacting protein kinase 3 are target genes of this miRNA cluster; thus, a link between these genes, their regulatory miRNAs, and T-cell leukemia has recently been established.[167]

Murine Fli-3 is the integration site of Friend murine leukemia virus (F-MuLV)−induced erythroleukemia; its human homolog, C13orf25, includes a region encoding the miR-17-92 miRNA cluster and is the target gene of 13q31 chromosomal amplification in human B-cell lymphomas. Erythroleukemias resulting from either insertional activation or amplification of Fli-3 express higher levels of the primary or mature miRNAs derived from miR-17-92. Ectopic expression of Fli-3 in an erythroblastic cell line also switches erythropoietin (Epo)-induced differentiation to Epo-induced proliferation through activation of the Ras and PI3K pathways. *Fli-3* encoding miR-17-92 is implied strongly in hematopoiesis and the development of erythroleukemia.[168]

Collectively, the considerable amount of data generated for leukemia indicates that miRNA expression profiles may become useful biomarkers in cancer diagnostics. In addition, miRNA therapy could also be a powerful tool for cancer prevention and therapeutics.

2.5 CONCLUSIONS AND FUTURE PERSPECTIVES

The general mechanisms underlying leukemic transformation with stem cells or their progenitors being targets for genetic rearrangements are begining to be elucidated. Current chemotherapeutic agents are thought to have limited efficacies in leukemia since they target both the leukemic and normal stem cell populations. The expression of drug transporters is also enhanced in LSCs, making this disease drug resistant. It thus becomes imperative to develop new strategies to target the LSC population specifically and preferentially while sparing normal stem cells. Some drugs that are lethal for LSCs are currently in preclinical testing, including parthenolide and TDZD8.[169] The existing correlation of specific surface markers and cancer-initiating cells coupled with a leukemogenic marker also provides an excellent prognostic tool to track postchemotherapeutic minimal residual disease, which at a later stage leads to a stage of relapse. However, such strategies have yet to be developed at the clinical level.

Targeting the fusion products of chromosomal translocations is another current approach. These proteins not only retain the cells in an immature state, but also contribute significantly to the development of a relative insensitivity to treatment.[170] In CML, the bcr-abl translocation product is an activated tyrosine kinase that activates myelocyte cells; this activation may be inhibited by imatinib mesylate (Gleevec, STI-571), which prevents phosphorylation, resulting in cell differentiation and apoptosis. In APL, the PML-RARα fusion product blocks the activity of the promyelocytic protein required for formation of the granules of promyelocytes and prevents further differentiation. Retinoic acids bind to the RARα component of the fusion product, resulting in degradation of the fusion protein by ubiquitinization. This allows normal PML to participate in granule formation and differentiation of the promyelocytes. Leukemias resulting from *FLT3* mutations that result in constitutive activation of the IL-3 receptor may be blocked by agents that inhibit farnesyl transferase. Thus, specific inhibition of the genetically altered activation molecules of leukemic cells could induce differentiation and apoptosis. Use of optimal combinations of specific inhibitors that act on the effects of different specific genetic lesions promises to result in more effective and permanent treatment.

In conclusion, a number of similarities are evident between normal and leukemic stem cells, yet distinct differences also exist. The latter could be exploited to provide a therapeutic window in which specially designed small molecule inhibitors could specifically target LSCs in a clinical setting.

REFERENCES

1. Becker AJ, McCulloch EA, Till JE. Cytological demonstration of the clonal nature of spleen colonies derived from transplanted mouse marrow cells. *Nature*. 1963; 197:452–454.
2. Bonnet D, Dick JE. Human acute myeloid leukemia is organized as a hierarchy that originates from a primitive hematopoietic cell. *Nat Med*. 1997; 3:730–737.
3. Eaves CJ, Sutherland HJ, Udomsakdi C, et al. The human hematopoietic stem cell invitro and invivo. *Blood Cells*. 1992; 18:301–307.
4. Dick JE. Human stem cell assays in immune-deficient mice. *Curr Opin Hematol*. 1996; 3:405–409.
5. Weissman IL. Stem cells: units of development, units of regeneration, and units in evolution. *Cell*. 2000; 100:157–168.
6. Clarke MF, Dick JE, Dirks PB, et al. Cancer stem cells: perspectives on current status and future directions. AACR Workshop on Cancer Stem Cells. *Cancer Res*. 2006; 66:9339–9344.
7. Wright DE, Wagers AJ, Gulati AP, Johnson FL, Weissman IL. *Science*. 2001; 294:1933–1936.

8. Morrison SJ, Weissman IL. The long-term repopulating subset of hematopoietic stem cells is deterministic and isolatable by phenotype. *Immunity.* 1994; 1:661–673.

9. Morrison SJ, Wright DE, Weissman IL. Identification of a lineage of multipotent hematopoietic progenitors. *Development.* 1997; 124:1929–1939.

10. Kondo M, Weissman IL, Akashi K. Identification of clonogenic common lymphoid progenitors in mouse bone marrow. *Cell.* 1997; 91:661–672.

11. Akashi K, Traver D, Miyamoto T, Weissman IL. A clonogenic common myeloid progenitor that gives rise to all myeloid lineages. *Nature.* 2000; 404:193–197.

12. Traver D, Akashi K, Manz M, et al. Development of CD8alpha-positive dendritic cells from a common myeloid progenitor. *Science.* 2000; 290:2152–2154.

13. Manz MG, Traver D, Miyamoto T, Weissman IL, Akashi K. Dendritic cell potentials of early lymphoid and myeloid progenitors. *Blood.* 2001; 97:3333–3341.

14. Jamieson CH, Weissman IL, Passegué E. Chronic versus acute myelogenous leukemia: a question of self-renewal. *Cancer Cell.* 2004; 6:531–533.

15. Domen J. The role of apoptosis in regulating hematopoietic stem cell numbers. *Apoptosis.* 2001; 6:239–252.

16. Morishita K. Leukemogenesis of the *EVI1/MEL1* gene family. *Int J Hematol.* 2007; 85:279–286.

17. Eaves C, Cashman J, Eaves A. Defective regulation of leukemic hematopoiesis in chronic myeloid leukemia. *Leuk Res.* 1998; 12:1085–1096.

18. Vaziri H, Dragowska W, Allsopp RC, Thomas TE, Harley CB, Lansdorp PM. Evidence for a mitotic clock in human hematopoietic stem cells: loss of telomeric DNA with age. *Proc Natl Acad Sci U S A.* 1994; 91:9857–9860.

19. Allsopp RC, Chang E, Kashefi-Aazam M, et al. Telomere shortening is associated with cell division invitro and invivo. *Exp Cell Res.* 1995; 220:194–200.

20. Albanell J, Han W, Mellado B, et al. Telomerase activity is repressed during differentiation of maturation-sensitive but not resistant human tumor cell lines. *Cancer Res.* 1996; 56:1503–1508.

21. Drummond MW, Hoare SF, Monaghan A, et al. Dysregulated expression of the major telomerase components in leukaemic stem cells. *Leukemia.* 2005; 19:381–389.

22. Morishita K. Leukemogenesis of the *EVI1/MEL1* gene family. *Int J Hematol.* 2007; 85:279–286.

23. Sirard C, Lapidot T, Vormoor J, et al. Normal and leukemic SCID-repopulating cells (SRC) coexist in the bone marrow and peripheral blood from CML patients in chronic phase, whereas leukemic SRC are detected in blast crisis. *Blood.* 1996; 87:1539–1548.

24. Nowell P, Hungerford D. A minute chromosome in human chronic granulocytic leukemia. *Science.* 1960; 132:1497.

25. Graham SJ, Jorgensen HG, Allan E, et al. Primitive, quiescent, Philadelphia-positive stem cells from patients with chronic myeloid leukemia are insensitive to STI571 invitro. *Blood.* 2002; 99:319–325.

26. Goldman JM, Melo JV. Chronic myeloid leukemia: advances in biology and new approaches to treatment. *N Engl J Med.* 2003; 349:1451–1464.

27. Chomel JC, Brizard F, Veinstein A, et al. Persistence of *BCR-ABL* genomic rearrangement in chronic myeloid leukemia patients in complete and sustained cytogenetic remission after interferon-alpha therapy or allogeneic bone marrow transplantation. *Blood.* 2000; 95:404–408.

28. Bose S, Deninger M, Gora-Tybor J, Goldman JM, Melo JV. The presence of typical and atypical *BCR-ABL* fusion genes in leukocytes of normal individuals: biologic significance and implications for the assessment of minimal residual disease. *Blood*. 1998; 92:3362–3367.

29. Holyoake TL, Jiang X, Drummond MW, Eaves AC, Eaves CJ. Elucidating critical mechanisms of deregulated stem cell turnover in the chronic phase of chronic myeloid leukemia. *Leukemia*. 2002; 16:549–558.

30. Jaiswal S, Traver D, Miyamoto T, Akashi K, Lagasse E, Weissman IL. Expression of *BCR/ABL* and *BCL-2* in myeloid progenitors leads to myeloid leukemias. *Proc Natl Acad Sci U S A*. 2003; 100:10002–10007.

31. Sawyers CL. Chronic myeloid leukemia. *N Engl J Med*. 1999; 340:1330–1340.

32. Passegue E, Jamieson CHM, Ailles LE, Weissman IL. Normal and leukemic hematopoiesis: Are leukemias a stem cell disorder or a reacquisition of stem cell characteristics? *Proc Natl Acad Sci U S A*. 2003; 100:11842–11849.

33. Manz MG, Miyamoto T, Akashi K, Weissman IL. Prospective isolation of human clonogenic common myeloid progenitors. *Proc Natl Acad Sci U S A*. 2002; 99:11872–11877.

34. Jamieson CHM, Ailles LE, Dylla SJ, et al. Granulocyte–macrophage progenitors as candidate leukemic stem cells in blast-crisis CML. *N Engl J Med*. 2004; 351:657–667.

35. Traver D, Akashi K, Weissman IL, Lagasse E. Mice defective in two apoptotic pathways in the myeloid lineage develop acute myeloblastic leukemia. *Immunity*. 1998; 9:47–57.

36. Rego EM, Ruggero D, Tribioli C, et al. Leukemia with distinct phenotypes in transgenic mice expressing PML/RAR alpha, PLZF/RAR alpha or NPM/RAR alpha. *Oncogene*. 2006; 25:1974–1979.

37. Vincent AM, Cawley JC, Burthem J. Integrin function in chronic lymphocytic leukemia. *Blood*. 1996; 87:4780–4788.

38. Keating MJ. Progress in CLL, chemotherapy, antibodies and transplantation. *Biomed Pharmacother*. 2001; 55:524–528.

39. Hervé M, Xu K, Ng YS, et al. Unmutated and mutated chronic lymphocytic leukemias derive from self-reactive B cell precursors despite expressing different antibody reactivity. *J Clin Invest*. 2005; 115:1636–1643.

40. Dürig J, Naschar M, Schmücker U, et al. CD38 expression is an important prognostic marker in chronic lymphocytic leukaemia. *Leukemia*. 2002; 16:30–35.

41. Schattner EJ. CD40 ligand in CLL pathogenesis and therapy. *Leuk Lymphoma*. 2000; 37:461–472.

42. Nakagawa R, Soh JW, Michie AM. Subversion of protein kinase C alpha signaling in hematopoietic progenitor cells results in the generation of a B-cell chronic lymphocytic leukemia-like population invivo. *Cancer Res*. 2006; 66:527–534.

43. Möhle R, Failenschmid C, Bautz F, Kanz L. Overexpression of the chemokine receptor CXCR4 in B cell chronic lymphocytic leukemia is associated with increased functional response to stromal cell–derived factor-1 (SDF-1). *Leukemia*. 1999; 13:1954–1959.

44. Duke VM, Gandini D, Sherrington PD, et al. V(H) gene usage differs in germline and mutated B-cell chronic lymphocytic leukemia. *Haematologica*. 2003; 88:1259–1271.

45. Scalzulli PR, Musto P, Dell'Olio M, Carotenuto M. CFU-GM are not increased in peripheral blood of patients with chronic lymphocytic leukaemia. *Eur J Haematol.* 1996; 57:180–181.

46. Cuní S, Pérez-Aciego P, Pérez-Chacón G, et al. A sustained activation of PI3K/NF-kappaB pathway is critical for the survival of chronic lymphocytic leukemia B cells. *Leukemia.* 2004; 18:1391–1400.

47. Giné E, Bosch F, Villamor N, et al. Simultaneous diagnosis of hairy cell leukemia and chronic lymphocytic leukemia/small lymphocytic lymphoma: a frequent association? *Leukemia.* 2002; 16:1454–1459.

48. Dearden CE. T-cell prolymphocytic leukemia. *Med Oncol.* 2006; 23:17–22.

49. Ravandi F, Kantarjian H, Jones D, Dearden C, Keating M, O'Brien S. Mature T-cell leukemias. *Cancer.* 2005; 104:1808–1818.

50. Assaf C, Hummel M, Dippel E, et al. Common clonal T-cell origin in a patient with T-prolymphocytic leukaemia and associated cutaneous T-cell lymphomas. *Br J Haematol.* 2003; 120:488–491.

51. Weng AP, Lau A. Notch signaling in T-cell acute lymphoblastic leukemia. *Future Oncol.* 2005; 1:511–519.

52. Aster JC. Deregulated Notch signaling in acute T-cell lymphoblastic leukemia/lymphoma: new insights, questions, and opportunities. *Int J Hematol.* 2005; 82:295–301.

53. Kastan MB, Schlaffer E, Russo JE, Colvin OM, Civin CI, Hilton J. Direct demonstration of elevated aldehyde dehydrogenase in human hematopoietic progenitor cells. *Blood.* 1990; 75:1947–1950.

54. Pearce DJ, Taussig D, Simpson C, et al. Characterization of cells with a high aldehyde dehydrogenase activity from cord blood and acute myeloid leukemia samples. *Stem Cells.* 2005; 23:752–760.

55. Cheung AM, Wan TS, Leung JC, et al. Aldehyde dehydrogenase activity in leukemic blasts defines a subgroup of acute myeloid leukemia with adverse prognosis and superior NOD/SCID engrafting potential. *Leukemia.* 2007; 21:1423–1430.

56. Lapidot T, Sirard C, Vormoor J, et al. A cell initiating human acute myeloid leukaemia after transplantation into SCID mice. *Nature.* 1994; 367:645–648.

57. Cobaleda C, Gutiérrez-Cianca N, Pérez-Losada J, et al. A primitive hematopoietic cell is the target for the leukemic transformation in human Philadelphia-positive acute lymphoblastic leukemia. *Blood.* 2000; 95:1007–1013.

58. Jordan CT, Upchurch D, Szilvassy SJ, et al. The interleukin-3 receptor alpha chain is a unique marker for human acute myelogenous leukemia stem cells. *Leukemia.* 2000; 14:1777–1784.

59. Florian S, Sonneck K, Hauswirth AW, et al. Detection of molecular targets on the surface of $CD34^+/CD38^-$ stem cells in various myeloid malignancies. *Leuk Lymphoma.* 2006; 47:207–222.

60. Peeters SD, van der Kolk DM, de Haan G, et al. Selective expression of cholesterol metabolism genes in normal $CD34^+CD38^-$ cells with a heterogeneous expression pattern in AML cells. *Exp Hematol.* 2006; 34:622–630.

61. Jin L, Hope KJ, Zhai Q, Smadja-Joffe F, Dick JE. Targeting of CD44 eradicates human acute myeloid leukemic stem cells. *Nat Med.* 2006; 12:1167–1174.

62. Hosen N, Park CY, Tatsumi N, et al. CD96 is a leukemic stem cell–specific marker in human acute myeloid leukemia. *Proc Natl Acad Sci U S A.* 2007; 104:11008–11103.

63. Fuchs A, Colonna M. The role of NK cell recognition of nectin and nectin-like proteins in tumor immunosurveillance. *Semin Cancer Biol.* 2006; 16:359–366.

64. Gramatzki M, Ludwig WD, Burger R, et al. Antibodies TC-12 ("unique") and TH-111 (CD96) characterize T-cell acute lymphoblastic leukemia and a subgroup of acute myeloid leukemia. *Exp Hematol.* 1998; 26:1209–1214.

65. Wang L, O'Leary H, Fortney J, Gibson LF. Ph^+/VE^- cadherin$^+$ identifies a stem cell like population of acute lymphoblastic leukemia sustained by bone marrow niche cells. *Blood.* 2007; 110:3334–3344.

66. Walkley CR, Olsen GH, Dworkin S, et al. A Microenvironment-induced myeloproliferative syndrome caused by retinoic acid receptor γ deficiency. *Cell.* 2007; 129:1097–1110.

67. Walkley CR, Shea JM, Sims NA, Purton LE, Orkin SH. Rb regulates interactions between hematopoietic stem cells and their bone marrow microenvironment. *Cell.* 2007; 129:1081–1095.

68. Shimamoto T, Ohyashiki K, Toyama K, Takeshita K. Homeobox genes in hematopoiesis and leukemogenesis. *Int J Hematol.* 1998; 67:339–350.

69. Sengupta A, Banerjee D, Chandra S, et al. Deregulation and cross talk among Sonic Hedgehog, Wnt, Hox and Notch signaling in chronic myeloid leukemia progression. *Leukemia.* 2007; 21:949–955.

70. Ysebaert L, Chicanne G, Demur C, et al. Expression of beta-catenin by acute myeloid leukemia cells predicts enhanced clonogenic capacities and poor prognosis. *Leukemia.* 2006; 20:1211–1216.

71. Lu D, Zhao Y, Tawatao R, et al. Activation of the Wnt signaling pathway in chronic lymphocytic leukemia. *Proc Natl Acad Sci U S A.* 2004; 101:3118–3123.

72. Liu TX, Becker MW, Jelinek J, et al. Chromosome 5q deletion and epigenetic suppression of the gene encoding alpha-catenin (CTNNA1) in myeloid cell transformation. *Nat Med.* 2007; 13:78–83.

73. Zheng X, Beissert T, Kukoc-Zivojnov N, et al. Gamma-catenin contributes to leukemogenesis induced by AML-associated translocation products by increasing the self-renewal of very primitive progenitor cells. *Blood.* 2004; 103:3535–3543.

74. Müller-Tidow C, Steffen B, Cauvet T, et al. Translocation products in acute myeloid leukemia activate the Wnt signaling pathway in hematopoietic cells. *Mol Cell Biol.* 2004; 7:2890–2904.

75. Mikesch JH, Steffen B, Berdel WE, Serve H, Müller-Tidow C. The emerging role of Wnt signaling in the pathogenesis of acute myeloid leukemia. *Leukemia.* 2007; 21:1638–1647.

76. Tickenbrock L, Schwäble J, Wiedehage M, et al. Flt3 tandem duplication mutations cooperate with Wnt signaling in leukemic signal transduction. *Blood.* 2005; 105:3699–3706.

77. Simon M, Grandage VL, Linch DC, Khwaja A. Constitutive activation of the Wnt/beta-catenin signalling pathway in acute myeloid leukaemia. *Oncogene.* 2005; 24:2410–2420.

78. De Toni F, Racaud-Sultan C, Chicanne G, et al. A crosstalk between the Wnt and the adhesion-dependent signaling pathways governs the chemosensitivity of acute myeloid leukemia. *Oncogene.* 2006; 25:3113–3122.

79. Román-Gómez J, Cordeu L, Agirre X, et al. Epigenetic regulation of Wnt-signaling pathway in acute lymphoblastic leukemia. *Blood.* 2007; 109:3462–3469.

80. Suzuki R, Onizuka M, Kojima M, et al. Preferential hypermethylation of the Dickkopf-1 promoter in core-binding factor leukaemia. *Br J Haematol.* 2007; 138:624–631.

81. Román-Gómez J, Jiménez-Velasco A, Agirre X, et al. Transcriptional silencing of the Dickkopfs-3 (*Dkk-3*) gene by CpG hypermethylation in acute lymphoblastic leukaemia. *Br J Cancer.* 2004; 91:707–713.

82. Mazieres J, You L, He B, et al. Inhibition of Wnt16 in human acute lymphoblastoid leukemia cells containing the t(1;19) translocation induces apoptosis. *Oncogene.* 2005; 24:5396–5400.

83. Nygren MK, Døsen G, Hystad ME, Stubberud H, Funderud S, Rian E. Wnt3A activates canonical Wnt signalling in acute lymphoblastic leukaemia (ALL) cells and inhibits the proliferation of B-ALL cell lines. *Br J Haematol.* 2007; 136:400–413.

84. Khan NI, Bradstock KF, Bendall LJ. Activation of Wnt/beta-catenin pathway mediates growth and survival in B-cell progenitor acute lymphoblastic leukaemia. *Br J Haematol.* 2007; 138:338–348.

85. Liang H, Chen Q, Coles AH, et al. Wnt5a inhibits B cell proliferation and functions as a tumor suppressor in hematopoietic tissue. *Cancer Cell.* 2003; 5:349–360.

86. Radtke F, Wilson A, Mancini SJ, MacDonald HR. Notch regulation of lymphocyte development and function. *Nat Immunol.* 2004; 5:247–253.

87. Vercauteren SM, Sutherland HJ. Constitutively active Notch4 promotes early human hematopoietic progenitor cell maintenance while inhibiting differentiation and causes lymphoid abnormalities in vivo. *Blood.* 2004; 104:2315–2322.

88. O'Neil J, Grim J, Strack P, et al. Clurman BE, Look AT. FBW7 mutations in leukemic cells mediate Notch pathway activation and resistance to gamma-secretase inhibitors. *J Exp Med.* 2007; 204:1813–1824.

89. Nemoto N, Suzukawa K, Shimizu S, et al Identification of a novel fusion gene MLL-MAML2 in secondary acute myelogenous leukemia and myelodysplastic syndrome with inv(11)(q21q23). *Genes Chromosomes Cancer.* 2007; 46:813–819.

90. Ma X, Renda MJ, Wang L, et al. Rbm15 modulates Notch-induced transcriptional activation and affects myeloid differentiation. *Mol Cell Biol.* 2007; 27:3056–3064.

91. Grabher C, von Boehmer H, Look AT. Notch 1 activation in the molecular pathogenesis of T-cell acute lymphoblastic leukaemia. *Nat Rev Cancer.* 2006; 6:347–359.

92. Zhu YM, Zhao WL, Fu JF, et al. Notch1 mutations in T-cell acute lymphoblastic leukemia: prognostic significance and implication in multifactorial leukemogenesis. *Clin Cancer Res.* 2006; 12:3043–3049.

93. Palomero T, Sulis ML, Cortina M, et al. Mutational loss of PTEN induces resistance to Notch1 inhibition in T-cell leukemia. *Nat Med.* 2007; 10:1203–1210.

94. Nie L, Wu H, Sun XH. Ubiquitination and degradation of Tal1/SCL is induced by Notch signaling and depends on Skp2 and CHIP. *J Biol Chem.* 2008; 283(2):684–692.

95. Sharma VM, Draheim KM, Kelliher MA. The Notch1/c-Myc pathway in T cell leukemia. *Cell Cycle.* 2007; 6:927–930.

96. Chan SM, Weng AP, Tibshirani R, Aster JC, Utz PJ. Notch signals positively regulate activity of the mTOR pathway in T-cell acute lymphoblastic leukemia. *Blood.* 2007; 110:278–286.

97. Zweidler-McKay PA, He Y, Xu L, et al. Notch signaling is a potent inducer of growth arrest and apoptosis in a wide range of B-cell malignancies. *Blood.* 2005; 106:3898–3906.

98. Vacca A, Felli MP, Palermo R, et al. Notch3 and pre-TCR interaction unveils distinct NF-kappaB pathways in T-cell development and leukemia. *EMBO J.* 2006; 25:1000–1008.

99. Allman D, Karnell FG, Punt JA, et al. Separation of Notch1 promoted lineage commitment and expansion/transformation in developing T cells. *J Exp Med.* 2001; 194:99–106.

100. Felli MP, Vacca A, Calce A, et al. PKC theta mediates pre-TCR signaling and contributes to Notch3-induced T-cell leukemia. *Oncogene.* 2005; 24:992–1000.

101. Zweidler-McKay PA, Pear WS. Notch and T cell malignancy. *Semin Cancer Biol.* 2004; 14:329–340.

102. Vilimas T, Mascarenhas J, Palomero T, et al. Targeting the NF-kappaB signaling pathway in Notch1-induced T-cell leukemia. *Nat Med.* 2007; 13:70–77.

103. Murata-Ohsawa M, Tohda S, Kogoshi H, Nara N. The Notch ligand, Delta-1, reduces TNF-alpha-induced growth suppression and apoptosis by decreasing activation of caspases in U937 cells. *Int J Mol Med.* 2004; 14:861–866.

104. Kumano K, Chiba S, Shimizu K, et al. Notch1 inhibits differentiation of hematopoietic cells by sustaining GATA-2 expression. *Blood.* 2001; 98:3283–3289.

105. Jang MS, Miao H, Carlesso N, et al Notch-1 regulates cell death independently of differentiation in murine erythroleukemia cells through multiple apoptosis and cell cycle pathways. *J Cell Physiol.* 2004; 199:418–433.

106. Isufi I, Seetharam M, Zhou L, et al. Transforming growth factor-beta signaling in normal and malignant hematopoiesis. *J Interferon Cytokine Res.* 2007; 27:543–252.

107. Møller GM, Frost V, Melo JV, Chantry A. Upregulation of the TGFbeta signalling pathway by Bcr-Abl: implications for haemopoietic cell growth and chronic myeloid leukaemia. *FEBS Lett.* 2007; 581:1329–1334.

108. Chu S, Li L, Singh H, Bhatia R. BCR-tyrosine 177 plays an essential role in *Ras* and *Akt* activation and in human hematopoietic progenitor transformation in chronic myelogenous leukemia. *Cancer Res.* 2007; 67:7045–7053.

109. Dorrell C, Takenaka K, Minden MD, Hawley RG, Dick JE. Hematopoietic cell fate and the initiation of leukemic properties in primitive primary human cells are influenced by *Ras* activity and farnesyltransferase inhibition. *Mol Cell Biol.* 2004; 16:6993–7002.

110. Bunting KD. STAT5 signaling in normal and pathologic hematopoiesis. *Front Biosci.* 2007; 12:2807–2820.

111. Tazzari PL, Cappellini A, Ricci F, et al. Multidrug resistance–associated protein 1 expression is under the control of the phosphoinositide 3 kinase/Akt signal transduction network in human acute myelogenous leukemia blasts. *Leukemia.* 2007; 21:427–438.

112. Kato Y, Iwama A, Tadokoro Y, et al. Selective activation of STAT5 unveils its role in stem cell self-renewal in normal and leukemic hematopoiesis. *J Exp Med.* 2005; 202:169–179.

113. Buick RN, Minden MD, McCulloch EA. Self-renewal in culture of proliferative blast progenitor cells in acute myeloblastic leukemia. *Blood.* 1979; 54:95–104.

114. Lessard J, Sauvageau G. Bmi-1 determines the proliferative capacity of normal and leukaemic stem cells. *Nature.* 2003; 423:255–260.

115. Wang GG, Pasillas MP, Kamps MP. Persistent transactivation by meis1 replaces hox function in myeloid leukemogenesis models: evidence for co-occupancy of meis1-pbx and hox-pbx complexes on promoters of leukemia-associated genes. *Mol Cell Biol.* 2006; 26:3902–3916.

116. Yilmaz OH, Valdez R, Theisen BK, et al. Pten dependence distinguishes haematopoietic stem cells from leukaemia-initiating cells. *Nature.* 2006; 441(7092):475–482.

117. Cheung AM, Mak TW. PTEN in the haematopoietic system and its therapeutic indications. *Trends Mol Med.* 2006; 12:503–505.

118. Rossi DJ, Weissman IL. Pten, tumorigenesis, and stem cell self-renewal. *Cell.* 2006; 125:229–231.

119. Fukuda R, Hayashi A, Utsunomiya A, et al. Alteration of phosphatidylinositol 3-kinase cascade in the multilobulated nuclear formation of adult T cell leukemia/lymphoma (ATLL). *Proc Natl Acad Sci U S A.* 2005; 102:15213–15218.

120. Román-Gómez J, Jiménez-Velasco A, Agirre X, et al. Epigenetic regulation of *PRAME* gene in chronic myeloid leukemia. *Leuk Res.* 2007; 31(11):1521–1528.

121. Corn PG, Kuerbitz SJ, van Noesel MM, et al. Transcriptional silencing of the *p73* gene in acute lymphoblastic leukemia and Burkitt's lymphoma is associated with 5′ CpG island methylation. *Cancer Res.* 1999; 59:3352–3356.

122. Flotho C, Paulun A, Batz C, Niemeyer CM. *AKAP12*, a gene with tumor suppressor properties, is a target of promoter DNA methylation in childhood myeloid malignancies. *Br J Haematol.* 2007; 138:644–650.

123. Campo Dell'Orto M, Banelli B, Giarin E, et al. Down-regulation of DLX3 expression in MLL-AF4 childhood lymphoblastic leukemias is mediated by promoter region hypermethylation. *Oncol Rep.* 2007; 18:417–423.

124. Motiwala T, Majumder S, Kutay H, et al. Methylation and silencing of protein tyrosine phosphatase receptor type O in chronic lymphocytic leukemia. *Clin Cancer Res.* 2007; 13:3174–3181.

125. Ying J, Gao Z, Li H, et al. Frequent epigenetic silencing of protocadherin 10 by methylation in multiple haematologic malignancies. *Br J Haematol.* 2007; 136:829–832.

126. Fazi F, Zardo G, Gelmetti V, et al. Heterochromatic gene repression of the retinoic acid pathway in acute myeloid leukemia. *Blood.* 2007; 109:4432–4440.

127. Cheng Q, Cheng C, Crews KR, et al. Epigenetic regulation of human gamma-glutamyl hydrolase activity in acute lymphoblastic leukemia cells. *Am J Hum Genet.* 2006; 79:264–274.

128. Shi H, Guo J, Duff DJ, et al. Discovery of novel epigenetic markers in non-Hodgkin's lymphoma. *Carcinogenesis.* 2007; 28:60–70.

129. Mertens D, Wolf S, Tschuch C, et al. Allelic silencing at the tumor-suppressor locus 13q14.3 suggests an epigenetic tumor-suppressor mechanism. *Proc Natl Acad Sci U S A.* 2006; 103:7741–7746.

130. Mulero-Navarro S, Carvajal-Gonzalez JM, Herranz M, et al. The dioxin receptor is silenced by promoter hypermethylation in human acute lymphoblastic leukemia through inhibition of Sp1 binding. *Carcinogenesis.* 2006; 27:1099–1104.

131. Agrelo R, Setien F, Espada J, et al. Inactivation of the lamin A/C gene by CpG island promoter hypermethylation in hematologic malignancies, and its association with poor survival in nodal diffuse large B-cell lymphoma. *J Clin Oncol.* 2005; 23:3940–3947.

132. Raval A, Lucas DM, Matkovic JJ, et al. TWIST2 demonstrates differential methylation in immunoglobulin variable heavy chain mutated and unmutated chronic lymphocytic leukemia. *J Clin Oncol.* 2005; 23:3877–3885.

133. Tsuji T, Sugahara K, Tsuruda K, et al. Clinical and oncologic implications in epigenetic down-regulation of CD26/dipeptidyl peptidase IV in adult T-cell leukemia cells. *Int J Hematol.* 2004; 80:254–260.

134. Liu TC, Lin SF, Chang JG, Yang MY, Hung SY, Chang CS. Epigenetic alteration of the *SOCS1* gene in chronic myeloid leukaemia. *Br J Haematol.* 2003; 123:654–661.

135. Raval A, Tanner SM, Byrd JC, et al. Downregulation of death-associated protein kinase 1 (DAPK1) in chronic lymphocytic leukemia. *Cell.* 2007; 129:879–890.

136. Watt PM, Kumar R, Kees UR. Promoter demethylation accompanies reactivation of the HOX11 proto-oncogene in leukemia. *Genes Chromosomes Cancer.* 2000; 29:371–377.

137. Taylor KH, Pena-Hernandez KE, Davis JW, et al. Large-scale CpG methylation analysis identifies novel candidate genes and reveals methylation hotspots in acute lymphoblastic leukemia. *Cancer Res.* 2007; 67:2617–2625.

138. Li J, Bench AJ, Vassiliou GS, Fourouclas N, Ferguson-Smith AC, Green AR. Imprinting of the human *L3MBTL* gene, a Polycomb family member located in a region of chromosome 20 deleted in human myeloid malignancies. *Proc Natl Acad Sci U S A.* 2004; 101:7341–7346.

139. Yang J, Chai L, Liu F, et al. *Bmi-1* is a target gene for SALL4 in hematopoietic and leukemic cells. *Proc Natl Acad Sci U S A.* 2007; 104:10494–10499.

140. Kn H, Bassal S, Tikellis C, El-Osta A. Expression analysis of the epigenetic methyltransferases and methyl-CpG binding protein families in the normal B-cell and B-cell chronic lymphocytic leukemia (CLL). *Cancer Biol Ther.* 2004; 3:989–994.

141. Ougolkov AV, Bone ND, Fernandez-Zapico ME, Kay NE, Billadeau DD. Inhibition of glycogen synthase kinase-3 activity leads to epigenetic silencing of nuclear factor kappaB target genes and induction of apoptosis in chronic lymphocytic leukemia B cells. *Blood.* 2007; 110:735–742.

142. Guenther MG, Jenner RG, Chevalier B, et al. Global and Hox-specific roles for the MLL1 methyltransferase. *Proc Natl Acad Sci U S A.* 2005; 102:8603–8608.

143. Ono R, Nosaka T, Hayashi Y. Roles of a trithorax group gene, *MLL*, in hematopoiesis. *Int J Hematol.* 2005; 81:288–293.

144. Yang XJ, Ullah M. MOZ and MORF, two large MYSTic HATs in normal and cancer stem cells. *Oncogene.* 2007; 26:5408–5419.

145. Liu TH, Raval A, Chen SS, Matkovic JJ, Byrd JC, Plass C. CpG island methylation and expression of the secreted *frizzled*-related protein gene family in chronic lymphocytic leukemia. *Cancer Res.* 2006; 66:653–658.

146. Rossetti S, Van Unen L, Touw IP, Hoogeveen AT, Sacchi N. Myeloid maturation block by AML1-MTG16 is associated with Csf1r epigenetic downregulation. *Oncogene.* 2005; 24:5325–5332.

147. Okada Y, Feng Q, Lin Y, et al. hDOT1L links histone methylation to leukemogenesis. *Cell.* 2005; 121:167–178.

148. Chen CZ, Lodish HF. MicroRNAs as regulators of mammalian hematopoiesis. *Semin Immunol.* 2005; 17:155–165.

149. Lawrie CH. MicroRNAs and haematology: small molecules, big function. *Br J Haematol.* 2007; 137:503–512.

150. Yu J, Wang F, Yang GH, et al. Human microRNA clusters: genomic organization and expression profile in leukemia cell lines. *Biochem Biophys Res Commun.* 2006; 349:59–68.

151. Debernardi S, Skoulakis S, Molloy G, Chaplin T, Dixon-McIver A, Young BD. MicroRNA miR-181a correlates with morphological sub-class of acute myeloid leukaemia and the expression of its target genes in global genome-wide analysis. *Leukemia.* 2007; 21:912–916.

152. Marton S, Garcia MR, Robello C, et al. Small RNAs analysis in CLL reveals a deregulation of miRNA expression and novel miRNA candidates of putative relevance in CLL pathogenesis. *Leukemia.* 2007 [Epub ahead of print].

153. Calin GA, Dumitru CD, Shimizu M, et al. Frequent deletions and down-regulation of micro-RNA genes miR15 and miR16 at 13q14 in chronic lymphocytic leukemia. *Proc Natl Acad Sci U S A.* 2002; 99:15524–15529.

154. Calin GA, Sevignani C, Dumitru CD, et al. Human microRNA genes are frequently located at fragile sites and genomic regions involved in cancers. *Proc Natl Acad Sci U S A.* 2004; 101:2999–3004.

155. Calin GA, Croce CM. Genomics of chronic lymphocytic leukemia microRNAs as new players with clinical significance. *Semin Oncol.* 2006; 33:167–173.

156. Calin GA, Pekarsky Y, Croce CM. The role of microRNA and other non-coding RNA in the pathogenesis of chronic lymphocytic leukemia. *Best Pract Res Clin Haematol.* 2007; 20:425–437.

157. Zhang B, Pan X, Cobb GP, Anderson TA. MicroRNAs as oncogenes and tumor suppressors. *Dev Biol.* 2007; 302:1–12.

158. Garzon R, Pichiorri F, Palumbo T, et al. MicroRNA gene expression during retinoic acid–induced differentiation of human acute promyelocytic leukemia. *Oncogene.* 2007; 26:4148–4157.

159. Pekarsky Y, Santanam U, Cimmino A, et al. Tcl1 expression in chronic lymphocytic leukemia is regulated by miR-29 and miR-181. *Cancer Res.* 2006; 66:11590–11593.

160. Costinean S, Zanesi N, Pekarsky Y, et al. Pre-B cell proliferation and lymphoblastic leukemia/high-grade lymphoma in E(mu)-miR155 transgenic mice. *Proc Natl Acad Sci U S A.* 2006; 103:7024–7029.

161. Fulci V, Chiaretti S, Goldoni M, Quantitative technologies establish a novel microRNA profile of chronic lymphocytic leukemia. *Blood.* 2007; 109:4944–4951.

162. Fazi F, Racanicchi S, Zardo G, et al. Epigenetic silencing of the myelopoiesis regulator microRNA-223 by the AML1/ETO oncoprotein. *Cancer Cell.* 2007; 12:457–466.

163. Fazi F, Rosa A, Fatica A, et al. A minicircuitry comprised of microRNA-223 and transcription factors NFI-A and C/EBPalpha regulates human granulopoiesis. *Cell.* 2005; 123:819–831.

164. Venturini L, Battmer K, Castoldi M, et al. Expression of the miR-17-92 polycistron in chronic myeloid leukemia (CML) CD34$^+$ cells. *Blood.* 2007; 109:4399–4405.

165. Fontana L, Pelosi E, Greco P, et al. MicroRNAs 17-5p-20a-106a control monocytopoiesis through AML1 targeting and M-CSF receptor upregulation. *Nat Cell Biol.* 2007; 9:775–787.

166. Zanette DL, Rivadavia F, Molfetta GA, Barbuzano FG, Proto-Siqueira R, Silva WA Jr. miRNA expression profiles in chronic lymphocytic and acute lymphocytic leukemia. *Braz J Med Biol Res.* 2007; 40:1435–1440.

167. Landais S, Landry S, Legault P, Rassart E. Oncogenic potential of the miR-106-363 cluster and its implication in human T-cell leukemia. *Cancer Res.* 2007; 67:5699–5707.

168. Cui JW, Li YJ, Sarkar A, et al. Retroviral insertional activation of the Fli-3 locus in erythroleukemias encoding a cluster of microRNAs that convert Epo-induced differentiation to proliferation. *Blood.* 2007; 110:2631–2640.

169. Jordan CT. The leukemic stem cell. *Best Pract Res Clin Haematol.* 2007; 20:13–18.

170. Jiang X, Smith C, Eaves A, Eaves C. The challenges of targeting chronic myeloid leukemia stem cells. *Clin Lymphoma Myeloma.* 2007; 7 Suppl 2:S71–S80.

3 Isolation and Characterization of Breast and Brain Cancer Stem Cells

MEERA SAXENA and ANNAPOORNI RANGARAJAN

3.1 INTRODUCTION

Even though the existence of functional hierarchies within tumors had been predicted over 40 years ago,[1] the first conclusive evidence came only in 1994, when Lapidot and his co-workers demonstrated the existence of a small subpopulation of tumor-initiating cells within acute myeloid leukemia (AML).[2,3] This raised further questions: Is the existence of such a rare subpopulation of tumor-initiating cells a rule for all kinds of tumors, or is AML an exception?[4] Do solid tumors, which comprise the bulk of adult tumors, also obey this rule? Fueled by research in the field of leukemic stem cells, the hunt began for the identification and isolation of a subset of cells within solid tumors capable of initiating and maintaining tumors. However, the field of cancer stem cell (CSC) biology had to wait almost a decade more before answers to these questions were forthcoming. In 2003, researchers in the laboratory of Michael Clarke elegantly demonstrated the existence of a small subpopulation of cells within breast cancers that alone had the potential to initiate new tumors.[5] These were considered to be putative breast cancer stem cells. This landmark study quickly paved the way for the identification and isolation of CSCs in several other solid cancers, with brain tumors following suit immediately. In this chapter, we look at how researchers succeeded in isolating breast and brain cancer stem cells.

Cancer Stem Cells: Identification and Targets, Edited by Sharmila Bapat
Copyright © 2009 John Wiley & Sons, Inc.

3.2 BREAST CANCER STEM CELLS

3.2.1 Mammary Gland Architecture and Cell Types

The development of the mammary gland or breast is a dynamic process that responds to changes in age, menstrual cycle, and the pregnancy status of the individual.[6] The mammary gland is a well-architectured tubuloalveolar organ made up of ducts and lobules. An apt way to appreciate the architecture of the mammary gland is to compare it with an upturned tree. The secretory alveoli or the *terminal end buds* could be considered to represent the leaves on this tree and they synthesize and secrete milk. Many of these alveoli are grouped together into *lobules*, which drain milk into the short branchlike *intralobular ducts*. The terminal end buds and the intralobular ducts together comprise the structural and functional unit of the breast called as the *terminal ductal lobular unit* (TDLU). The intralobular ducts, in turn, drain milk into the major branchlike *interlobular ducts*. The lobules are organized into 15 to 20 *lobes*, each of which empties into separate *lactiferous sinuses*, and from there into *lactiferous ducts*, which open into the trunklike *nipple* (Fig. 3.1). The mammary gland is supported by interspersed interstitial and intralobular adipose and connective tissue, comprising mainly fibroblasts, endothelial cells, blood cells, and nerve cells.

The mammary epithelium is made up of three cell types: the *myoepithelial cells*, which form the basal layer of ducts and alveoli; the *ductal epithelial cells*, which line the lumen of the ducts, and the *alveolar epithelial cells*, which synthesize the milk proteins. Together, the ductal and alveolar cells constitute the luminal epithelium[7] (Fig. 3.1).

The ability of the mammary gland to replenish itself following cycles of menstruation, pregnancy, and lactation suggested long ago the existence of a repopulating pool of cells within the mammary gland. However, the exact identification and isolation of such cells has proven difficult. It is now well established that tissue-specific adult stem cells represent the parental undifferentiated cells within an organ, which maintain the cell pool through their ability to both self-renew and differentiate into the various cell types of the tissue. Very recently, in the mouse mammary gland, epithelial stem cells were identified as bearing a Lin$^-$CD29^{hi}CD24^{low} cell surface marker profile since an entire functional mammary gland could be derived from a single cell of this phenotype in transplantation experiments.[8] Although a definite breast epithelial stem cell in human mammary gland is yet to be identified, Dontu et al. developed an in vitro system to isolate and propagate normal human breast epithelial stem cells. When grown in nonadherent, suspension conditions, in serum-free media, this research group showed that human mammary epithelial cells generated floating spheres that they termed *mammospheres*. These nonadherent mammospheres contained cells that retained their undifferentiated, multipotent, and proliferative state, suggesting that mammospheres are enriched in cells with functional characteristics of stem or progenitor cells.[9]

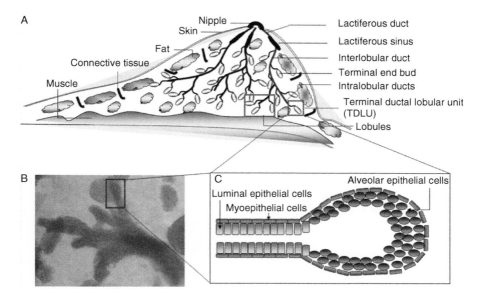

FIGURE 3.1 Mammary gland architecture and cell types: (A) parts of the mammary gland; (B) phase-contrast microscopy of organoid obtained upon mechanical and enzymatic treatment of breast tissue [the boxed area shows the terminal ductal lobular unit (TDLU)]; (C) TDLU and cell types of mammary gland.

3.2.2 Breast Cancer

Breast cancer today is the most prevalent cancer among women, affecting approximately 1 million women worldwide. The most common type of breast cancer is cancer of the ducts, which originates in the terminal end bud and is called invasive ductal carcinoma (IDC). The in situ form of this type of tumor is referred to as ductal carcinoma in situ (DCIS). Other less common types of breast tumors are invasive lobular carcinoma (ILC), which progresses from lobular carcinoma in situ (LCIS). A few examples of rare types of breast cancer are tubular carcinoma, medullary carcinoma, and metaplastic carcinoma. Phylloides tumor and sarcoma are tumors that arise in the connective tissue of the breast.

3.2.3 Identification of Breast Cancer Stem Cells

Al-Hajj and co-workers, in the laboratory of Michael Clarke,[5] identified and isolated putative breast cancer stem cells based on the differential cell surface expression of adhesion markers using flow cytometry (Fig. 3.2A). They sorted the primary breast tumor–derived cancer cells or metastatic pleural effusions, based on the expression of cell adhesion molecules CD44 and CD24 [cluster of

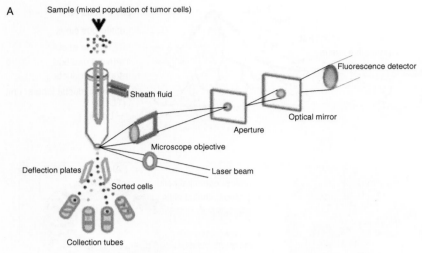

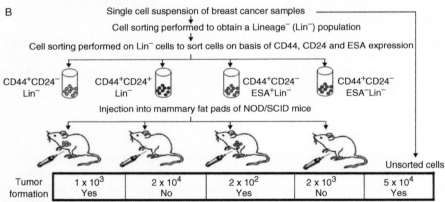

FIGURE 3.2   (A) Principle of flow cytometry to sort cells. The sample of cell suspension is introduced in the center of a narrow, rapidly flowing stream of sheath fluid. Light of a single wavelength is focused onto this stream of cells. Each fluorescent antibody-labeled cell passing through the laser beam may get excited and emit light at a longer wavelength which is detected by a number of detectors. A vibrating mechanism causes the stream of cells to break into individual droplets. Based on the fluorescence signal, an electric charge is placed onto the drop containing a cell which then gets deflected and sorted into different collection tubes by an electrostatic deflection system. (B) Strategy used by Al-Hajj et al. to identify tumorigenic breast cancer cells.

differentiation (CD) cell surface markers]. Often, cells isolated from a primary tissue are contaminated by cells of hematopoietic lineage. Such contaminating hematopoietic cells were eliminated by sorting out for cells expressing the lineage markers CD2, CD3, CD10, CD16, CD18, CD31, and CD64, which are associated with leukocyte, endothelial cell, mesothelial cell, and fibroblast cell populations.

The resulting Lineage$^-$ (Lin$^-$) breast cancer cell fraction was sorted further using the CD44/CD24 parameter. This yielded several expressions, including CD44^{high}, CD44^{low}, CD24^{high}, CD24$^{-/low}$, and combinations thereof, which were collected after sorting and individually grafted into the mammary fat pad of nanobese diaketic/severely compromised immunodeficient (NOD/SCID) mice to assess their tumorigenicity potential [10] (Fig. 3.2B).

3.2.4 Putative Breast Cancer Stem Cells that Exhibit the CD44$^+$ CD24$^{-/low}$Lin$^-$ Marker Profile

Of the Lin$^-$ cells obtained after cell sorting, the authors demonstrated that only the CD44$^+$ CD24$^{-/low}$ fraction of cells represented the prospective breast cancer–initiating cells as assessed by their capacity to generate tumors in NOD/SCID mice; the alternate cell populations sorted on the basis of other combination of markers failed to form detectable tumors. This clearly indicated that the tumor-initiating property resided within the CD44$^+$ CD24$^{-/low}$ cell populations. Yet another support for this was the fact that whereas at least 5×10^4 unsorted breast cancer cells were required to initiate tumors, as few as 10^3 CD44$^+$ CD24$^{-/low}$Lin$^-$ cells consistently generated tumors, indicating that these markers are enriched in cells with tumor-forming potential in NOD/SCID mice. Among the various breast tumors and pleural effusions analyzed, 11 to 35% of the Lin$^-$ cells were CD44$^+$ CD24$^{-/low}$, indicating that the tumor-initiating potential rested within only a small subpopulation of the tumor cells, while the majority of the tumor bulk failed to do so.

3.2.5 ESA$^+$ Subpopulation of CD24$^{-/low}$Lin$^-$ Cells Enriched by Tumorigenicity

On using another cell surface marker, ESA (epithelial surface antigen, which is expressed in epithelial cancers), the ESA$^+$ subpopulation of CD44$^+$ CD24$^{-/low}$Lin$^-$ cells became further enriched in their ability to form tumors when injected into NOD/SCID mice. As few as 200 of the CD44$^+$ CD24$^{-/low}$Lin$^-$ ESA$^+$ cells formed tumors consistently, whereas 10 times CD44$^+$ CD24$^{-/low}$Lin$^-$ ESA$^-$ cells (2000 cells), or 20,000 CD44$^+$ CD24$^+$ cells, failed to form tumors. Thus, the ability of such a small number of CD44$^+$ CD24$^{-/low}$Lin$^-$ ESA$^+$ cells to regenerate a complete tumor is similar to the ability of normal stem cells to regenerate an entire tissue.

3.2.6 Tumorigenic Breast Cells Displaying Properties of Stem Cells

The two hallmark properties of stem cells are self-renewal and the ability to regenerate all the other cell types of the tissue. To test whether the tumorigenic subpopulation of breast cancers indeed behaved like stem cells, they were screened for an expression of the properties of self-renewal and an ability to generate phenotypically diverse cells. Tumors generated in NOD/SCID mice through the engraftment of 200 CD44$^+$ CD24$^{-/low}$Lin$^-$ ESA$^+$ cells were dissociated and subjected to flow cytometry after immunostaining for specific surface markers. These cells exhibited heterogeneity with respect to the expression of CD44, CD24, and ESA surface molecules, thereby recapitulating the diversity of cell repertoire within the parent tumor. Thus, in addition to the phenotype of the cells injected (i.e., CD44$^+$ CD24$^-$ ESA$^+$), cells of other phenotypes were detected in the newly generated tumor. This small subpopulation of breast cancer cells indeed had the ability to propagate itself as well as to generate the phenotypic heterogeneity characteristic of the parent tissue, confirming their stem cell–like nature.

The extensive self-renewal and proliferation potential of these putative CSCs was assessed further by monitoring them through sequential cycles of sorting the specific subpopulation followed by tumor development in vivo in mice. The tumorigenic CD44$^+$ CD24$^{-/low}$Lin$^-$ breast cancer cells could thus be serially passaged through a minimum of four such cycles of sorting and tumor formation, thus demonstrating their extensive self-renewal and proliferation capacity.

In conclusion, to modify the classical definition of normal stem cells, the two properties required to qualify a cell to gain recognition as a cancer stem cell are (1) its ability to self-renew and proliferate extensively, and (2) its ability to generate the various cell types of the parent tumor in order to recapitulate the original tumor heterogeneity. Since very few of the CD44$^+$ CD24$^{-/low}$Lin$^-$ ESA$^+$ cells were able to be passaged serially in vivo, mimicking the extensive cell proliferation ability of stem cells and also give rise to a heterogeneous tumor recapitulating the parent tumor, these cells have been termed breast cancer–initiating or breast cancer stem cells, the first of their kind to be identified in solid tumors.[10]

3.2.7 In Vitro Propagation of Breast Cancer Stem Cells as Mammospheres

In an effort to understand the role and importance of CSCs, and the molecular and cellular events that occur in them during seeding of cancer, one must first harvest appreciable numbers of these rare cells. For this it is essential to isolate and propagate such CSCs in long-term in vitro culture. Toward meeting this demand, advances made in the field for culturing normal breast stem/progenitor cells as floating mammospheres came to the rescue.[9] In this strategy, cells are seeded in low-attachment plates in serum-free media supplemented with growth factors that provide a stimulus of activation of the stem cells, and included bFGF, EGF and insulin. While a vast majority of cells die, owing to their inability to attach (i.e., grow in an anchorage-independent manner), the few surviving cells generate floating spheres termed *mammospheres*. Mammospheres generated by normal breast tissue were found to contain cells capable of self-renewal and could be maintained in an undifferentiated state for several passages, and also demonstrated multilineage differentiation potential on receiving the appropriate differentiation cues from the environment. This strategy thus allowed the isolation and propagation of stem and progenitor cells in vitro (Fig. 3.3A).

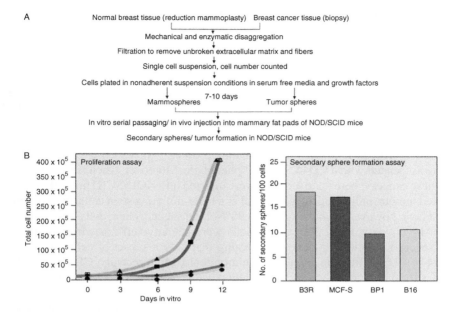

FIGURE 3.3 (A) Method of in vitro propagation of normal and cancer mammospheres; (B) proliferation and sphere formation efficiency of breast cancer precursor cell lines: ▲, B3R, grade 3 carcinoma; ■, MCF-S, breast carcinoma cell line; ♦, BP1, grade 2 carcinoma; ●, B16, grade 1 carcinoma. (Adapted from ref. 11.)

Breast Cancer–Derived Cells that Generate Cancer Spheres   To further address whether breast cancer stem cells would also have the potential to generate mammospheres, Ponti and co-workers subjected primary breast cancer–derived cells to suspension culture.[11] The investigators observed that of the 16 different primary cancers that were subjected to this assay, three had the capacity to give rise to long-term cultures. These cancer-derived mammospheres were composed of undifferentiated, self-renewing tumor cells which could be differentiated into epithelial-like cells expressing markers of the mature mammary gland epithelium, thus demonstrating that they were endowed with stem/progenitor cell properties. In addition, the authors observed that the three tumors from which these cells were derived were all estrogen receptor–positive lesions, which suggested a possible correlation between hormonal cues and maintenance of stemness.[10,11]

Sphere-Forming Efficiency Correlated with Tumor Grade   Another interesting observation made by this group was that the primary culture established from aggressive grade III breast carcinomas had more proliferating cells and correspondingly higher sphere-forming efficiency than those of a grade II breast carcinoma; the latter in turn, had more proliferating cells and higher sphere-forming potential than those of the least invasive grade I breast carcinoma (Fig. 3.3B). Since sphere formation efficiency is indicative of the number of self-renewing stem cells or proliferating progenitors in a cell pool, a higher sphere formation efficiency observed in cells obtained from a more aggressive tumor suggests the presence of more self-renewing cells in that tumor.

Breast Cancer Mammosphere–Derived Cells that Are CD44+ CD24− and that Initiate Tumors In Vivo   As Al-Hajj et al. had demonstrated that the subpopulation of breast cancer cells capable of behaving as if CSCs had a CD44+ CD24− phenotype,[5] Ponti et al. subjected the cancer spheres derived from breast cancer to flow cytometry to assess for their marker profile.[11] Indeed, the cancer spheres were associated with a CD44+ CD24− phenotype. Moreover, when small numbers of these cancer sphere–derived cells were injected into NOD/SCID mice, they generated tumors, indicating their potential as cancer-initiating stem cells. On a discordant note, however, even though 95 to 98% of the cancer sphere cells were CD44+ CD24−/low, only 10 to 20% of the cells were actually self-renewing, as shown by sphere formation efficiency, suggesting that only a subset of cells within the CD44+ CD24− fraction are actually tumorigenic. More recently, CD44+ CD24− CSCs from individual human breast tumors were shown to be clonally related but not always identical, suggesting that few additional genetic changes may be further rendered for the progression of CD44+ CD24− CSCs into the various other phenotypes that contribute to the heterogeneity within a tumor.[12]

3.3 BRAIN CANCER STEM CELLS

3.3.1 Brain Architecture and Cell Types

The human brain, although just 2% of human body weight, is the most complex of all organs. It is the prime site for integration of all the physiological functions of a human body. The brain consists of extremely convoluted neural tissue that is supported and protected by the *skin*, the cranial bones of the *skull*, the three layers of connective tissue membranes called the *meninges*, and the *cerebrospinal fluid* (CSF). The CSF flows between the spaces of the meninges, providing nourishment to the brain tissue and also functioning as a shock absorber against physical stresses.

Anatomically, the brain can be divided broadly into three parts: the *cerebrum, brain stem*, and *cerebellum*. The functional and structural units of the brain are the *nerve cells* or *neurons* and the *glial cells* (Fig. 3.4). The brain contains approximately 100 billion neurons that receive, process, and transmit information as electrochemical signals to and from the brain. Glial cells make up 90% of all brain cells and are called the "nerve glue," as they hold the nerve cells together, support, and nourish them. The three main types of glial cells found in the central nervous system are the *astrocytes* (star-shaped cells that provide physical and nutritional support for neurons), *oligodendrocytes* (which form the protective coatings called myelin around neurons), and the *microglial cells* (phagocytes that clear away the dead neurons).

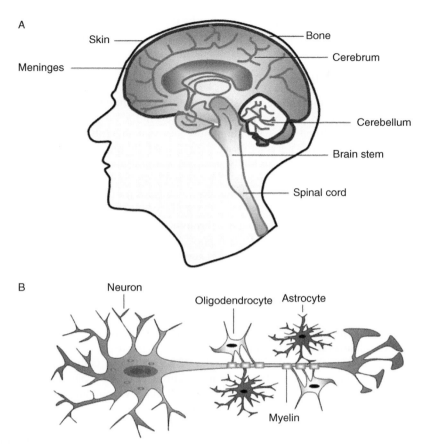

FIGURE 3.4 (A) Architecture of the brain; (B) brain cell types. (*See insert for color representation of figure.*)

3.3.2 Brain Cancers

Brain cancers are the leading cause of cancer-related mortality in children. The nomenclature of these cancers is based on the cell type or part of the brain in which they arise. The most common type of brain cancers arise in the glial cells and hence are called the *gliomas*. Within the gliomas, astrocytomas (arising from astrocytes) are the most common tumors. A type of grade 3 astrocytoma called anaplastic astrocytoma and a grade 4 astrocytoma called glioblastoma multiforme are both recognized to be highly malignant brain cancers. Other gliomas include brain stem glioma, ependymoma (arises in ependymal cells that line the brain ventricles), and the rare oligodendroglioma (arises in the oligodendrocytes). Some of the tumors that do not arise in the glial cells include the medulloblastoma (arises in cerebellum and is the most common brain tumor in children), pineal gland tumor (arises in the pineal gland), craniopharyngioma (arises at the base of the brain near the pituitary gland and most often afflicts children), and meningioma (arises in the meninges).

3.3.3 Brain Stem Cells

The long-held belief that the nervous tissue does not regenerate was shaken in the 1960s when few researchers discovered two regions in the rat brain that contained cells capable of dividing.[13–15] The dogmatic view that an adult brain is static and incapable of regenerating nervous tissue was discarded completely in the 1990s, when scientists showed that the adult brain contains undifferentiated stem cells that could generate the three major types of brain cells: the neurons, astrocytes, and oligodendrocytes. In 1992, Reynolds and Weiss isolated for the first time, neuronal stem cells from the subventricular zone (SVZ) of the mouse brain.[15] In fact, it is now known that neurogenesis occurs throughout adulthood in restricted areas of the brain such as the SVZ and the hippocampus (present in cerebral cortex), leading to production of as many as 30,000 new neuronal precursors or neuroblasts every day.[14–16] They developed for the first time an in vitro assay for propagation of neural stem cells as nonadherent neurospheres.[15] A *neurosphere* is a clone of cells formed through the self-renewal of one parent neural stem cell. Therefore, neurosphere generation is an in vitro assay indicative of the self-renewal capability of cells. In fact, it is neurosphere cultures that pioneered the strategy for in vitro propagation of several types of stem cells as floating spheres.

Soon after, Uchida et al. identified CD133-expressing cells derived from human neural tissues as candidate human central nervous system stem cells (hCNS-SC).[17] As mentioned earlier, each cell type within an organ can be identified and distinguished from other cell types on the basis of a characteristic cell surface marker profile. In their study, Uchida et al. screened 50 different monoclonal antibodies against dissociated fetal brain tissue to identify specific hCNS-SC markers and found that the cell population expressing CD133 displayed attributes of stem cells.[17] These cells were able to self-renew, expand in neurosphere cultures, and differentiate in vitro into neurons and glial cells. Moreover, they also observed engraftment of the CD133$^+$ cells on transplantation into the brains of NOD/SCID mice.

3.3.4 Brain Cancer Stem Cells

While globally, researchers were revising their concepts with respect to the discovery that the adult nervous tissue can regenerate, there came a report in 2003 by Singh and co-workers from the laboratory of Peter Dirks claiming the identification of a CD133-expressing subpopulation of brain cancers as brain cancer stem cells, which they called *brain tumor stem cells* (BTSCs).[18] At that time, it was the only other report of the identification and isolation of cancer stem cells from a solid tumor following the identification of breast cancer stem cells. Unlike the case of breast cancer stem cells, evidence for the existence of BTSCs first came from in vitro experiments, which were subsequently supported by in vivo experiments.

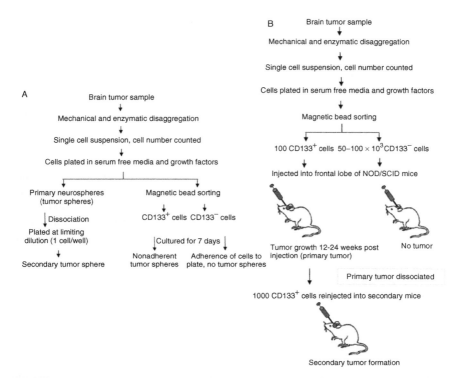

FIGURE 3.5 (A) Strategy for in vitro assays to identify brain tumor cancer stem cells; (B) strategy for in vivo tumorigenicity assays to identify CD133$^+$ brain tumor cancer stem cells. (Adapted from refs. 18 and 19.)

3.3.5 Brain Cancer–Derived Cells that Generate Tumor Spheres

Sheila Singh and co-workers in 2003 utilized the previously established technique of culturing normal neural stem cells as neurospheres to propagate brain tumor stem cells as nonadherent cancer spheres (Fig. 3.5A). They established cultures from 14 solid primary pediatric brain tumors and showed that tumor spheres generated from all 14 samples expressed nestin (an intermediate filament protein characteristic of normal neural stem cells and progenitors) and CD133 (a hematopoetic and neural stem cell marker). The tumor spheres did not express markers for differentiated neurons, astrocytes, or oligodendrocytes, such as βIII tubulin and glial fibrillary acidic protein (GFAP).[18]

Tumor Spheres That Show Stem Cell Properties   The ability of primary spheres to generate secondary spheres after single-cell dissociation has been used to demonstrate self-renewal. All such primary tumor–derived neurospheres generated secondary neurospheres when seeded at serial dilutions, revealing the presence of stem cell–like cells capable of self-renewal. Using a limiting dilution assay, the minimum number of cells required to generate at least one tumor-derived neurosphere per well in the case of medulloblastoma was found to be much lower (23 ± 17) than in pilocytic astrocytoma (98.25 ± 4.6). Additionally, when an equal number of cells (100 cells per well) of different tumor subtypes were seeded, a higher number of secondary tumor neurospheres were generated from cells derived from a more aggressive cancer (medulloblastoma) than from a less aggressive cancer (pilocytic astrocytoma). These experiments indicated that aggressive brain cancers would harbor a higher number of CSCs then would the less aggressive cancers.

On exposing the dissociated tumor neurosphere cells to differentiation-inducing conditions, the cell types generated recapitulated the same phenotype as the tumor from which they were derived. For example, on growing medulloblastoma-derived tumor sphere cells in the presence of FBS, the cells differentiated and expressed βIII tubulin (neuronal cell marker), while cells from pilocytic astrocytoma (cancer of astrocytes)–derived tumor spheres expressed GFAP, an astrocyte marker, thus mimicking the parent tumor phenotype. On differentiation, however, these cells failed to express the stem cell markers nestin and CD133.

Tumor Spheres Initiated Only by CD133$^+$ Cells   To characterize the BTSC population further on the basis of CD133 expression, Singh et al. used magnetic bead sorting to separate CD133$^+$ and CD133$^-$ populations and grew both populations in conditions that favor tumor neurosphere formation. They found that only the CD133$^+$ tumor cells grew as nonadherent tumor neurospheres, whereas the CD133$^-$ cells adhered to culture dishes, did not proliferate, and did not generate neurospheres. Using a limiting dilution assay, they further confirmed that only the CD133$^+$ cells (and not CD133$^-$ cells) could yield neurospheres and hence had stem cell activity. Thus, the CD133-expressing subpopulation of brain tumors represents a distinct class of brain tumor cells possessing stem cell activity.

CD133$^+$ Cells Behaving as Tumor-Initiating Cells (CSCs) In Vivo   The same group in 2004 reinforced their findings using more convincing in vivo assays. The hallmark properties of a cancer stem cell are its ability to (1) self-renew, (2) undergo multilineage differentiation, and (3) recapitulate the phenotype of the original heterogeneous tumor. To attribute the same characteristics to the CD133$^+$ putative BTSC, Singh et al. sorted the CD133$^+$ and CD133$^-$ fractions using magnetic sorting and then transplanted both types of sorted cells into the frontal lobes of NOD/SCID mice[19] (Fig. 3.5B). They showed that in contrast to the high cell numbers (10^5 to 10^6) of unsorted tumor cells or cell line–derived cells required to form tumors, as few as 100 CD133$^+$ cells were sufficient for the formation of brain tumors in NOD/SCID mice within 12 to 24 weeks of injection. Interestingly, even 50,000 to 100,000 CD133$^-$ cells were not able to form tumors in the brains of NOD/SCID mice, thereby indicating that the tumor-initiating potential was present only within the subpopulation of CD133$^+$ cells. Highly aggressive glioblastoma were found to harbor 11 to 19%, while the medulloblastoma harbor 6 to 21% of tumorigenic CD133$^+$ cells.

The authors went on to characterize the xenografts formed on injection of medulloblastoma and glioblastoma multiforme (GBM)–derived CD133$^+$ cells by immunohistochemistry. They found that all of the medulloblastoma and GBM xenografts had cells expressing nestin and vimentin, markers of neural stem and precursor cells. Also, a large majority of cells in a medulloblastoma xenografts expressed βIII tubulin, similar to the original patient tumor. A few of the cells were also shown to express GFAP, an astrocyte marker. While the majority of the cells in GBM xenografts expressed GFAP, a minority of the cells expressed a neuronal marker, microtubule-associated protein (MAP2), indicating that the CD133$^+$ cells derived from tumors were capable of multilineage differentiation. On immunostaining for Ki67 (a marker of proliferation), a high number of proliferating cells were found in the xenografts. These results indicated that CD133$^+$ cells were capable of proliferation and differentiation when injected into NOD/SCID mice, thus recapitulating the original tumor from which they were derived.

Serial passaging of tumor-derived and tumor-sorted cells is considered the gold standard to demonstrate and assess the long-term self-renewal capacity of stem cells. To show beyond doubt that CD133$^+$ cells could self-renew and hence were authentic BTSC, the authors performed in vivo serial passaging (retransplantation) of the primary xenografts obtained from a pediatric and an adult GBM case. It was observed that the secondary brain tumors generated recapitulated the phenotype of the original patient tumor. This experiment provided direct evidence for the self-renewal capacity of this population. None of the mice injected with CD133$^-$ tumor cells developed brain tumors. Thus, CD133$^+$ brain tumor–derived cells are concluded to represent the tumorigenic subpopulation that show stem cell properties such as self-renewal, extensive proliferation, multilineage differentiation, and the ability to recapitulate the original tumor, thereby qualifying as BTSCs.

3.4 CONCLUSIONS AND FUTURE PERSPECTIVES

The stem cell concept of tumors seems to be shifting the focus of cancer biology from treating the bulk of the tumor to a rare subpopulation within it that alone has the potential to regenerate an entire new tumor.[20] Since these rare stem cells also have enhanced radio and chemoresistant properties, it is possible that the continuing presence of these cells causes tumor relapse following treatment. Therefore, it seems logical to design novel therapeutic strategies that specifically target this minority population. Thus, research in the field of breast and brain cancer stem cells has provided evidence for the existence of functional heterogeneity within solid tumors. However, this area of research still has several unanswered questions, such as:

1. Do the CD44$^+$ CD24$^{-/\text{low}}$Lin$^-$ subpopulation of breast cancer and CD133$^+$ subpopulation of brain tumors represent true CSCs, or do they simply

represent markers that promote better engraftment in mice models of xenograft assays?[21,22]

2. Are mammo- and neurospheres actually generated by proliferation of single stem cells, or does cell aggregation or clumping contribute to the process? Thus, can sphere-forming efficiency be considered to be a true reflection of the number of stem cells present in the starting tissue? [23]

3. The cell type of origin of these CSCs within the normal organ and their evolution from normalcy to a state of transformation is not yet clear. Based on their similarities to normal stem cells, it is predicted that CSCs may originate through the accumulation of mutations in normal stem or progenitor cells within the adult tissue; however, concrete evidence is yet to emerge.[20,22−25]

4. What are the unique tumorigenic mechanisms operating within the CSCs that make them distinct from the tumor bulk? [26]

Cancer stem cell research no doubt promises to provide exciting answers to several of these questions in the years to come.

REFERENCES

1. Bergsagel DE, Valeriote FA. Growth characteristics of a mouse plasma cell tumor. *Cancer Res.* 1968; 28:2187−2196.

2. Lapidot T, Sirard C, Vormoor J, et al. A cell initiating human acute myeloid leukaemia after transplantation into SCID mice. *Nature.* 1984; 367:645−648.

3. Huntly BJP, Gary D. Gilliland leukaemia stem cells and evolution of cancer stem cell research. *Nat Rev Cancer.* 2005; 5:311−321.

4. Wang JC, Dick JE. Cancer stem cells: lessons from leukaemia. *Trends Cell Biol.* 2005; 15:494−501.

5. Al-Hajj M, Wicha MS, Benito-Hernandez A, Morrison SJ, Clarke MF. Prospective identification of tumorigenic breast cancer cells. *Proc Natl Acad Sci U S A.* 2003; 100:3983−3988.

6. Bissell MJ, Radisky DC, Rizki A, Weaver VM, Petersen OW. The organizing principle: microenvironmental influences in the normal and malignant breast. *Differentiation.* 2002; 70:537−546.

7. Smalley M, Ashworth A. Stem cells and breast cancer: a field in transit. *Nat Rev Cancer.* 2003; 3:832−834.

8. Shackleton M, Francois V, Simpson KJ, et al. Generation of a functional mammary gland from a single stem cell. *Nature.* 2006; 439:84−88.

9. Dontu G, Abdallah WM, Foley JM, et al. In Vitro propagation and transcriptional profiling of human mammary stem/progenitor cells. *Genes Dev.* 2003; 17:1253−1270.

10. Ponti D, Zaffaroni N, Capelli C, Daidone M. Breast cancer stem cells: an overview. *Eur J Cancer.* 2006; 42:1219−1224.

11. Ponti D, Costa A, Zaffaroni N, et al. In Vitro propagation of breast tumorigenic cancer cells with stem/progenitor cell properties. *Cancer Res.* 2005; 65:5506−5511.

12. Shipitsin M, Campbell LL, Argani P, et al. Molecular definition of breast tumor heterogeneity. *Cancer Cell.* 2007; 11:259–273.

13. Weiss S. Is there a neural stem cell in the mammalian forebrain? *Trends Neurosci.* 1996; 19:387–393.

14. Luskin MB. Restricted proliferation and migration of postnatally generated neurons derived from the forebrain subventricular zone. *Neuron.* 1993; 11:173–189.

15. Reynolds BA, Weiss S. Generation of neurons and astrocytes from isolated cells of the adult mammalian central nervous system. *Science.* 1992; 255:1707–1710.

16. Galli R, Gritti A, Bonfanti L, Vescovi AL. Neural stem cells: an overview. *Circ Res.* 2003; 92:598–608.

17. Uchida N, Buck DW, He D, et al. Direct isolation of human central nervous system stem cells. *Proc Natl Acad Sci U S A.* 2000; 97:15720–15725.

18. Singh SK, Clarke ID, Terasaki M, et al. Identification of a cancer stem cell in human brain tumors. *Cancer Res.* 2003; 63:5821–5828.

19. Singh SK, Hawkins C, Clarke ID, et al. Identification of human brain tumor initiating cells. *Nature.* 2004; 432:396–401.

20. Reya T, Morrison SJ, Clarke MF, Weissman IL. Stem cells, cancer, and cancer stem cells. *Nature.* 2001; 414:105–111.

21. Fillmore C, Kuperwasser C. Human breast cancer stem cell markers CD44 and CD24: enriching for cells with functional properties in mice or in man? *Breast Cancer Res.* 2007; 9:303.

22. Calabrese C, Poppleton H, Kocak M, et al. A perivascular niche for brain tumor stem cells. *Cancer Cell.* 2007; 11:69–82.

23. Singec I, Knoth R, Meyer RP, et al. Defining the actual sensitivity and specificity of the neurosphere assay in stem cell biology. *Nat Methods.* 2006; 3(10): 801–809.

24. Polyak K, Hahn WC. Roots and stems: stem cells in cancer. *Nat Med.* 2005; 11:296–300.

25. Pardal R. Applying the principles of stem-cell biology to cancer. *Nat Rev Cancer.* 2003; 3:895–902.

26. Linheng Li, Neaves WB. Normal stem cells and cancer stem cells: the niche matters. *Cancer Res.* 2006; 66(9): 4553–4557.

4 Cancer Stem Cell Side Populations

DANUTA BALICKI and RAYMOND BEAULIEU

4.1 INTRODUCTION

The cancer stem cell (CSC) hypothesis represents a paradigm shift in oncology.[1,2] It unifies our clinical observations into a single cohesive cancer model. In the classic stochastic model, all cancer cells have tumorigenic potential. In contrast, the cancer stem cell model suggests that only a small proportion of cancer cells have intrinsic tumorigenic properties.[2] In addition to their multipotential capacity to differentiate into progenitor cells of different lineages, stem cells self-renew. CSCs are cells capable of cancer initiation, expansion, and progression. Mutated normal stem cells and progenitors are sources of cancer stem cells.[3,4] CSCs are quiescent but have a high level of proliferative potential under specific environmental stimuli of the stem cell niche. These cells may evade the asymmetric cell division of adult stem cells, giving rise to their self-renewal while generating differentiated daughter cells.[5,6] Instead, clonal expansion of malignant cells occurs, contributing to the malignant phenotype (Fig. 4.1). The fate of stem cells, including their capacity to self-renew and differentiate, may also be programmed by the stem cell niche: that is, the cellular microenvironment providing the necessary support and stimuli to sustain self-renewal and differentiation, in addition to cellular adhesion, division, and survival. [4,7−13]

Conventional chemotherapy eradicates cycling differentiated cancer cells but is ineffective against quiescent CSCs. Under the influence of the stem cell niche, these cells not only thrive but also escape cell cycle and proliferation checkpoints by conventional therapies, leading to tumor recurrence and metastases. The mechanisms of resistance of CSCs include high levels of expression of transporter molecules, which export therapeutic agents.[14] Thus, it is necessary to isolate CSCs and identify their biological properties so that they can subsequently be eliminated by targeted therapies. Stem cell side populations (SPs) are recognized to be a useful tool for the isolation and analysis of stem cells, including CSCs.

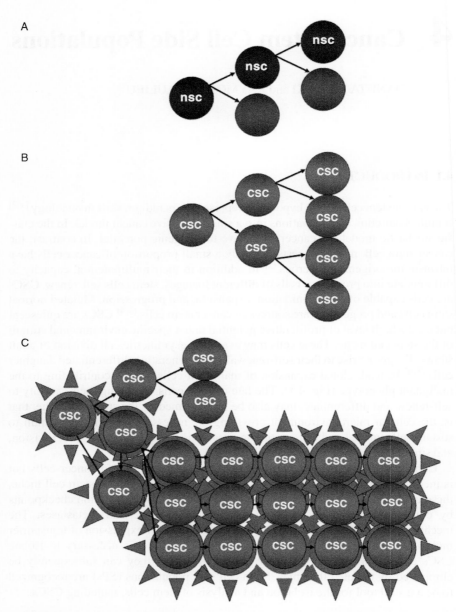

FIGURE 4.1   Comparative physiology of (A) normal stem cells (NSCs), (B) cancer stem cells (CSCs), and (C) cancer stem cells within their niche.

4.2 STEM CELL SIDE POPULATIONS

SPs represent cell populations enriched in primitive and undifferentiated cells.[15,16] The isolation of SPs is based on the technique initially described in 1996 by Goodell et al.[17] During a study of the cell cycle distribution of whole bone marrow cells using Hoechst 33342 vital dye staining, they discovered that the display of Hoechst fluorescence simultaneously at two emission wavelengths (blue 450 nm and red 675 nm) localizes a small, yet distinct nonstained cell population that expresses stem cell markers (Sca1$^+$ Lin$^{neg/low}$). SPs are localized in the left lower quadrant of a fluorescence activated cell sorter (FACS) profile (Fig. 4.2). Furthermore, the degree of efflux is correlated with the maturation state such that cells exhibiting the highest efflux are the most primitive or least restricted with regard to their differentiation potential.[18] Hoechst 33342 stoichiometrically binds to the A-T-rich areas of the minor DNA groove, and the intensity of its fluorescence is an index of DNA content, chromatin structure and conformation, and cell cycle.[18] It has been proposed that the exclusion of Hoechst 33342 by SP cells is an active process involving multidrug resistance transporter 1 (MDR1), a member of ATP-binding cassette (ABC) transporter transmembrane proteins. To support this hypothesis, the MDR1 inhibitor verapamil was used to demonstrate that Hoechst 33342 exclusion by SP cells decreases when the MDR1 pump is blocked with calcium channel blockers. However, verapamil is a nonspecific inhibitor of MDR1 transport. Furthermore, the MDR1 antibody does not identify SP cells exclusively, as MDR1 is expressed by 65% of bone marrow cells and SP represent only 0.1% of these cells.[17] Thus, MDR1 cannot be taken as a single marker to identify and isolate SP cells, and additional transporters should be analyzed. For example, Zhou et al.[19] have demonstrated that the breast cancer resistance protein (BCRP) is a marker of the SP phenotype in breast tissues; further, the expression of integrin β_3 is also correlated with the properties of quiescent stem cells possessing the SP phenotype.[20]

In addition to their identification in a broad spectrum of normal cells,[18,21–30] SPs have also been identified in cancer cell lines and tumors (e.g., neuroblastoma,[31] melanoma,[32] ovarian,[33] hepatocellular,[34] and glioma cell lines.[35]). SP cells have been utilized to isolate and identify malignant cell markers and properties, where their ability to expel nuclear dyes correlates with multidrug resistance (i.e., the expulsion of cytotoxic drugs.[31]). The transporters associated with this functionality, ABCB1 (MDR1) and ABCG2 (BCRP), can also be studied using SPs since these transporters bring about active exclusion of chemotherapeutic agents (e.g., ABCB1 expels vinblastin[36] and paclitaxel,[37] while ABCG2 expels imatinib mesylate,[38] topotecan, and methotrexate[39]). SPs and ABC transporters have been identified concurrently in certain cell lines: for example, ABCA3 and ABCG2 in human neuroblastoma cell lines,[31] ABCB1 and ABCG2 in human gastrointestinal cell lines,[40] and Mdr1a in the mouse muscle cell line C2C12.[41]

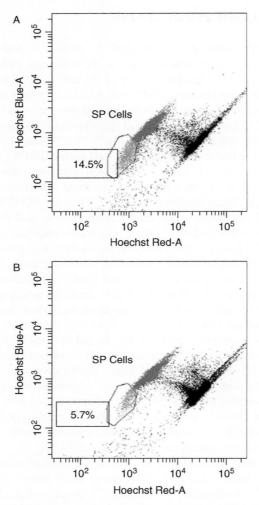

FIGURE 4.2   Presence of a side population in the human MCF-7 breast adenocarcinoma cell line. Cells labeled with Hoechst 33342 reveal the presence of a small low-staining SP population (A). When treated with 50 μL verapamil, an inhibitor of SP transporters, this population diminishes in size (B). The cells were counterstained with 1 μg/mL propidium iodide. (*See insert for color representation of figure.*)

Although there is a certain degree of disagreement regarding the phenotypes that best characterize stem cells, similarities between stem cells and SPs are also evident. In two independent studies[31,41] it has been demonstrated that (1) hematopoietic stem cells are also present in the non-SP compartment, and (2) mammary-repopulating cells capable of reconstituting a cleared mammary fat pad are independent of the SP phenotype.[37,38] Given that SPs are composed of stem

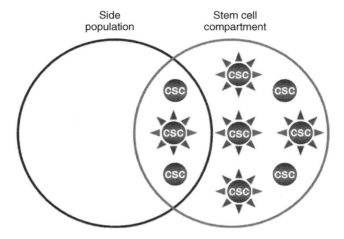

FIGURE 4.3   Overlap between a side population and a stem cell compartment.

and nonstem cells, and that some stem cells do not reside within SPs, it may be surmised that the SP fraction of cells within any normal tissue or tumors does not completely overlap with the stem cell compartment (Fig. 4.3).

Despite the conclusion above, it has been reported that a majority of SP cells within tumors can be characterized by their immature, poorly differentiated, and highly tumorigenic nature. At the genetic level, comparative gene expression profiles confirm that SP cells are less differentiated than non-SP cells.[16] They also express high levels of stem cell markers and low levels of differentiation markers.[40] There remains a controversy regarding the differential capacity of SP cells to divide asymmetrically, as is the case with adult stem cells. In some reports, only SPs are able to generate SPs and non-SPs, whereas non-SPs are unable to do so[31,35,40]; others have shown that both SPs and non-SPs are able to generate either SP or non-SPs.[41] It thus seems most likely that SPs are heterogeneous populations, and only a fraction of these populations has the capacity to undergo asymmetric division. Moreover, SPs and non-SPs differ in terms of their tumorigenicity in vivo and their potential for multilineage differentiation, SPs being more malignant than non-SPs and more likely to differentiate.[35]

To characterize this property further, the SP cell population in the tail region (of flow cytometry profiles) has been divided into different compartments according to their capacity to efflux dyes. It has been demonstrated that cells at the tip of the SP tail (which represent the highest Hoechst efflux activity) compared to the distal portion,[43] also have the highest progenitor activity. Thus, under defined conditions and circumstances, the level of Hoechst exclusion may be a direct indication of cellular differentiation and presumably also have higher transporter levels. Since the SP phenotype has also often been correlated with ABCG2 expression, which is dynamic during progenitor commitment,[44] it may be concluded that the subpopulations found in the SP compartment reflect various degrees of cellular differentiation. Thus, in specific cases and under defined experimental conditions, the isolation of cells with the highest dye efflux activity within the SPs could lead to a highly enriched stem cell population.

4.3 SIDE POPULATIONS IN NORMAL TISSUE

Since the SP phenotype is correlated with dye exclusion (ABC transporter) activity, several studies have been undertaken to establish a specific association between the increased expression of a given transporter in specific tissues. For example, since expression of the ABCG2/Bcrp1 transporter is the most frequent association with the SP phenotype, it has undergone intense study. High levels of expression of this transporter has been reported in skeletal muscle,[45] bone marrow,[46] skin,[47,48] and cornea.[49–51] Low to complete absence of SP expression has been observed in Bcrp1 knockout mice in the lungs,[29] skeletal muscle, and bone marrow.[46] However, both ABCG2/Bcrp1 and MDR1/Mdr1 transporter expression have been described in the mouse mammary gland[52] and human pancreatic islets of Langerhans.[53] Taken together, these results also suggest the tissue-dependent distribution of these transporters.

Gene expression profile studies of SP cells using an in-house microarray assay platform[15] or Affymetrix 430 2.0 array chips[52] have shown that several genes are up-regulated. Notably, the transcription factor ID2, translation initiation factor EIF5, and growth factor NRG1[15] were identified at significantly higher levels. The up-regulation of transcription- and translation-associated genes have also been reported, while genes regulating the immune response and cell adhesion are down-regulated.[54] The latter suggests that SPs contain a particular cell fraction whose survival is not exclusively or entirely dependent on cell–cell interactions.[55] In addition, cell cycle checkpoints/repression genes have been reported to be up-regulated in SPs, suggesting that SP cells will arrest at a particular phase of the cell cycle. Correspondingly, G0/G1 cell cycle arrest has been reported within SPs, which supports the relationship between SP and stem cells. It is thus concluded that SP and stem cells are both slow-cycling cells that may reside in a quiescent state.[52,56,57]

Universal stem cell markers have been proposed for characterizing SPs. For example, the stem cell antigen (Sca1$^+$) is present in heart, bone marrow, mammary gland, and pituitary cells of mouse tissue [58–61]; proteins associated with the Notch-1 and Wnt self-renewal pathways[62,63] are often detected in SP fractions. Tissue-dependent stem cell markers have also been described in association with SPs. For example, CD34$^+$ has been identified in hematopoietic SP,[64] keratin 14 (K14), and keratin 19 (K19) in skin SPs. This suggests that SPs derived from specific tissues may express a given pattern of tissue-dependent markers in addition to the common functionality of dye exclusion. However, certain exceptions to this principle exist in the literature. For example, nestin is a neural stem cell marker reported initially in nervous system tissues,[65] but is now identified in several other tissue- derived SP fractions, including those from the pancreas.[53] Despite the lack of agreement as to whether or not SPs express a universal or specific stem cell marker, there is a consensus as to the presence of an immature cellular population within the SP fraction. While immature cell markers are expressed at high levels within SPs, they also express low levels of differentiation markers [e.g., epithelial membrane antigen (EMA) in mouse mammary epithelial cells].[66] Thus, it is not surprising that stem cells can be identified within SPs, as they contain immature and undifferentiated cell populations.

There is also a heightened controversy regarding the expression of SP markers in tissue samples compared with cell lines, with a multitude of contradictory results. A pertinent example is provided by estrogen receptor (ER) expression within the mammary gland. ERs have variably been reported to be either absent in SPs[67] or present at high levels in SPs,[68] or may be expressed in SPs and non-SPs at similar levels.[66] Factors that have been identified to explain these absolutely contradictory results include the effects of tissue dissociation in the preparation of single-cell suspensions, cell counting, Hoechst concentration, and SP gate selection.[69] Age is yet another confusing factor that yields mixed results. It has been demonstrated that SP size may increase[70] or decrease[71] with aging, whereas Triel et al.[48] have concluded from their studies that SP size is age independent. Although SPs may be an enriched source of stem cells throughout the lifetime of a given organism, a thorough characterization of this cell population will elucidate the origins, markers, and functions of its constituent cells.

4.4 SIDE POPULATIONS IN TUMORS

Two general approaches are currently being used to identify and characterize CSCs. First, molecular markers have been proposed: for example, as described elsewhere in the book, $CD44^+/CD24^-$ for breast cancer stem cells,[72,73] $CD34^+/CD38^-$ for leukemia stem cells,[74] and $CD44^+/_{\alpha2}\beta_1^{hi}/CD133^+$ for prostate cancer stem cells.[75] Second, SP cells isolated from tumors have proven to be an attractive alternative strategy to the study of CSCs, especially in cases where specific surface marker associations with normal stem cells of the organ of origin have not been identified. This stems from the proposition that since SPs isolated from normal tissues are enriched in normal stem cells, the same population from tumors may be enriched in CSCs. For example, Grichnik et al.[32] have isolated SPs from human melanoma samples and have shown that SP cells have overlapping properties common with normal stem cells, including the finding that they are small, possess the capacity to give rise to larger cells, and have the greatest ability to expand in culture.

Since then, SP-enriched cells that express stem cell characteristics have been identified in a wide range of cancers, including retinoblastoma,[76] bone marrow from patients with acute myeloid leukemia (AML),[77] and neuroblastoma.[31] Miletti-Gonzalez et al. have demonstrated that CD44 and MDR1/P-gp expression are correlated in several carcinoma cell lines.[78] Since MDR1 is one of the ABC transporters responsible for the SP phenotype, a relationship between SPs and

CD44 could be speculated at least in some types of cancers. Furthermore, as an SP is defined by ABC transporter activity, the expression of these transporters has been analyzed in various malignancies. For example, cells from patients with acute lymphoblastic leukemia (ALL) express high levels of MRP1 to MRP6 (ABCC)[79] members, gastrointestinal stromal tumors express high expression levels of ABCB,1[80] and neuroblastoma cells express high levels of ABCG2 and ABCA.3[31]

4.5 OVERCOMING SIDE POPULATION LIMITATIONS

Given that SPs represent a small fraction of the entire stem cell cohort, additional approaches are required to enrich this source of stem cells. The concomitant study of molecular markers of stem cells is mandatory to achieve their complete characterization. For example, the $Sca1^+$ Lin^- $Thy1^-$ c-kit^+ phenotype of hematopoietic stem cells is commonly accepted[39] as being indicative of stem cells. However, in no single instance has a complete correlation with specific markers been achieved for the SP compartments in either normal or tumor tissues.

The second limitation of SP studies relates to the requirement of an ultraviolet (UV) laser to excite Hoechst 33342. Whereas flow cytometers equipped with UV sources are not common because of the cost of both the laser and optics that can transmit UV light, violet laser sources are inexpensive and are now common fixtures on flow cytometers. In comparison to Hoechst 33342, the DyeCycle Violet (DCV) reagent has emission characteristics, but with a longer wavelength excitation maximum (369 nm). When this dye is loaded into hematopoietic cells, a side population is sharply resolved that is similar in appearance to that seen with Hoechst 33342, except that DCV-SP is similar in appearance to both violet and UV excitation.[81] DCV-SP can be inhibited with fumitremorgin C and has the same membrane pump specificity as Hoechst 33342. The simultaneous immunophenotyping with stem cell markers in mouse bone marrow demonstrated that DCV-SP was restricted to the Lin^- $Sca1^-$ c-kit^- population, as is Hoechst SP and that they identify similar stem cell populations.

Yet another limitation of SPs as a stem cell source is the toxicity of Hoechst 33342,[63] the fluorescent dye employed for their isolation. As Hoescht dyes are toxic to many cells, a comparison of the biological and functional properties of SPs and non-SPs is difficult. For example, using the SP phenotype of mobilized human peripheral blood (mPB), Fischer et al.[42] have demonstrated that Hoechst staining depleted mobilized human peripheral blood of short-term repopulating cells with myeloid potential. Thus, additional studies to optimize this technique and define the conditions associated with minimum toxicity are required. An alternative strategy is to develop and utilize a less or nontoxic reagent to detect SPs via FACS. Rhodamine 123 is a mitochondrial dye that is less toxic than Hoechst in SP applications. The amount of fluorescence emitted by a cell stained with this dye reflects either its mitochondrial content or its kinetics. It has been used alone or in combination with Hoechst, and its associated SPs may be enriched in stem cells in certain cases compared with Hoechst SPs.[82]

Due to the growing need for new markers to identify and isolate stem cells, an assay based on aldehyde dehydrogenase isoenzyme 1 (ALDH1) activity has been proposed and identified as a promising alternative. Storms et al. designed a novel system using a non-UV-light-excitable fluorochrome named Aldefluor [BODIPY-conjugated aminoacetaldehyde (BAAA)], which is metabolized by both murine and human cells.[83] They combined inhibition of multidrug resistance using verapamil with Aldefluor staining of human umbilical cord blood mononuclear cells and obtained a population highly enriched for Lin^- $CD34^+$ $CD38^{lo/-}$ cells and other hematopoeitic progenitor cells, suggesting that ALDH1 is a reliable marker for the isolation of stem cells. In contrast to Hoescht staining, Aldefluor-stained cells are viable; hence cell separation based on Aldefluor staining can subsequently be utilized successfully in in vitro and in vivo experimental protocols, as has recently been described with mammary stem cells.[84]

4.6 CONCLUSIONS AND FUTURE PERSPECTIVES

While stem cells will undoubtedly continue to play pivotal roles in normal physiology and cancer, research in this area is limited by a lack of universally accepted markers. Despite some limitations, the advantage of SPs is that they entail no specific markers. SPs are well documented as an enriched source of stem cells in specific settings and remain a valid and promising tool for the identification, isolation, and characterization of stem cells, particularly when this approach is combined with other cell markers. Thus, it is a good starting point for stem cell studies, especially with newer reagents that are less toxic to cells.[85] The enrichment of stem cells in SPs is best demonstrated through the functional assay of in vivo serial passage and multilineage differentiation, which confirms the stem cell properties of self-renewal and differentiation.

Acknowledgments

Annamaria Hadnagy and Saad Mansour are gratefully acknowledged for Fig. 4.2, generated from their results in our laboratory. D.B. is a scholar of Fonds de la recherche en santé du Québec (FRSQ) and principal investigator of the National Cancer Institute of Canada's Clinical Trials Group (NCIC CTG). This study was also made possible through a Canadian Institutes for Health Research (CIHR) operating grant to D.B. The authors thankfully acknowledge the editorial assistance of Ovid Da Silva, Editor, Research Support Office, CRCHUM, as well as the expert assistance of Rhyna Salinas (CRCHUM) in the preparation of this chapter.

REFERENCES

1. Massard C, Deutsch E, Soria JC. Tumor stem cell–targeted treatment: elimination or differentiation. *Ann Oncol.* 2006; 17:1620–1624.

2. Wicha MS, Liu S, Dontu G. Cancer stem cells: an old idea—a paradigm shift. *Cancer Res.* 2006; 66:1883–1890.

3. Jordan CT, Guzman ML, Noble M. Cancer stem cells. *N Engl J Med.* 2006; 355:1253–1261.

4. Krivtsov AV, Twomey D, Feng Z, et al. Transformation from committed progenitor to leukaemia stem cell initiated by MLL-AF9. *Nature.* 2006; 442:818–822.

5. Dontu G, El Ashry D, Wicha MS. Breast cancer, stem/progenitor cells and the estrogen receptor. *Trends Endocrinol Metab.* 2004; 15:193–197.

6. Clevers H. Stem cells, asymmetric division and cancer. *Nat Genet.* 2005; 37:1027–1028.

7. Dontu G, Liu S, Wicha MS. Stem cells in mammary development and carcinogenesis: implications for prevention and treatment. *Stem Cell Rev.* 2005; 1:207–213.

8. Balicki D, Leyland-Jones B, Wicha MS. Breast cancer stem cells and their niche: lethal seeds in lethal soil. In: Leyland-Jones B, ed. *Pharmacogenetics of Breast Cancer: Towards the Individualization of Therapy.* New York: Informa Healthcare; 2008; 259–289.

9. Arai F, Hirao A, Suda T. Regulation of hematopoietic stem cells by the niche. *Trends Cardiovasc Med.* 2005; 15:75–79.

10. Scadden DT. The stem-cell niche as an entity of action. *Nature.* 2006; 441:1075–1079.

11. Dontu G, Abdallah WM, Foley JM, et al. In vitro propagation and transcriptional profiling of human mammary stem/progenitor cells. *Genes Dev.* 2003; 17:1253–1270.

12. Yang ZJ, Wechsler-Reya RJ. Hit 'em where they live: targeting the cancer stem cell niche. *Cancer Cell.* 2007; 11:3–5.

13. Moore KA, Lemischka IR. Stem cells and their niches. *Science.* 2006; 311:1880–1885.

14. Donnenberg VS, Donnenberg AD. Multiple drug resistance in cancer revisited: the cancer stem cell hypothesis. *J Clin Pharmacol.* 2005; 45:872–877.

15. Larderet G, Fortunel NO, Vaigot P, et al. Human SP keratinocytes exhibit long-term proliferative potential, a specific gene expression profile, and can form a pluristratified epidermis. *Stem Cells.* 2006; 24:965–974.

16. Decraene C, Benchaouir R, Dillies MA, et al. Global transcriptional characterization of SP and MP cells from the myogenic C2C12 cell line: effect of FGF6. *Physiol Genomi.* 2005; 23(2): 132–149.

17. Goodell MA, Brose K, Paradis G, Conner AS, Mulligan RC. Isolation and functional properties of murine hematopoietic stem cells that are replicating in vivo. *J Exp Med.* 1996; 183:1797–1806.

18. Challen GA, Little MH. A side order of stem cells: the SP phenotype. *Stem Cells.* 2006; 24:3–12.

19. Zhou S, Schuetz JD, Bunting KD, et al. The ABC transporter Bcrp1/ABCG2 is expressed in a wide variety of stem cells and is a molecular determinant of the side-population phenotype. *Nat Med.* 2001; 7:1028–1034.

20. Umemoto T, Yamato M, Shiratsuchi Y, et al. Expression of integrin beta3 is correlated to the properties of quiescent hemopoietic stem cells possessing the side population phenotype. *J Immunol.* 2006; 177:7733–7739.

21. Redvers RP, Li A, Kaur P. Side population in adult murine epidermis exhibits phenotypic and functional characteristics of keratinocyte stem cells. *Proc Natl Acad Sci USA.* 2006; 103:13168–13173.

22. Reynolds SD, Shen H, Reynolds P, et al. Molecular and functional properties of lung side population cells. *Am J Physiol Lung Cell Mol Physiol.* 2007; 92:L972–L983.

23. Yoshida S, Shimmura S, Nagoshi N, et al. Isolation of multipotent neural crest–derived stem cells from the adult mouse cornea. *Stem Cells.* 2006; 24:2714–2722.

24. Iohara K, Zheng L, Ito M, Tomokiyo A, Matsushita K, Nakashima M. Side population cells isolated from porcine dental pulp tissue with self-renewal and multipotency for dentinogenesis, chondrogenesis, adipogenesis, and neurogenesis. *Stem Cells* 2006; 24:2493–2503.

25. Uezumi A, Ojima K, Fukada S, et al. Functional heterogeneity of side population cells in skeletal muscle. *Biochem Biophys Res Commun.* 2006; 341:864–873.

26. Poliakova L, Pirone A, Farese A, MacVittie T, Farney A. Presence of nonhematopoietic side population cells in the adult human and nonhuman primate pancreas. *Transplant Proc.* 2004; 36:1166–1168.

27. Asakura A, Seale P, Girgis-Gabardo A, Rudnicki MA. Myogenic specification of side population cells in skeletal muscle. *J Cell Biol.* 2002; 159:123–134.

28. Wulf GG, Luo KL, Jackson KA, Brenner MK, Goodell MA. Cells of the hepatic side population contribute to liver regeneration and can be replenished with bone marrow stem cells. *Haematologica.* 2003; 88:368–378.

29. Summer R, Kotton DN, Sun X, Ma B, Fitzsimmons K, Fine A. Side population cells and Bcrp1 expression in lung. *Am J Physiol Lung Cell Mol Physiol.* 2003; 285:L97–L104.

30. Hierlihy AM, Seale P, Lobe CG, Rudnicki MA, Megeney LA. The post-natal heart contains a myocardial stem cell population. *FEBS Lett.* 2002; 530:239–243.

31. Hirschmann-Jax C, Foster AE, Wulf GG, et al. A distinct "side population" of cells with high drug efflux capacity in human tumor cells. *Proc Natl Acad Sci U S A.* 2004; 101:14228–14233.

32. Grichnik JM, Burch JA, Schulteis RD, et al. Melanoma, a tumor based on a mutant stem cell? *J Invest Dermatol.* 2006; 126:142–153.

33. Szotek PP, Pieretti-Vanmarcke R, Masiakos PT, et al. Ovarian cancer side population defines cells with stem cell–like characteristics and Mullerian inhibiting substance responsiveness. *Proc Natl Acad Sci U S A.* 2006; 103:11154–11159.

34. Chiba T, Kita K, Zheng YW, et al. Side population purified from hepatocellular carcinoma cells harbors cancer stem cell–like properties. *Hepatology.* 2006; 44:240–251.

35. Kondo T, Setoguchi T, Taga T. Persistence of a small subpopulation of cancer stem-like cells in the C6 glioma cell line. *Proc Natl Acad Sci U S A.* 2004; 101:781–786.

36. Morita Y, Ema H, Yamazaki S, Nakauchi H. Non-side-population hematopoietic stem cells in mouse bone marrow. *Blood.* 2006; 108:2850–2856.

37. Shackleton M, Vaillant F, Simpson KJ, et al. Generation of a functional mammary gland from a single stem cell. *Nature.* 2006; 439:84–88.

38. Stingl J, Eirew P, Ricketson I, et al. Purification and unique properties of mammary epithelial stem cells. *Nature*. 2006; 439:993–997.

39. Wang XW. Microinjection technique used to study functional interaction between p53 and hepatitis B virus X gene in apoptosis. *Mol Biotechnol*. 2001; 18:169–177.

40. Haraguchi N, Utsunomiya T, Inoue H, et al. Characterization of a side population of cancer cells from human gastrointestinal system. *Stem Cells*. 2005; 24:506–513.

41. Benchaouir R, Rameau P, Decraene C, et al. Evidence for a resident subset of cells with SP phenotype in the C2C12 myogenic line: a tool to explore muscle stem cell biology. *Exp Cell Res*. 2004; 294:254–268.

42. Fischer M, Schmidt M, Klingenberg S, Eaves CJ, Von Kalle C, Glimm H. Short-term repopulating cells with myeloid potential in human mobilized peripheral blood do not have a side population (SP) phenotype. *Blood*. 2006; 108:2121–2123.

43. Parmar K, Sauk-Schubert C, Burdick D, Handley M, Mauch P. Sca$^+$CD34$^-$ murine side population cells are highly enriched for primitive stem cells. *Exp Hematol*. 2003; 31:244–250.

44. Scharenberg CW, Harkey MA, Torok-Storb B. The ABCG2 transporter is an efficient Hoechst 33342 efflux pump and is preferentially expressed by immature human hematopoietic progenitors. *Blood*. 2002; 99:507–512.

45. Meeson AP, Hawke TJ, Graham S, et al. Cellular and molecular regulation of skeletal muscle side population cells. *Stem Cells*. 2004; 22:1305–1320.

46. Zhou S, Morris JJ, Barnes Y, Lan L, Schuetz JD, Sorrentino BP. Bcrp1 gene expression is required for normal numbers of side population stem cells in mice, and confers relative protection to mitoxantrone in hematopoietic cells in vivo. *Proc Natl Acad Sci U S A* 2002; 99:12339–12344.

47. Tamaki T, Akatsuka A, Okada Y, Matsuzaki Y, Okano H, Kimura M. Growth and differentiation potential of main- and side-population cells derived from murine skeletal muscle. *Exp Cell Res*. 2003; 291:83–90.

48. Triel C, Vestergaard ME, Bolund L, Jensen TG, Jensen UB. Side population cells in human and mouse epidermis lack stem cell characteristics. *Exp Cell Res*. 2004; 295:79–90.

49. de Paiva CS, Pflugfelder SC, Li DQ. Cell size correlates with phenotype and proliferative capacity in human corneal epithelial cells. *Stem Cells*. 2005; 24:368–375.

50. Budak MT, Alpdogan OS, Zhou M, Lavker RM, Akinci MA, Wolosin JM. Ocular surface epithelia contain ABCG2-dependent side population cells exhibiting features associated with stem cells. *J Cell Sci*. 2005; 118:1715–1724.

51. Du Y, Funderburgh ML, Mann MM, SundarRaj N, Funderburgh JL. Multipotent stem cells in human corneal stroma. *Stem Cells*. 2005; 23:1266–1275.

52. Behbod F, Xian W, Shaw CA, Hilsenbeck SG, Tsimelzon A, Rosen JM. Transcriptional profiling of mammary gland side population cells. *Stem Cells*. 2005; 24:1065–1074.

53. Lechner A, Leech CA, Abraham EJ, Nolan AL, Habener JF. Nestin-positive progenitor cells derived from adult human pancreatic islets of Langerhans contain side population (SP) cells defined by expression of the ABCG2 (BCRP1) ATP-binding cassette transporter. *Biochem Biophys Res Commun*. 2002; 293:670–674.

54. Liadaki K, Kho AT, Sanoudou D, et al. Side population cells isolated from different tissues share transcriptome signatures and express tissue-specific markers. *Exp Cell Res*. 2005; 303:360–374.

55. Dontu G, Wicha MS. Survival of mammary stem cells in suspension culture: implications for stem cell biology and neoplasia. *J Mammary Gland Biol Neoplasia.* 2005; 10:75–86.

56. de Paiva CS, Chen Z, Corrales RM, Pflugfelder SC, Li DQ. ABCG2 transporter identifies a population of clonogenic human limbal epithelial cells. *Stem Cells.* 2005; 23:63–73.

57. Woodward WA, Chen MS, Behbod F, Rosen JM. On mammary stem cells. *J Cell Sci.* 2005; 118:3585–3594.

58. Tsinkalovsky O, Rosenlund B, Laerum OD, Eiken HG. Clock gene expression in purified mouse hematopoietic stem cells. *Exp Hematol.* 2005; 33:100–107.

59. Welm BE, Tepera SB, Venezia T, Graubert TA, Rosen JM, Goodell MA. Sca-1(pos) cells in the mouse mammary gland represent an enriched progenitor cell population. *Dev Biol.* 2002; 245:42–56.

60. Chen J, Hersmus N, Van DV, Caesens P, Denef C, Vankelecom H. The adult pituitary contains a cell population displaying stem/progenitor cell and early embryonic characteristics. *Endocrinology.* 2005; 146:3985–3998.

61. Martin CM, Meeson AP, Robertson SM, et al. Persistent expression of the ATP-binding cassette transporter, Abcg2, identifies cardiac SP cells in the developing and adult heart. *Dev Biol.* 2004; 265:262–275.

62. Patrawala L, Calhoun T, Schneider-Broussard R, Zhou J, Claypool K, Tang DG. Side population is enriched in tumorigenic, stem-like cancer cells, whereas ABCG2$^+$ and ABCG2$^-$ cancer cells are similarly tumorigenic. *Cancer Res.* 2005; 65:6207–6219.

63. Liu BY, McDermott SP, Khwaja SS, Alexander CM. The transforming activity of Wnt effectors correlates with their ability to induce the accumulation of mammary progenitor cells. *Proc Natl Acad Sci U S A.* 2004; 101:4158–4163.

64. Uchida N, Fujisaki T, Eaves AC, Eaves CJ. Transplantable hematopoietic stem cells in human fetal liver have a CD34(+) side population (SP) phenotype. *J Clin Invest.* 2001; 108:1071–1077.

65. Ernst C, Christie BR. The putative neural stem cell marker, nestin, is expressed in heterogeneous cell types in the adult rat neocortex. *NeuroScience.* 2005; 138:183–188.

66. Alvi AJ, Clayton H, Joshi C, et al. Functional and molecular characterisation of mammary side population cells. *Breast Cancer Res.* 2003; 5:R1–R8.

67. Clayton H, Titley I, Vivanco M. Growth and differentiation of progenitor/stem cells derived from the human mammary gland. *Exp Cell Res.* 2004; 297:444–460.

68. Clarke RB, Spence K, Anderson E, Howell A, Okano H, Potten CS. A putative human breast stem cell population is enriched for steroid receptor–positive cells. *Dev Biol.* 2005; 277:443–456.

69. Montanaro F, Liadaki K, Schienda J, Flint A, Gussoni E, Kunkel LM. Demystifying SP cell purification: viability, yield, and phenotype are defined by isolation parameters. *Exp Cell Res.* 2004; 298:144–154.

70. Pearce DJ, Anjos-Afonso F, Ridler CM, Eddaoudi A, Bonnet D. Age dependent increase in SP distribution within hematopoiesis: implications for our understanding of the mechanism of aging. *Stem Cells.* 2006; 25:828–835.

71. Yano S, Ito Y, Fujimoto M, Hamazaki TS, Tamaki K, Okochi H. Characterization and localization of side population cells in mouse skin. *Stem Cells.* 2005; 23:834–841.

72. Al Hajj M, Wicha MS, Benito-Hernandez A, Morrison SJ, Clarke MF. Prospective identification of tumorigenic breast cancer cells. *Proc Natl Acad Sci U S A.* 2003; 100:3983–3988.

73. Ponti D, Costa A, Zaffaroni N, et al. Isolation and in vitro propagation of tumorigenic breast cancer cells with stem/progenitor cell properties. *Cancer Res.* 2005; 65:5506–5511.

74. Raaijmakers MH, de Grouw EP, Heuver LH, et al. Breast cancer resistance protein in drug resistance of primitive CD34$^+$38$^-$ cells in acute myeloid leukemia. *Clin Cancer Res.* 2005; 11:2436–2444.

75. Collins AT, Berry PA, Hyde C, Stower MJ, Maitland NJ. Prospective identification of tumorigenic prostate cancer stem cells. *Cancer Res.* 2005; 65:10946–10951.

76. Seigel GM, Campbell LM, Narayan M, Gonzalez-Fernandez F. Cancer stem cell characteristics in retinoblastoma. *Mol Vis.* 2005; 11:729–737.

77. Wulf GG, Wang RY, Kuehnle I, et al. A leukemic stem cell with intrinsic drug efflux capacity in acute myeloid leukemia. *Blood.* 2001; 98:1166–1173.

78. Miletti-Gonzalez KE, Chen S, Muthukumaran N, et al. The CD44 receptor interacts with P-glycoprotein to promote cell migration and invasion in cancer. *Cancer Res.* 2005; 65:6660–6667.

79. Plasschaert SL, de Bont ES, Boezen M, et al. Expression of multidrug resistance–associated proteins predicts prognosis in childhood and adult acute lymphoblastic leukemia. *Clin Cancer Res.* 2005; 11:8661–8668.

80. Theou N, Gil S, Devocelle A, et al. Multidrug resistance proteins in gastrointestinal stromal tumors: site-dependent expression and initial response to imatinib. *Clin Cancer Res.* 2005; 11:7593–7598.

81. Telford WG, Bradford J, Godfrey W, Robey RW, Bates SE. Side population analysis using a violet-excited cell permeable DNA binding dye. *Stem Cells.* 2006; 25:1029–1036.

82. Lo KC, Brugh VM III, Parker M, Lamb DJ. Isolation and enrichment of murine spermatogonial stem cells using rhodamine 123 mitochondrial dye. *Biol Reprod.* 2005; 72:767–771.

83. Storms RW, Trujillo AP, Springer JB, et al. Isolation of primitive human hematopoietic progenitors on the basis of aldehyde dehydrogenase activity. *Proc Natl Acad Sci USA.* 1999; 96:9118–9123.

84. Ginestier C, Hur MH, Charafe-Jauffret M, et al. ALDH1 is a marker of normal and malignant human mammary stem cells and a predictor of poor clinical outcome. *Cell Stem Cell.* 2007; 1:555–567.

85. Hadnagy A, Gaboury L, Beaulieu R, Balicki D. SP analysis may be used to identify cancer stem cell populations. *Exp Cell Res.* 2006; 312:3701–3710.

5 Evidence for Cancer Stem Cells in Retinoblastoma

GAIL M. SEIGEL

5.1 INTRODUCTION

Retinoblastoma (RB) is the most common intraocular tumor of childhood. It involves mutations in the retinoblastoma gene(*RB*) and occurrs in both germline (40%) and sporadic (60%) forms.[1] Retinoblastoma treatment presents a clinical challenge with its incidence of metastatic or secondary tumors that affect life span and quality of life. Survivors of germline retinoblastoma have an increased frequency of secondary malignant neoplasms[2] at a rate of about 2% per year.[3] Secondary malignancies are believed to result from a number of possible causes, including extraretinal manifestations of the *RB* mutation, radiation-induced neoplasms,[4] or invasion and seeding of metastatic tumor cells. The prognosis for metastatic retinoblastoma is reported to be very poor.[5] Advances in the treatment of RB and accompanying metastases will require an intimate knowledge of RB tumor progression, including a better understanding of the divergent cell types involved in the establishment and progression of the disease.

5.2 ELUSIVE ORIGINS OF RETINOBLASTOMA

The *RB* gene encodes for a nuclear phosphoprotein that regulates the proliferation and development of a normal retinoblast cell[6–8] and exhibits tumor suppressor activity.[9] Loss of tumor suppression and ensuing RB tumor growth occur when both alleles of the *RB* gene on human chromosome 13 are inactivated within the same retinoblast cell.[10,11] More than 20 years ago it was proposed that RB tumors originate from a primitive neuroectodermal cell.[12,13] More recently, the cancer stem cell theory in retinoblastoma has been suggested,[14] which may involve a naturally occurring, death-resistant retinal stem/precursor cell.[15] Approaches that have been used to show the evidence of cancer stem cells (CSCs) in other tumors

Cancer Stem Cells: Identification and Targets, Edited by Sharmila Bapat
Copyright © 2009 John Wiley & Sons, Inc.

have provided newer approaches to exploring the origins of RB in a variety of models.

5.3 SOURCES OF RETINOBLASTOMA CELLS FOR STUDY

Ideally, fresh RB tumors from human surgical specimens are preferred for the most clinically relevant experiments. However, the number of new RB cases can be limited, as the clinical goal is to spare the eye and thereby preserve visual acuity without enucleation or surgical intervention. Due to the scarcity of fresh surgical human specimens, it is necessary to examine retinal tissues from appropriate mouse models of RB. One example is the long-studied LH-betaTAG mice, created by expression of a viral oncogene (simian virus 40 T-antigen) in the retina of transgenic mice, resulting in heritable ocular tumors with histological, ultrastructural, and immunohistochemical characteristics comparable to those of human RB.[16] In addition, newer and improved preclinical mouse models have been developed over the last couple of years[17,18] that more closely mimic the underlying genetic basis of RB.

In addition to the human and mouse models of RB, there are two very well characterized human retinoblastoma cell lines that are most frequently used in such studies, Y79[19] and WERI-RB27,[20] derived from genetically related individuals.[21] Established RB cell lines provide a pliant system for experimentation since they are easy to grow and can be used to work out experimental conditions for comparisons with in vivo models. The Y79 and WERI-RB27 human cell lines have been in culture for many years and have retained many features of RB[12], including the ability to form intraocular tumors in vivo,[22] but are not identical to RB tumors in situ. Another consideration regarding human versus mouse RB models is that mouse RB (but not human RB) requires inactivation of both *RB* and *p*107. [15,23] Since all RB models are not created equal, our studies have included multiple RB models, including cell lines, mouse tumors, and human tumors in order to seek out both commonalities and differences between the models.

5.4 PRECEDENT FOR CANCER STEM CELLS

As seen in previous chapters, some cancers appear to contain stemlike cells that are slowly dividing, chemoresistant, and capable of tumor progression (for reviews, see refs. 24 and 25). The existence of CSCs was first shown in acute myeloid leukemia[26,27] and has since been demonstrated in other cancers, including breast cancer,[28] human brain tumors,[29] and even the C6 glioma cell line, which has been maintained in culture over many years.[30] Therefore, there is a precedent for the persistence and association of CSCs expressing unique stem cell–like properties with specific cancers.

5.5 SIDE POPULATIONS IN RETINOBLASTOMA

Recently, expression of ABCG2 (or BCRP), a cell surface drug-resistance marker has been utilized to identify stem cells. Specifically, ABCG2 expression confers upon cells the ability to exclude Hoechst dye 33342 in a verapamil-sensitive manner[31–34] as well as resistance to at least 20 different chemotherapeutic agents, including methotrexate, doxorubicin, indolcarbazole, and others (for a review, see ref. 35). This dye-excluding side population (SP) phenotype has been used in a variety of tissues to isolate presumptive stem cells, including stem cells from hematopoetic populations,[32] bone marrow, skeletal muscle,[33] mammary gland,[36] lung,[37] and developing retina.[38] Recently, our group has detected side populations in dissociated mouse RB tumors as well as in human RB cell lines.[39] Correspondingly, we have visualized ABCG2-immunopositive cells that are colocalized as Hoechst-dim by fluorescence microscopy. The percentage of these sp cells in RB generally ranges between 0.1 and 0.4% of the total population. This percentage is similar to that seen in other tissues and tumor cell lines. In six other tumor cell lines—C6 (rat glioma), MCF-7 (breast cancer), U-20 and SaOS-2 (osteosarcoma), B104 (rat neuroblastoma), and HeLa (adenocarcinoma, cultured over 50 years)—Kondo et al. found that all but the U-20 and SaOS-2 cells contained side population cells.[30] Despite the many years since they were first established and propagated in culture, Y79 and WERI-Rb27 cells appear to have retained a degree of heterogeneity in terms of a side population and immunoreactivity to ABCG2. The presence of side populations and small numbers of ABCG2-bright/Hoechst-dim cells is significant for RB cell lines, as they were previously thought to be more homogeneous in nature.

5.6 IMMUNOREACTIVITY TO STEM CELL MARKERS IN RETINOBLASTOMA

In addition to the ABCG2 transporter, we have also detected immunoreactivity to other classical stem cell markers in RB,[39] including Sca1 (mouse), ALDH1, Oct3/4, and Nanog(human). The presence of these markers in small subpopulations of RB is another sign of heterogeneity. A summary of these markers is shown in Table 5.1. It has, however, not been established whether all of these stem cell markers colocalize to the same cell(s) or represent nonoverlapping subpopulations.

In addition to the markers shown in Table 5.1, we have detected immunore-activity to the marker MCM2,[39] a minichromosome maintenance gene reported to label neural stem cells.[40] An example of MCM2 immunoreactivity in human RB is shown in Fig. 5.1. Expression of MCM2 in RB was later confirmed in a large-scale study in which MCM2- and ABCG2-positive cells were detected in over 50% of RB tumor samples examined; with increased levels of ABCG2 and MCM2 expression correlating with more highly invasive tumors.[41] This suggests that quantitation of some stem cell markers may be a useful parameter that can be applied in developing improved classification schemes for RB tumors.

TABLE 5.1   Summary of Stem Cell Characteristics in RB

	Human Cell Lines	Mouse Tumor	Human Tumor
Side population	+	+	N.D.[a]
ABCG2	+	+	+
SCA-1 (mouse)	−	+	−
ALDH1	+	+	+
Oct3/4 (human)[b]	+	−	+
Nanog (human)[b]	+	−	+

[a]Not done.
[b]Embryonic stem cell markers.[42]

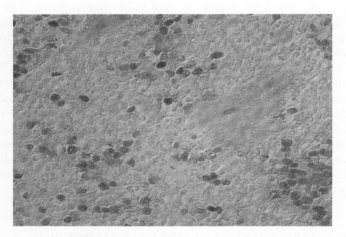

FIGURE 5.1   MCM2 immunoreactivity in human RB. An archival paraffin-embedded human retinoblastoma tumor was de-paraffinized and immunostained with MCM2 antibody (Santa Cruz Biochemicals, N-19). MCM2 immunoreactive cells were labeled with brown DAB reaction product. (*See insert for color representation of figure.*)

In work by ourselves and by Mohan et al.[41], a few cells within the normal retina of RB-containing eyes were found to be immunopositive for stem cell markers (Fig. 5.2). These stemlike cells in normal areas of RB retina could represent "seeding" for initiation of new retinal tumors or simply be an interesting characteristic of RB-susceptible retinal tissue. This differs from the normal postmitotic adult mammalian retina, in which retinal progenitors of unknown function have been detected in the ocular ciliary body and not in the neural retina.[38]

Another interesting and recent finding is the immunoreactivity of human embryonic stem cell markers in RB[42] (Table 5.1). Oct 3/4 and Nanog are human embryonic transcription factors that maintain stem cell pluripotency and self-renewal.[43] The presence of these human embryonic markers in RB is interesting, considering that RB is a tumor of early childhood, with the majority of RB tumors diagnosed

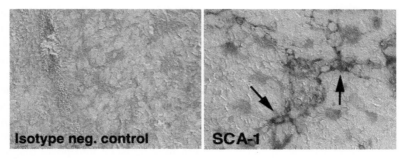

FIGURE 5.2   SCA-1 immunoreactivity in mouse RB. A frozen mouse RB tumor was immunostained with Sca-1 antibody (BD Biosciences, 557403). Arrows indicate areas of immunopositive cells. An isotype control panel is shown for comparison. (Adapted from ref. 39, with permission of the authors.) (*See insert for color representation of figure.*)

by age 2.[1] It is intriguing to speculate whether the presence of these embryonic markers could represent remnants carried over from embryonic retinal development. However, further experiments will be needed to determine whether there is either a postnatal persistence or de novo expression of human embryonic markers in RB, and what impact this may have on RB tumor progression and differentiation potential.[44,45]

5.7 CONCLUSIONS AND FUTURE PERSPECTIVES

Further evidence is needed to confirm the presence of cancer stem cells in retinoblastoma   The definition of a cancer stem cell is that it is a cell that is able to give rise to a new tumor that recapitulates the tumor of origin. A critical future experiment will involve the isolation of putative RB CSCs for further examination in terms of tumorigenicity. If an RB subpopulation of interest is indeed a CSC, it will preferentially form new RB tumors in animals, while the non–stem cell component of the cell population, lacking stem cell characteristics, may survive or engraft, but not form tumors. In this way, the existence of an RB-CSC population can be shown more definitively.

More recently, tumorigenic retinal stemlike cells have been identified in RB lesions as undifferentiated cells capable of extensive proliferation as clonal nonadherent spheres that have a capacity to differentiate into different retinal cells in vitro.[46] Cultured cells from retinoblastomas express several retinal development-related genes, including *Nestin, CD133, Pax6, Chx10*, and *Rx*, as well as *Bmi1*, a gene associated with self-renewal and proliferation of stem and progenitor cells. Further, in the classical severe compromised immune deficiency (SCID) mouse model, these cells gave rise to new tumors with histomorphological features and immunophenotypes similar to those of the parental tumors. This recent study provides a new insight into retinoblastoma tumors.

As of this writing, it is clear that mouse and human RB tumor cells contain a small subpopulation of cells that exhibit some CSC characteristics. Especially significant is the expression of ABCG2, a chemoresistance protein. Previous studies have shown other multidrug resistance markers in pathology specimens of retinoblastoma.[47] RB heterogeneity and drugresistance are important to consider in the context of RB treatment strategies. No matter whether these particular RB subpopulations eventually fit within the generally accepted realm of cancer stem cells, chemoresistant subpopulations in RB are clinically significant in terms of future targets for therapies that will promote long-term survival and improve the quality of life in RB patients. This, in turn, may translate into novel diagnostic means for determining RB prognosis and intervention.

Acknowledgments

Dr. Seigel's work was supported by Fight for Sight as well as an unrestricted departmental grant from Research to Prevent Blindness. Dr. Seigel is also supported by the Sybil Harrington Special Scholar Award from Research to Prevent Blindness.

REFERENCES

1. Augsburger JJ, Oehlschlager U, Manzitti JE. Multinational clinical and pathologic registry of retinoblastoma. Retinoblastoma International Collaborative Study Report 2. *Graefe's Arch Clin Exp Ophthalmol.* 1995; 233:469–475.
2. Gallie BL, Dunn JM, Chan HS, Hamel PA, Phillips RA. The genetics of retinoblastoma: relevance to the patient. *Pediatr Clin North Am.* 1991; 38:299–315.
3. Abramson DH, Melson MR, Dunkel IJ, Frank CM. Third (fourth and fifth) nonocular tumors in survivors of retinoblastoma. *Ophthalmology.* 2001; 108:1868–1876.
4. Wong FL, Boice JD Jr, Abramson DH, et al. Cancer incidence after retinoblastoma: radiation dose and sarcoma risk. *JAMA.* 1997; 278:1262–1267.
5. Gunduz K, Muftuoglu O, Gunalp I, Unal E, Tacyildiz N. Metastatic retinoblastoma: clinical features, treatment, and prognosis. *Ophthalmology.* 2006; 113:1558–1566.
6. Zhang J, Gray J, Wu L, et al. Rb regulates proliferation and rod photoreceptor development in the mouse retina. *Nat Genet.* 2004; 36:351–360.
7. Donovan SL, Dyer MA. Developmental defects in Rb-deficient retinae. *Vision Res.* 2004; 44:3323–3333.
8. MacPherson D, Sage J, Kim T, Ho D, McLaughlin ME, Jacks T. Cell type–specific effects of Rb deletion in the murine retina. *Genes Dev.* 2004; 18:1681–1694.
9. Madreperla SA, Whittum-Hudson JA, Prendergast RA, Chen PL, Lee WH. Intraocular tumor suppression of retinoblastoma gene–reconsituted retinoblastoma cells. *Cancer Res.* 1991; 51:6381–6384.
10. Knudson AG Jr. Mutation and cancer: statistical study of retinoblastoma. *Proc Natl Acad Sci U S A.* 1971; 68:820–823.
11. Kyritsis AP, Tsokos M, Triche TJ, Chader GJ. Retinoblastoma: origin from a primitive neuroectodermal cell? *Nature.* 1984; 307:471–473.

12. Campbell M, Chader GJ. Retinoblastoma cells in tissue culture. *Ophthalmic Paediatr Genet.* 1988; 9:171–199.

13. Schlesinger HR, Rorke L, Jamieson R, Hummeler K. Neuronal properties of neuroectodermal tumors in vitro. *Cancer Res.* 1981; 41:2573–2575.

14. Dyer MA, Bremner R. The search for the retinoblastoma cell of origin. *Nat Rev Cancer.* 2005; 5:91–101.

15. Chen D, Livne-Bar I, Vanderluit JL, Slack RS, Agochiya M, Bremner R. Cell-specific effects of RB or RB/p107 loss on retinal development implicate an intrinsically death-resistant cell-of-origin in retinoblastoma. *Cancer Cell.* 2004; 5:539–551.

16. Windle JJ, Albert DM, O'Brien JM, et al. Retinoblastoma in transgenic mice. *Nature.* 1990; 343:665–669.

17. Zhang J, Schweers B, Dyer MA. The first knockout mouse model of retinoblastoma. *Cell Cycle.* 2004; 3:952–959.

18. Dyer MA, Rodriguez-Galindo C, Wilson MW. Use of preclinical models to improve treatment of retinoblastoma. *PLoS Med.* 2005; 2:971–976.

19. Reid TW, Albert DM, Rabson AS, et al. Characteristics of an established cell line of retinoblastoma. *J Natl Cancer Inst.* 1974; 53:347–360.

20. Sery TW, Lee EY, Lee WH, et al. Characteristics of two new retinoblastoma cell lines: WERI-Rb24 and WERI-Rb27. *J Pediatr Ophthalmol Strabismus.* 1990; 27:212–217.

21. Madreperla SA, Bookstein R, Jones OW, Lee WH. Retinoblastoma cell lines Y79, RB355 and WERI-Rb27 are genetically related. *Ophthalmic Paediatr Genet.* 1991; 12:49–56.

22. del Cerro M, Seigel GM, Lazar E, et al. Transplantation of Y79 cells into rat eyes: an in vivo model of human retinoblastomas. *Invest Ophthalmol Vis Sci.* 1993; 34:3336–3346.

23. Robanus-Mandag EC, Van der Valk M, Vlaar M, et al. Developmental rescue of an embryonic lethal mutation in the retinoblastoma gene in chimeric mice. *EMBO J.* 1994; 13:4260–4268.

24. Pardal R, Clarke MF, Morrison SJ. Applying the principles of stem-cell biology to cancer. *Nat Rev Cancer.* 2003; 3:895–902.

25. Reya T, Morrison SJ, Clarke MF, Weissman IL. Stem cells, cancer, and cancer stem cells. *Nature.* 2001; 414:105–111.

26. Bonnet D, Dick JE. Human acute myeloid leukemia is organized as a hierarchy that originates from a primitive hematopoietic cell. *Nat Med.* 1997; 3:730–737.

27. Lapidot T, Sirard C, Vormoor J, et al. A cell initiating human acute myeloid leukaemia after transplantation into SCID mice. *Nature.* 1994; 367:645–648.

28. Alvi AJ, Clayton H, Joshi C, et al. Functional and molecular characterisation of mammary side population cells. *Breast Cancer Res.* 2003; 5:R1–R8.

29. Singh SK, Clarke ID, Terasaki M, et al. Identification of a cancer stem cell in human brain tumors. *Cancer Res.* 2003; 63:5821–5828.

30. Kondo T, Setoguchi T, Taga T. Persistence of a small subpopulation of cancer stem-like cells in the C6 glioma cell line. *Proc Natl Acad Sci U S A.* 2004; 101:781–786.

31. Kim M, Turnquist H, Jackson J, et al. The multidrug resistance transporter ABCG2 (breast cancer resistance protein 1) effluxes Hoechst 33342 and is overexpressed in hematopoietic stem cells. *Clin Cancer Res.* 2002; 8:22–28.

32. Scharenberg CW, Harkey MA, Torok-Storb B. The ABCG2 transporter is an efficient Hoechst 33342 efflux pump and is preferentially expressed by immature human hematopoietic progenitors. *Blood*. 2002; 99:507–512.

33. Zhou, S, Schuetz JD, Bunting KD, et al. The ABC transporter Bcrp1/ABCG2 is expressed in a wide variety of stem cells and is a molecular determinant of the side-population phenotype. *Nat Med*. 2001; 7:1028–1034.

34. Goodell M, Brose K, Paradis G, Conner AS, Mulligan RC. Isolation and functional properties of murine hematopoetic stem cells that are replicating in vivo. *J Exp Med*. 1996; 183:1797–1806.

35. Doyle LA, Ross DD. Multidrug resistance mediated by the breast cancer resistance protein BCRP (ABCG2). *Oncogene*. 2003; 22:7340–7358.

36. Welm BE, Tepera SB, Venezia T, Graubert TA, Rosen J, Goodell M. Sca1 positive cells in the mouse mammary gland represent an enriched progenitor cell population. *Dev Biol*. 2002; 245:42–56.

37. Summer R, Kotton DN, Sun X, Ma B, Fitzsimmons K, Fine A. Side population cells and Bcrp1 expression in lung. *Am J Physiol Lung Cell Mol Physiol*. 2003; 285:L97–L104.

38. Bhattacharya S, Jackson JD, Das AV, et al. Direct identification and enrichment of retinal stem cells/progenitors by Hoechst dye efflux assay. *Invest Ophthalmol Vis Sci*. 2003; 44:2764–2773.

39. Seigel GM, Campbell LM, Narayan M, Gonzalez-Fernandez F. Cancer stem cell characteristics in retinoblastoma. *Mol Vision*. 2005; 11:729–737.

40. Maslov AY, Barone TA, Plunkett RJ, Pruitt SC. Neural stem cell detection, characterization, and age-related changes in the subventricular zone of mice. *J Neurosci*. 2004; 24:1726–1733.

41. Mohan A, Kandalam M, Ramkumar HL, Gopal L, Krishnakumar S. Stem cell markers: ABCG2 and MCM2 expression in retinoblastoma. *Br J Ophthalmol*. 2006; 90:889–893.

42. Seigel GM, Hackam AS, Ganguly A, Mandell LM, Gonzalez-Fernandez F. Human embryonic and neuronal stem cell markers in retinoblastoma. *Mol Vision*. 2007; 13:823–832.

43. Rodda DJ, Chew JL, Lim LH, et al. Transcriptional regulation of nanog by Oct4 and Sox2. *J Biol Chem*. 2005; 280:24731–24737.

44. Seigel GM, Notter MF. Differentiation of Y79 retinoblastoma cells induced by succinylated concanavalin A. *Cell Growth Differ*. 1993; 4:1–7.

45. Seigel GM, Tombran-Tink J, Becerra SP, et al. Differentiation of Y79 retinoblastoma cells with pigment epithelial-derived factor and interphotoreceptor matrix wash: effects on tumorigenicity. *Growth Factors*. 1994; 10:289–297.

46. Zhong X, Li Y, Peng F, et al. Identification of tumorigenic retinal stem–like cells in human solid retinoblastomas. *Int J Cancer*. 2007; 121(10): 2125–2131.

47. Krishnakumar S, Mallikarjuna K, Desai N, et al. Multidrug resistant proteins: P-glycoprotein and lung resistance protein expression in retinoblastoma. *Br J Ophthalmol*. 2004; 88:1521–1526.

6 Ovarian Stem Cell Biology and the Emergence of Ovarian Cancer Stem Cells

ANJALI KUSUMBE and SHARMILA BAPAT

6.1 INTRODUCTION

Ovarian cancer is widely recognized as an asymptomatic disease and progresses rapidly, delaying the diagnosis to an advanced stage when prognosis is extremely poor.[1,2] Survival rates at the early stage of ovarian carcinoma range from 70 to 90%, compared with 20 to 25% at the advanced stage of the disease.[3] Recently, cancer has been viewed as being propagated by a minority of cells that display stem cell–like characteristics and are termed *cancer stem cells* (CSCs). Unfortunately, research in this facet of ovarian cancer is at present limited, due to the paucity of understanding of the stem cell population residing within normal ovarian tissue. Unlike the hematopoietic system or organs such as breast and brain, where normal stem cell populations have been identified and their commitment and differentiation along various lineages dissected out precisely, ovarian stem cell lineages and markers for their identification remain unrevealed.

6.2 OVERVIEW OF THE HUMAN OVARY

6.2.1 Histological Landmarks

The surface of the adult ovary is covered by a layer of epithelium termed *ovarian surface epithelium* (OSE) and is continuous with the mesothelium of the ovarian ligament (mesovarium) and peritoneum. The OSE proliferates under the influence of pituitary gonadotrophins, ovarian steroids, and other growth factors to repair the postovulation trauma. Beneath the OSE lies the dense connective tissue capsule termed the *ovarian cortical interstitium* or *tunica albuginea*. As in many other organs, ovarian tissue is also organized into an outer cortex and an inner medulla.

Cancer Stem Cells: Identification and Targets, Edited by Sharmila Bapat
Copyright © 2009 John Wiley & Sons, Inc.

The cortex consists of a very cellular connective tissue component termed *stroma* in which the follicles are embedded; the loose connective tissue comprising the medulla contains blood vessels and nerves. Each follicle consists of an immature ovum surrounded by one or more layers of granulosa cells. Follicles at the various stages of development—primordial (resting) follicles, primary follicles, secondary follicles, and tertiary (graffian) follicles—coexist in the ovary[10] (Fig. 6.1C).

6.2.2 Ovarian Development: The Story Before Birth

Human ovarian tissue consists of two cell lineages: somatic cells and germ cells (GCs). The somatic cells include the OSE, which forms the outermost layer of the ovary; the tunica albugenia cells located beneath the OSE; and the granulosa cells, which surround the GCs. Analogous to the epithelia of urogenital system and adrenal cortex, the OSE derives its origin from the mesoderm. The process is initiated by the segregation of mesoderm into pluripotent mesenchyme and coelomic epithelium (also termed the peritoneal mesothelium). The latter invaginates over the genital ridges at 10 weeks of embryonic age to differentiate into the OSE; simultaneously, it proliferates to give rise to epithelial cords that penetrate the ovarian cortex.[4] At about the fifth month these cords give rise to the granulosa cells, which surround the oogonia, leading to development of the primordial follicles, which are separated from the overlying OSE by stromal cells.[5] Besides being the precursor of ovarian tissue, the coelomic epithelium also gives rise to linings of Mullerian ducts (i.e., the oviduct, endometrium, and endocervix). Thus, the epithelial linings of all these organs share a close developmental relationship.[5] The tunica albugenia is believed to originate through epithelial-to-mesenchymal transition of OSE cells during early development (Fig. 6.1A).[6]

The GCs arise from diploid embryonic precursors termed *primordial germ cells* (PGCs), found in the endoderm of the dorsal wall of the yolk sac. In the third week postfertilization they start proliferating and subsequently migrate through the endoderm, hindgut, and mesentery to reach the final destination, the genital ridges.[7] Further settlement of PGCs occurs at the end of the fifth week postfertilization, when these cells reach the gonadal primordium and colonize the superficial areas of the developing ovaries [8] (Fig. 6.1B). Despite a great deal of scientific interest in the origin of GCs that has extended over several decades, it is remarkable that the precise stage at which these cells segregate from the somatic cell lineage in the developing embryo has not yet been identified.

During further embryonic development, the PGCs differentiate into diploid oogonia, which divide rapidly to provide the repertoire of future oocytes. These oogonia, surrounded by the squamous granulosa cells, are termed primordial follicles. They begin a partial meiosis and can remain dormant (for up to four to five decades) at the prophase (diplotene) stage until they undergo the primordial-to-primary follicle transition. During this transition the primary follicles develop a cuboidal morphology, proliferate, and sequester theca from the surrounding stroma to mature into a graffian follicle. The primordial-to-primary

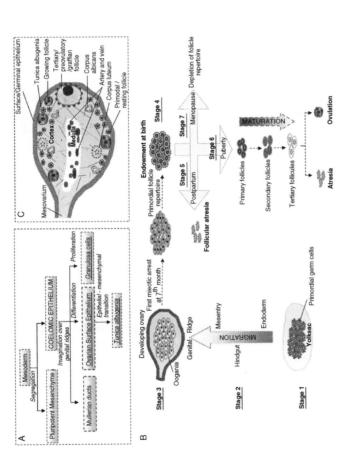

FIGURE 6.1 (A) Origin of somatic cells in the ovary. Early during embryonic development, mesoderm gives rise to mesenchyme and coelomic epithelium. Coelomic epithelium differentiates further into OSE and proliferates to generate granulose cells in the primordial follicles. Besides being the precursor of ovarian tissue, the coelomic epithelium also gives rise to linings of Mullerian ducts (i.e., the oviduct, endometrium, and endocervix). The tunica albugenia is also believed to originate through epithelial-to-mesenchymal transition of OSE cells during early development (B) Life history of ovarian follicles: primordial germ cells to oocytes. (C) Histology of the human ovary. The cross section through the human ovary shows two distinct compartments: outer cortex and the inner medulla. Cortex is surrounded by the epithelial lining termed OSE. Cortex encloses the follicles in the various stages of maturation, while the blood vessels are located within the medulla.

follicle transition is a nonreversible event. Maturation of the primary follicle continues until ovulation; alternatively, it is destroyed through atresia.[9] Despite loss of the majority of the follicles prior to birth, due to a wave of atresia, the two human ovaries together harbor approximately 700,000 oocytes at birth. By puberty the number of follicles declines further to about 400,000, due to continuing atresia.

6.2.3 The Mammalian Oogenesis Dogma

A basic doctrine of mammalian reproductive biology is that females are endowed with a preset reserve of nonrenewing GCs at the time of the birth, after which the production of new GCs (i.e., neo-oogenesis) ceases. In contrast, males retain germline stem cells (GSCs) for spermatogenesis throughout their adult life. The number of oocytes keeps on declining throughout postnatal life through ongoing atresia as described above, leading to GC depletion at around the fifth decade of life, thereby driving menopause (Fig. 6.1B). Thus, according to conventional dogma, the process of neo-oogenesis leading to de novo generation of GSCs does not exist in the postnatal mammalian ovaries.[11]

6.3 STEM/PROGENITOR CELLS IN THE ADULT MAMMALIAN OVARY

6.3.1 Historical Perspective

The debate over the issue of absence versus existence of GSCs in the postnatal mammalian ovaries exploded in the early twentieth century. As early as 1870, Waldeyer defined the OSE as a germinal epithelium, based on his observations that GCs arise from this layer of cells.[12] Allen (1923) supported this concept of germinal epithelium,[13] although earlier, Pearl and Schoppe stated that "during the life of the individual there neither is nor can be any increase in the number of primary oocytes."[14] Using differential follicle counting studies, Zuckerman also demonstrated oocyte production to be limited to prenatal life.[15] However, predictive studies by Vermande-Van Eck based on the incidence of atresia and the time taken for the atretic follicles to be cleared from the ovaries in a monkey model suggested that 90% of the GC reserve would be exhausted within the first two years of life.[16] In that case, neo-oogenesis would be indispensable to sustain the GC pool throughout the reproductive age of the animal. Nevertheless, the concept of neo-oogenesis in the postnatal ovaries was put in limbo until 1962, when through tritiated thymidine labeling of mouse oocyte nuclei it was demonstrated that the premeiotic S (synthesis) phase in the mouse occurs during the prenatal period. Thus, oocytes in juvenile and adult ovaries are the descendants of fetal GCs.[17,18] Thereafter, arguments regarding postnatal neo-oogenesis rested firmly with establishment of the dogma of mammalian females being endowed with a preset reserve of nonrenewing GCs.[11]

6.3.2 The Oogenesis Dogma Revisited

Despite the evidence supporting the absence of postnatal neo-oogenesis, from a phylogenetic perspective it seemed rather paradoxical that mammalian females would evolve a retrogressive mechanism that necessitates maintenance of their fetal GC pool throughout postnatal life. Conversely, neo-oogenesis has been observed in the adult ovaries of some prosimian primates.[19,20] An additional drawback of the oogenesis dogma was that preservation for over several decades would render GCs susceptible to spontaneous and environmentally induced genetic insults. These discrepancies led to the initiation of a series of studies in the field. In 2003, the first report of an in vitro differentiation of mice embryonic stem cells into oocytes was documented.[21] This was followed by several others (Table 6.1), suggesting the postnatal existence of GSCs. The key findings supporting this concept of postnatal neo-oogenesis are reviewed below.

Ovarian Surface Epithelium as the Source of Germline Stem Cells The first investigation to decipher the prevalence of neo-oogenesis in postnatal mammalian ovaries[22] was initiated with the observation that the total pool of healthy follicles in juvenile mice begins to decline at about 30 days of age, due to follicular atresia, and continues until the age of 4 months. Since these follicles have a brief turnover of about 4 days, this loss could be predicted to result in complete exhaustion of the follicular pool within a few weeks. Actual follicular monitoring, however, identifies the decline in the total number of healthy follicles within mice to be minimal, indicative of follicular renewal in the postnatal ovaries. The following data were presented in this study to support the foregoing notion:

1. Histological analysis of juvenile and adult mouse ovaries revealed the presence of germlike cells in the OSE, identified through the expression of Vasa (a conserved germ-cell marker). BrdU incorporation demonstrated the cyclical nature of these cells, indicative of a proliferative capacity.

2. Oocyte formation involves meiosis; this was confirmed through the expression of meiosis-specific proteins [i.e., synaptonemal complex protein 3 (Scp3), endonuclease Spo11, and the recombinase Dmc1].

3. Treating mice with the toxic drug busulfan (which targets only GSCs, not postmeotic GCs) rapidly depleted the pool of young follicles, indicating the involvement of neo-oogenesis in maintenance of the follicular pool.

4. Ovarian fragments from adult wild-type (WT) mice were grafted into transgenic mice with ubiquitous GFP (green fluorescent protein) expression. Confocal microscopic analysis of the ovaries from these transgenic mice later revealed the presence of hybrid follicles comprising GFP-expressing oocytes in the WT granulosa cells and ovarian fragments. This indicated infiltration of the WT graft with transgenic GFP-positive GCs to generate hybrid follicles (Fig. 6.2).

Collectively, the data above build a strong case in favor of the existence of GSCs, which are located within the OSE and function to replenish the GC pool in the postnatal mammalian ovary.

TABLE 6.1   Recent Reports Supporting the Persistence of Neo-oogenesis in Postnatal Ovaries of Mammalian Females

Year	Species	Cell Markers Used	Approach and Techniques	Significant Findings	Ref.
2004	Mouse	Vasa, Scp3, Spo11, Dmc1	Expression analysis of meiosis-specific proteins and the germ cell marker Vasa in adult ovaries. BrdU incorporation assay to determine the cyclical nature of cells in ovaries. Busulfan treatment to specifically target the GSCs. Engrafment of adult wild-type ovarian fragments in GFP$^+$ adult mice.	Adult ovaries are endowed with GSCs which seems to be residing within the OSE.	22
2004	Human	Cytokeratin, MAPK	Immunohistochemical analysis of adult ovaries to investigate whether the mesenchymal cells in the tunica albugenia function as bipotent progenitors with commitment for both GC and granulosa cells.	Bipotent mesenchymal progenitors from tunica albugenia function to replenish the GC reserve during postnatal life.	6
2005	Mouse and human	Oct4, Mvh, Dazl, Stella, and Fragilis	Expression profiling of BM for GC markers. Transplantation of BM from adult WT mice into chemotherapy-treated mice and congenitally sterile Atm mice.	BM and peripheral blood–derived GSCs contribute to neo-oogenesis in adult ovaries.	26
2005	Human	Cytokeratin 5, 6, 8,17,18	Cells from the surface of the ovaries were cultured to determine the possibility of their differentiation into oocytes and granulose cells.	Confirms OSE to be the source of GSCs during adult life.	27
2006	Porcine	Oct4, GDF9b, DAZL, Vasa, Scp3, c-Mos	Stem cells isolated from the skin were induced towards oocyte differentiation to investigate their germline potential.	Stem cells from fetal porcine skin demonstrate in vitro potential of germline differentiation.	28

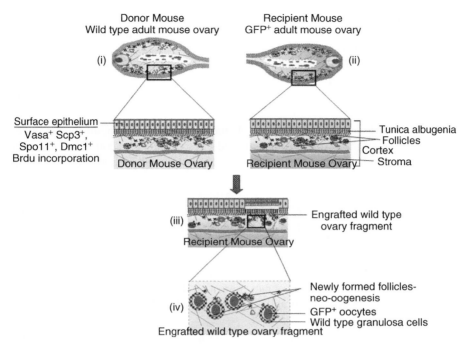

FIGURE 6.2   In vivo evidence of neo-oogenesis in postnatal mouse ovaries. Ovarian fragments from adult wild-type mice (i) were grafted into transgenic mice (ii) with ubiquitous GFP (green fluorescent protein) expression (iii). Subsequent to engraftment, ovaries from these transgenic mice revealed the presence of hybrid follicles. Hybrid follices were made up of GFP-expressing oocytes in wild-type granulosa cells and ovarian fragments (iv), thus indicating the infiltration of the wild-type graft with transgenic GFP-positive GCs to generate hybrid follicles.

Bone Marrow as a Putative Germline Stem Cell Reservoir   Although in accordance with the research, OSE serves to replenish the GC pool in the postnatal ovary, the regenerative potential of OSE is known to decline rapidly after mice attain puberty. This suggested the existence of an alternative source of GSCs to sustain oocyte production further until menopause. On the grounds mentioned below, bone marrow (BM) seemed to be the most probable candidate as an alternative source. During embryogenesis, hematopoietic stem cells (HSCs) and PGCs both originate from the same site (the proximal epiblast).[23] These two cell types migrate along a common path until the HSCs colonize the fetal liver.[24] During postnatal life HSCs not only sustain hematopoiesis but also have a capacity for multipotent differentiation.[25] Concurrently, PGCs have been identified with the potential to generate HSCs in vitro. Deriving a lead from this evidence, the BM has recently been identified as an alternative source of GCs in the postnatal mouse ovary.[26] The highlights of this study include:

1. Treatment of mice with doxorubicin (which results in the extensive destruction of primordial and early-growing follicles) followed by a recovery period evidenced spontaneous regeneration of the GC pool.

2. Expression profiling of freshly isolated and cultured BM for GC molecular markers, including Oct4, Mvh, Dazl, Stella, and Fragilis demonstrated the presence of all these markers in the BM, with significant variation in their expression levels during the mouse estrous cycle.

3. Fresh BM was transplanted into chemotherapy sterilized adult female mice. Later analyses of the ovaries from sterilized mice exihibited a complete spectrum of immature and mature oocyte-containing follicles and corpora lutea in these ovaries (indicative of the resumption of normal ovulatory cycles). Such a follicular pool remained evident for up to 11 months after transplantation.

4. BM transplants from WT female mice were further performed in congenitally sterile Atm (ataxia telangiectasia–mutated) mice, which lack the ability to produce mature GCs. Posttransplantation, oocyte-containing follicles could be identified in the ovaries of mutant females.

5. Peripheral blood cell transplantation may also be able to contribute to oocyte production in female recipients, a fact confirmed through detection of the markers Oct4, Mvh, Dazl, Stella, and Fragilis in the peripheral blood of the recipients.

The findings above suggest the BM to be a putative source of GSCs. Unfortunately, certain anomalies are distinctly evident in this work: for example, the specificity of the markers used for screening, the lack of evidence relating to the capacity of BM to regenerate oocytes in vitro (as achieved with embryonic stem cells), competence of oocytes thus produced for fertilization, and embryonic development and generation of viable offspring. Extensive study is thus required to confirm the functional competency of the BM and peripheral blood-derived oocytes.

Bipotent Progenitors for Germ Cells and Granulosa Cells   The findings above supporting postnatal neo-oogenesis were extended to humans by Bukovsky et al., who showed that adult human ovaries harbor a population of bipotent progenitor cells with commitment for both GCs and granulosa cells.[6] This study indicated that neo-oogenesis may be initiated when cytokeratin-negative mesenchymal cells of tunica albugenia differentiate into cytokeratin positive cells. The cyokeratin-expressing mesenchymal cells function as bipotent progenitors and undergo a mesenchymal-to-epithelial transition to revert back to the more primitive OSE cells. Being germinal in nature, the latter can differentiate into either granulosa cells or GCs upon receiving appropriate stromal/environmental cues (Fig. 6.3). Yet the follicular renewal ceases after a certain age, consequently driving the onset of the natural menopause or premature ovarian failure. The same group of researchers also demonstrated neo-oogenesis in vitro, wherein the OSE cells from adult human ovaries were grown and differentiated into oocytes and granulosa cells.[27] Thus, this study confirmed the earlier observations that in adult

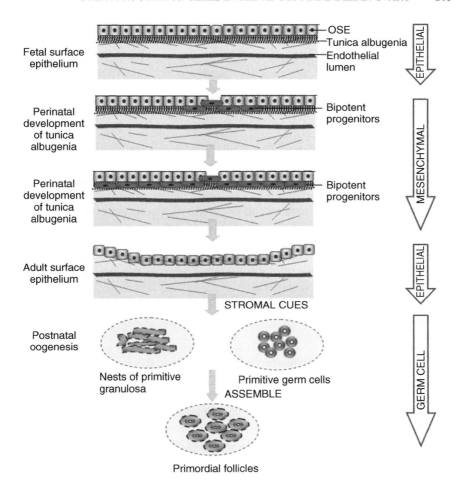

FIGURE 6.3 Four-step transition during oogenesis from bipotent progenitors. During early embryonic development the ovarian surface epithelium undergoes epithelial-to-mesenchymal transition to generate tunica albugenia. The tunica albugenia cells serve as the bipotent progenitors for the granulosa and germ cells during postnatal oogenesis. (*See insert for color representation of figure.*)

human ovaries, OSE may serve as a bipotent source for oocytes and granulosa cells.

Germline Potential of Stem Cells from Fetal Skin   More recently, it has been demonstrated that under specific in vitro growth conditions (medium supplemented with follicular fluid) stem cells isolated from the fetal porcine can undergo differentiation to give rise to oocytes.[28] In this study GC differentiation was monitored through the expression of markers Oct4, GDF9b, DAZL, and Vasa. In addition to expression of the GC phenotype, these cells formed follicle like aggregates that secreted estradiol and progesterone and responded to the cues that normally stimulate GC response. Finally, some of these aggregates gave rise spontaneously to embryolike structures through parthenogenesis.

6.4 IS OVARIAN CANCER A STEM CELL DISEASE?

Recently escalating evidence has accumulated supporting the existence of self-renewing tumor-initiating stem cells, termed cancer stem cells (CSCs), which drive tumor pathogenesis. Thus, analogous to the normal adult organs, cancers seem to be maintained through a hierarchical cellular organization. In addition, the CSCs have been identified to exhibit similarities with the normal adult stem cells. Briefly, like normal adult stem cells, CSCs exhibit self-renewal potential, establishment of cellular hierarchy, and slow cell cycling times.[29] In a broad sense a tumor can be considered as an aberrant organ initiated by CSCs; this facilitates an understanding of tumor biology based on the principles of normal stem cell biology and organogenesis.

Based on its cell of origin, ovarian cancer is classified into three broad histological types:

1. *Stromal tumors.* These account for only 5% of ovarian tumors and arise in the stromal cells located within the ovarian cortex.

2. *GC tumors.* These account for approximately 5% of all ovarian tumors and originate in the GCs. Based on their cellular composition, these are further classified as differentiated and undifferentiated tumors. The latter have a capacity to lead to a malignant state. Prior to the widely prevalent use of embryonic stem cells, undifferentiated germ cell tumor cell lines were used for cell development and differentiation studies.

3. *Epithelial ovarian carcinoma (EOC).* This is most widespread form of the disease and accounts for about 90% of ovarian cancer cases. This histologic type originates from the OSE. EOC is further divided into several subtypes, depending on the nature of cellular differentiation within the tumor. The most common subtype is *serous adenocarcinoma*, which resembles fallopian tube differentiation, while *clear cell* and *transitional cell (Brenner)* are rare subtypes and resemble gestational endometrium and urinary tract epithelium, respectively. Other subtypes include *mucinous* (resembling either the gastrointestinal tract or endocervix) and *endometrioid* type (resembles proliferative endometrium).[30]

6.4.1 Putative Role of Stem/Progenitor Cells in Ovarian Cancer

A cause–effect relationship between ovulation and EOC has been suspected for several years.[31] During the normal ovarian cycle, the OSE undergoes periodic degeneration and proliferative repair. Due to the location of OSE cells directly above the ovulation site, these are subjected to postovulatory genotoxins and oxidative damage. The continual cyclic wounding and repair processes, persisting for several decades until menopause, especially in cases that are uninterrupted by the long anovulatory rest periods during pregnancy and lactation, have been described as major predisposing factors for neoplasia. An observation that supports the hypothesis is that ovarian cancer is more common in human females than in other mammalian species, a fact attributed to a reduced frequency of pregnancies and lactation in the former during their reproductive lives.[32]

As described earlier, the existence of somatic stem cells and GSCs in the mammalian ovary has been suggested. Like any other adult stem cell, an ovarian stem cell would be expected to divide in response to stimulation for the requirement of tissue repair. Such stem cell divisions are known to be asymmetric, yielding one daughter cell that remains as a stem cell, while the other differentiates to function toward tissue repair (in case of ovarian stem cells, it would involve contributing to postovulation epithelial cell layer repair). In such a scenario, loss of a balance between differentiation and self-renewal simultaneous with the acquisition of mutations at either the stem cell stage or in their immediate progenitors would ultimately lead to their transformation into a CSC. This could lead to the development of EOC that progresses rapidly toward malignancy. Similarly, GSCs subsequent to genetic insults might seed GC tumors.

6.4.2 Tumor as an Aberrant Organ Initiated by Cancer Stem Cells

Although it is not known whether CSCs arise from normal adult stem cells or more differentiated cells, many recent studies have shown that CSCs resemble the normal stem or progenitor cells of the corresponding tissue of origin (see Chapter 1). There is evidence demonstrating that cell surface markers, generally restricted to normal adult stem cells, are also expressed by cancer stem cells. This permits a cell surface phenotype–based isolation and identification scheme for winnowing these notorious CSCs from tumors.

The first report on the isolation and identification of stemlike cells in ovarian cancer originated from our lab.[33] This study combined in vitro cell cloning with extensive screening for cell surface markers and other stem cell characteristics, thus providing evidence for the involvement of CSCs in ovarian cancer pathogenesis. As a part of the work, a novel in vitro model system comprising of 19 clones derived from a malignant ascites of human ovarian cancer was established. Of these, two clones were further identified as being CSCs, whereas the remaining 17 were untransformed and plausibly represented the pre-tumorigenic stage.

All 19 clones showed the presence of cell surface markers such as c-kit and stem cell factor, which are also expressed by stem cells, and expressed several other markers, such as CD44, EGFR, and E-cadherin. The two CSC clones (A2 and A4-T) showed the capacity of anchorage-independent growth, as evinced through their ability to form spheroids in the suspension culture and form multicellular colonies when grown in soft agar (an in vitro assay for carcinogenicity). Spheroid formation by these clones was construed to be a differentiation event, as it was associated with an expression of markers normally expressed by multiple ovarian lineages (i.e., the OSE, granulosa cells, and GCs). On injecting the CSCs in nude mice subcutaneous and intraperitoneal tumors were formed that were histolopathologically similar to the original tumor in the patient. Moreover, these tumors could be serially passaged through at least three generations of mice; indicative of a capacity for in vivo self-renewal. Thus, the study above provides evidence of ovarian cancer stem cells with stem cell phenotype, self-renewal, and differentiation abilities.

6.4.3 Ovarian Cancer Stem Cells Isolated as a Side Population

A small subset of cells termed as *side population* (SP) has been identified and characterized using flow cytometric analysis in several mammalian tissues, including bone marrow, mammary gland, lung, skin, and the cancers of these organs (details are presented in Chapter 4). A recent report by Szotek et al. demonstrates these properties within cells isolated as an SP in mouse ovarian cancer cell lines.[34] These SP cells were shown to have a capacity for self-renewal and produce heterologous descendent non-SP populations (i.e., the bulk that constitutes the majority of cells within a tumor). In culture, the SP cells were predominantly arrested at the G1 phase of the cell cycle. On injecting an equal number of SP and non-SP cells in vivo, the former formed palpable tumors much earlier and at a higher frequency than did non-SP cells. These cancer-derived mouse SP cells exhibit properties ascribed to CSCs, suggesting that the SP from mouse ovarian cancer cells is enriched for CSCs. A similar subpopulation of SP cells could also be isolated from human ovarian cancer cell lines and primary human ascites cells from ovarian cancer patients, thus providing an approach to the detection and isolation of ovarian CSCs for further characterization.

Compared to non-SP cells, the SP cells were highly resistant to doxorubicin (a conventionally used chemotherapeutic drug) that correlates with their expression of ABCG2 (a drug transporter frequently associated with SP populations). Subsequently, the authors demonstrated suppression of SP cell growth in the presence of Mullerian inhibiting substance. This provides a potential therapy that could be effective in restricting tumor progression driven by resistant ovarian SP-CSCs. However, these findings are based on in vitro studies in mouse cell lines, and extrapolation of these results to the clinic would require further study.

6.4.4 New Challenge: Targeting Ovarian Cancer Stem Cells

The two reports described above plausibly demonstrate that the ovarian cancers are derived from CSCs, which are more resistant than the bulk cells in the tumor. Thus, targeting these cells would prove extremely successful in the management of ovarian cancer (Table 6.2). Although the reports above define ovarian cancer as a stem cell disease, in neither of the studies have markers for these CSCs been revealed, rendering this system tricky for therapeutic research. Moreover, the overall similarities between normal stem cells and CSCs put forward another stumbling block in designing a therapy that will selectively target CSCs without significant toxicity

TABLE 6.2   Evidence Suggesting the Existence of Stemlike Cells (Putative CSCs) Within Ovarian Cancers

Year	Species	Main Criteria for CSC Identification	Cell Markers Used	Markers Identified for Ovarian CSCs	Conclusion	Ref.
2005	Human	Clonogenicity and cell surface phenotype	CD44, EGFR, c-kit, vimentin, c-met, Snail, Slug, SCF, E-caderin, Nanog, Oct4, Nestin	None	Ovarian cancer is a stem cell disease.	32
2006	Mouse and human	Dye effluxing potential— side population	CD24, CD34, c-kit, CD44, CD45, CD90, CD105, CD133, Sca1	None	SP from the ovarian cancer cells is enriched for the CSCs.	33

to normal stem cells and their niche. However, recent reports have provided proof of principle of such specific targeting of leukemic stem cells with minimal toxicity to their normal counterparts (i.e., HSCs).[35] In principal, similar studies in the field of ovarian cancer may be valuable in designing a therapy that selectively targets ovarian CSCs. It thus becomes imperative for reproductive and ovarian cancer biologists to initiate new series of investigations so as to extend or challenge these basic studies and resolve mysteries in this field of research.

REFERENCES

1. Ovarian cancer. ACOG Educational Bulletin 250, August 1998. In: *1999 Compendium of Selected Publications*. Washington, DC: American College of Obstetricians and Gynecologists; 1999: 665.

2. Wikborn C, Pettersson F, Silfversward C, Moberg PJ. Symptoms and diagnostic difficulties in ovarian epithelial cancer. *Int J Gynaecol Obstet*. 1993; 42:261.

3. Boring CC, Squires TS, Tong T. Cancer statistics. *CA Cancer J Clin*. 1993; 43:7.

4. Byskov AG. Differentiation of mammalian embryonic gonad. *Physiol Rev*. 1986; 66:71.

5. Auersperg N, Wong AST, Choi KC, Kang SK, Leung PCK. Ovarian surface epithelium: biology, endocrinology, and pathology. *Endocrinol Rev*. 2001; 22:255.

6. Bukovsky A, Caudle MR, Svetlikova M, Upadhyaya NB. Origin of germ cells and formation of new primary follicles in adult human ovaries. *Reprod Biol Endocrinol*. 2004; 2:20.

7. Ginsburg M, Snow MH, McLaren A. Primordial germ cells in the mouse embryo during gastrulation. *Development*. 1990; 110:521.

8. Van Wagner G, Simpson M, eds. *Embryology of the Ovary and Testis*: Homo sapiens and Macaca mulatta. New Haven, CT: Yale University Press; 1965: 1–171.

9. Kezele P, Nilsson EE, Skinner MK. Cell–cell interactions in primordial follicle assembly and development. *Front Biosci.* 2002; 7:d1990.

10. Fortune JE, Cushman RA, Wahl CM, Kito S. The primordial to primary follicle transition. *Mol Cell Endocrinol.* 2000; 163:53.

11. Franchi LL, Mandl AM, Zuckerman S. The development of the ovary and the process of oogenesis. In: *The Ovary.* London: Academic Press; 1962.

12. Byskov AG, Faddy MJ, Lemmen JG, Andersen CY. Eggs forever. *Differentiation.* 2005; 73: 9–10.

13. Allen E. Ovogenesis during sexual maturity. *Am J Anat.* 1923; 31:439.

14. Pearl R, Schoppe WE. Studies on the physiology of reproduction in the domestic fowl. *J Exp Zool.* 1921; 34:101.

15. Zuckerman S. The number of oocytes in the mature ovary. *Recent Prog Horm Res.* 1951; 6:63.

16. Vermande-Van Eck GJ. Neo-ovogenesis in the adult monkey; consequences of atresia of ovocytes. *Anat Rec.* 1956; 125:207.

17. Peters H. Migration of gonocytes into the mammalian gonad and their differentiation. *Philos Trans R Soc Lond B.* 1970; 259:91.

18. Borum K. Oogenesis in the mouse: a study of meiotic prophase. *Exp Cell Res.* 1961; 24:495.

19. Duke KL. Ovogenetic activity of the fetal-type in the ovary of the adult slow loris, *Nycticebus coucang. Folia Primatol(Basel).* 1967; 7:150.

20. David GF, Anand TC, Baker TG. Uptake of tritiated thymidine by primordial germinal cells in the ovaries of the adult slender loris. *J Reprod Fertil.* 1974; 41:447.

21. Hubner K, Fuhrmann G, Christenson LK, et al. Derivation of oocytes from mouse embryonic stem cells. *Science.* 2003; 300:1251.

22. Johnson J, Canning J, Kaneko T, Pru JK, Tilly JL. Germline stem cells and follicular renewal in the postnatal mammalian ovary. *Nature.* 2004; 428:145.

23. Lawson KA, Hage WJ. Clonal analysis of the origin of primordial germ cells in the mouse. *Ciba Found Symp.* 1994; 182:68.

24. Medvinsky A, Dzierzak E. Definitive hematopoiesis is autonomously initiated by the AGM region. *Cell.* 1996; 86:897.

25. Grove JE, Bruscia E, Krause DS. Plasticity of bone marrow derived stem cells. *Stem Cells.* 2004; 22:487.

26. Johnson J, Bagley J, Skaznik-Wikiel M, et al. Oocyte generation in adult mammalian ovaries by putative germ cells in bone marrow and peripheral blood. *Cell.* 2005; 122:303.

27. Bukovsky A, Svetlikova M, Caudle MR. Oogenesis in cultures derived from adult human ovaries. *Reprod Biol Endocrinol.* 2005; 3:17.

28. Dyce WP, Lihua Wen, Julang Li. In vitro germline potential of stem cells derived from fetal porcine skin. *Nat Cell Biol.* 2006; 8:4.

29. Reya T, Morrison SJ, Clarke MF, Weissman IL Stem cells, cancer and cancer stem cells. *Nature.* 2001; 414:105.

30. Naora H. The heterogeneity of epithelial ovarian cancers: reconciling old and new paradigms. *Expert Rev Mol Med.* 2007; 9(13): 1–12.

31. Fathalla MF. Incessant ovulation: a factor in ovarian neoplasia? *Lancet*. 1971; 2:163.

32. Murdoch JM, McDonnel CA. Roles of the ovarian surface epithelium in ovulation and carcinogenesis. *Reproduction*. 2002; 123:743.

33. Bapat SA, Mali AM, Koppikar CB, Kurrey NK. Stem and progenitor-like cells contribute to the aggressive behavior of human epithelial ovarian cancer. *Cancer Res*. 2005; 5:3025.

34. Szotek PP, Vanmarcke PR, Masiakos PT, et al. Ovarian cancer side population defines cells with stem cell–like characteristics and Mullerian inhibiting substance responsiveness. *Proc Natl Acad Sci U S A*. 2006; 103:30.

35. Yilmaz OH, Valdez R, Theisen BK, et al. Pten dependence distinguishes haematopoietic stem cells from leukaemia-initiating cells. *Nature*. 2006; 441:475.

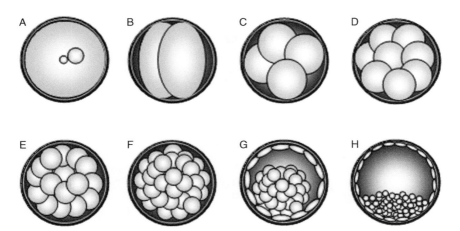

FIGURE 1.1 Developmental stages from normal early vertebrate embryo development to the blastocyst stage: (A) single-celled zygote; (B) two-celled embryo; (C) four-celled embryo; (D) early morula; (E) compacted morula; (F) late morula; (G) early blastocyst; (H) late blastocyst.

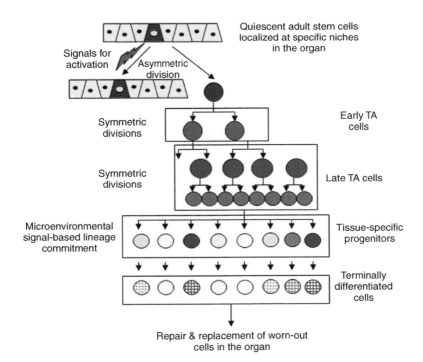

FIGURE 1.3 Stem cell hierarchies in normal tissue regeneration.

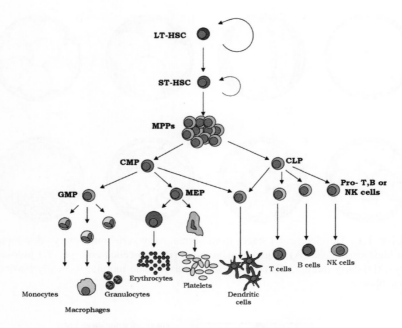

FIGURE 2.1 Normal hematopoietic cell hierarchy.

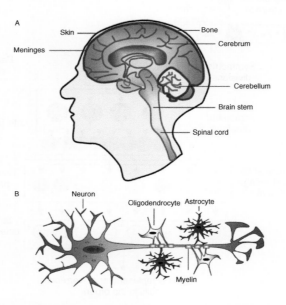

FIGURE 3.4 (A) Architecture of the brain; (B) brain cell types.

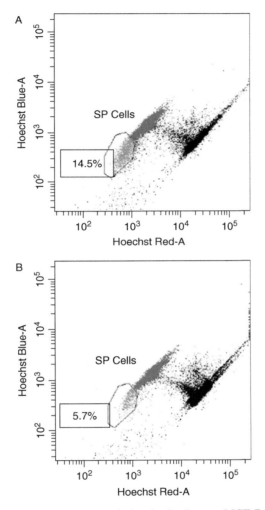

FIGURE 4.2 Presence of a side population in the human MCF-7 breast adenocarcinoma cell line. Cells labeled with Hoechst 33342 reveal the presence of a small low-staining SP population (A). When treated with 50 μL verapamil, an inhibitor of SP transporters, this population diminishes in size (B). The cells were counterstained with 1 μg/mL propidium iodide.

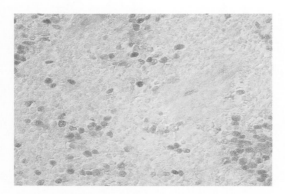

FIGURE 5.1 MCM2 immunoreactivity in human RB. An archival paraffin-embedded human retinoblastoma tumor was de-paraffinized and immunostained with MCM2 antibody (Santa Cruz Biochemicals, N-19). MCM2 immunoreactive cells were labeled with brown DAB reaction product.

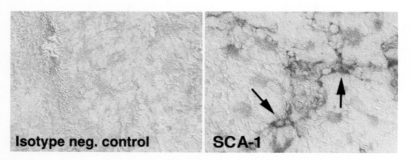

FIGURE 5.2 SCA-1 immunoreactivity in mouse RB. A frozen mouse RB tumor was immunostained with Sca-1 antibody (BD Biosciences, 557403). Arrows indicate areas of immunopositive cells. An isotype control panel is shown for comparison.

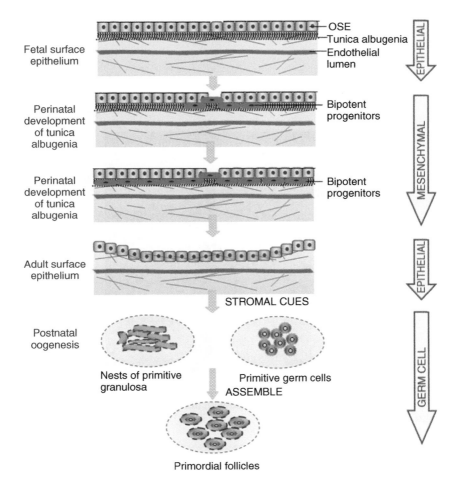

Fetal surface
epithelium

Perinatal
development
of tunica
albugenia

Perinatal
development
of tunica
albugenia

Adult surface
epithelium

Postnatal
oogenesis

OSE
Tunica albugenia
Endothelial
lumen

Bipotent
progenitors

Bipotent
progenitors

STROMAL CUES

Nests of primitive
granulosa

Primitive germ cells

ASSEMBLE

Primordial follicles

EPITHELIAL

MESENCHYMAL

EPITHELIAL

GERM CELL

FIGURE 6.3 Four-step transition during oogenesis from bipotent progenitors. During early embryonic development the ovarian surface epithelium undergoes epithelial-to-mesenchymal transition to generate tunica albugenia. The tunica albugenia cells serve as the bipotent progenitors for the granulosa and germ cells during postnatal oogenesis.

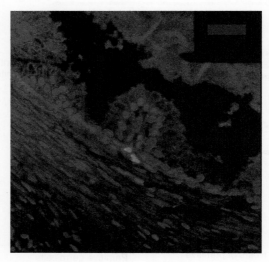

FIGURE 7.3 Expression of CD133 by a rare subset of basal cells in prostate epithelium. A paraffin section of prostatic acini is labeled with the nuclear stain DAPI (shown in blue) and anti-CD133 directly conjugated to PE (shown in red).

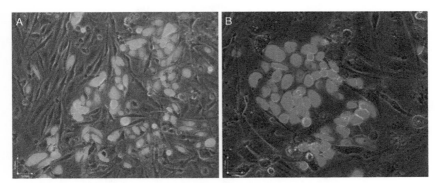

FIGURE 7.6 Transgenic prostate cancer stem cells cultured on feeder cells. Prostate cancer stem cells were isolated from clonal populations of P4E6 prostate cancer cells infected with a lentivirus delivering an (A) CMV-eGFP or (B) CMV-DsRed expression cassette, and were cultured on STO feeder cells. Pictures were taken after (A) 15 or (B) 10 days of culture. (A): 200 × magnification, (B): 400 × magnification.

FIGURE 8.1 Athymic nude mouse with melanoma induced by in vitro cultured CT-1413 cells. Melanoma developed within 10 days of subcutaneous inoculation cells.

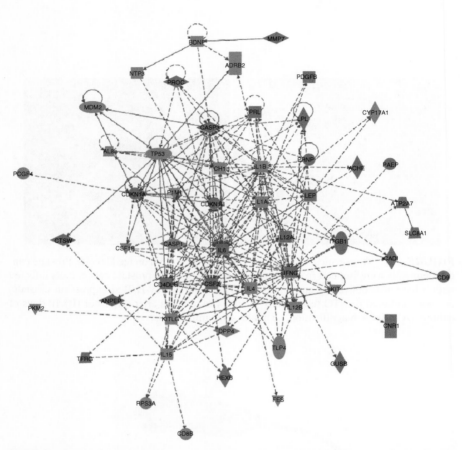

FIGURE 8.4 Graphical representation of the molecular relationship between proteins expressed in stem cell cat melanoma and network connections generated through the use of Ingenuity Pathway Analysis systems (www.ingenuity.com). Each protein is represented as a node, and the biological relationship between two nodes is represented by a line. All lines are supported by at least one reference from the literature, a textbook, or from canonical information stored in the Ingenuity Pathway knowledge base. The intensity of the node color indicates normalized quantities of expressed proteins [i.e., degree of expression, up- (red) or down- (green) regulation in melanoma cells] compared to those present in counterpart neuronal cells. Nodes are displayed using various shapes that represent the functional class of the protein. Fischer's exact test was used to calculate a p-value that determines the probability that each biological function assigned to that network is due to chance alone.

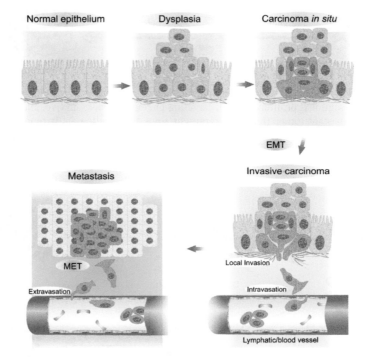

FIGURE 9.1 Metastasis is a multistep process. The major steps of the metastatic process are indicated: local invasion, involving basement membrane destruction and invasion into adjacent tissues; intravasation and survival in the bloodstream; and extravasation into distant organs and proliferation/survival in the new host organ. The specific steps where EMT (epithelial–mesenchymal transition) and the reverse MET (mesenchymal–epithelial transition) process are thought to occur are indicated.

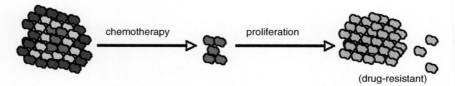

FIGURE 10.2 Based on several defining characteristics, cancer stem cells could probably resist conventional proliferative cell-targeting chemotherapies, survive, and repopulate a now drug-resistant tumor (red cells represent stem cells, yellow cells represent transit amplifying cells, and blue cells represent differentiated tumor cells). In the new tumor, the chemoresistant phenotype is passed on to stem cell progeny (gray).

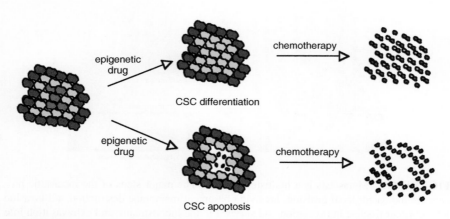

FIGURE 10.3 Based on the disruption of DNA methylation and/or histone deacetylation necessary for cancer stem cell (CSC)–initiated tumorigenesis, epigenetic therapies could induce CSC differentiation (top path) or apoptosis (lower path). Conventional chemotherapy could then eliminate the remaining non-stem cell population.

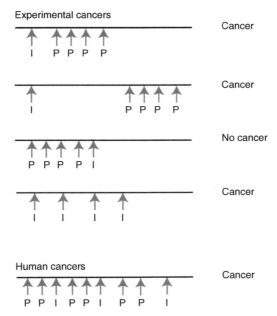

FIGURE 11.2 Tumor initiation and promotion. In experimental systems the adminis-tration of an initiator (mutagen) when followed by multiple administrations of a tumor promoter results in tumors. If the promoter is given first, no tumor results, although many administrations of the initiator can cause cancer. In the human system, chronic inflamma-tion and/or exposure to tissue-disrupting agents, combined with mutagen exposure, causes cancer.

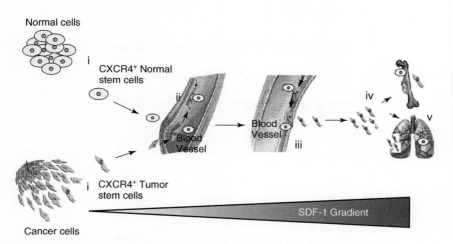

FIGURE 12.1 Migration of normal and cancer stem cells is a multistep process which involves: (i) Egress of CXCR4 expressing stem cells from the tissue; (ii) adhesion of stem cells to the endothelium; (iii) crossing the endothelial barrier; (iv) homing to the target tissues chemoattracted by the SDF-1 released; and (v) expansion of stem cells at the site that provides a favorable niche.

7 Prostate Cancer Stem Cells

STEFANIE HAGER, NORMAN J. MAITLAND, and ANNE COLLINS

7.1 INTRODUCTION

Causing about 200,000 deaths worldwide each year, prostate cancer is excelled only by lung cancer as the leading cause of cancer-related mortality among men. Improvements in detecting early-stage disease have been made over the past years, and treatment measures for localized prostate cancer have been refined. However, there remains little effective therapy for patients with locally advanced and/or metastatic disease. The majority of patients with advanced disease respond initially to androgen ablation therapy, but most go on to develop androgen-independent tumors that are inevitably fatal. The response to chemotherapeutic and radiotherapy treatments is similar.

Prostate epithelial stem cells have been assigned a role in prostate cancer. It is thought that these cells are residual after conventional androgen ablation therapy and that they represent the basis of commonly observed tumor regrowth following seemingly successful treatment. This hypothesis was strengthened by the identification and isolation of prostate cancer stem cells, based on surface expression of the markers integrin $\alpha_2\beta_1$ and CD133.[1] If prostate tumors are indeed derived from transformed stem cell we will need to reconsider the way we treat prostate cancer. To do this, we need to acquire a profound knowledge of these cells. In this chapter we discuss recent data as well as our current knowledge of prostate biology and pathology.

7.2 HUMAN PROSTATE BIOLOGY, GLAND ARCHITECTURE, AND PATHOLOGICAL ALTERATIONS

The human prostate gland is located in the pelvis and is surrounded posteriorly by the rectum and superiorly by the bladder. It is a nonessential organ that is supposedly involved in maintaining the viability of sperm and in semen production. Due to its compact, walnut-shaped structure, anatomical regions in the prostate are not

Cancer Stem Cells: Identification and Targets, Edited by Sharmila Bapat
Copyright © 2009 John Wiley & Sons, Inc.

easily discernible. A concept of anatomical zones which considers the site of origin of different pathologies is used instead.[2] The prostate is then described in terms of a peripheral, a central, and a transition zone as well as an anterior fibromuscular stroma. Benign prostate hyperplasia (BPH) originates mainly in the transition zone, whereas prostate carcinomas commonly arise in the peripheral zone of the prostate.[3]

Histologically, the prostate exhibits a tuboloalveolar gland architecture. An epithelial parenchyma is embedded within a connective tissue matrix, and epithelial cells are organized in glands that branch out from the urethra and terminate in secretory acini (Fig. 7.1). In adults, the prostate epithelium is organized as a bilayer consisting of a basal layer of flattened, undifferentiated cells attached to the basement membrane and a layer of terminally differentiated, columnar secretory cells residing on top of the basal cells and facing the gland lumen. This characteristic two-layered epithelial architecture develops during puberty under the influence of male sex hormones. Prior to this differentiation process, prostatic ducts and acini are lined with a multilayered epithelium consisting of immature cells.[4]

A third cell type discernible in adult prostate comprises the neuroendocrine cells, which are scattered throughout the basal compartment. They are characterized by the expression of neuropeptides such as chromogranin and serotonin, and are terminally differentiated androgen-insensitive cells.[5] Luminal cells are characterized by expression of the differentiation markers androgen receptor (AR), prostatic acid phosphatase (PAP), and prostate-specific antigen (PSA), which is currently used as a diagnostic marker for early prostate cancer.[6] Basal cells, in contrast, express the antiapoptotic gene *bcl-2*.[7,8] Both basal and luminal cells also express a characteristic set of cytokeratins, which commonly serve as specific

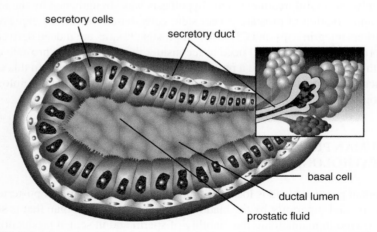

FIGURE 7.1 Organization of the prostate gland: cross section of the ductal region with labels indicating cell types present in prostatic ducts, including luminal secretory and basal cells. (From A.T. Collins and N.J. Maitland, Prostate cancer stem cells, *Eur. J. Cancer*, 2006; 42:1213–1218.)

markers of differentiation in epithelial tissues [9-13] (see Section 7.3.1). A further distinction between basal and luminal cells can be made on the basis of growth regulation: Basal cells are androgen-independent, whereas luminal cells are highly dependent on androgens for their growth and survival.[14]

In aging men, the prostate gland usually becomes hyperplastic. This alteration, termed *benign prostate hyperplasia* (BPH), arises mainly in the transition zone of the prostate. The state is nonmalignant and causes only minor disturbance of well-being.[3] In contrast, prostate intraepithelial neoplasia (PIN), is a premalignant lesion that is characterized by a progressive loss of the epithelial two-cell arrangement.[15]

Adenocarcinoma of the prostate accounts for 95% of all cancerous lesions in the prostate. It is the most commonly diagnosed neoplasm after skin cancer and is second only to lung cancer as the leading cause of cancer-related death among men in the Western world.[15] Prostate carcinoma arises primarily in the peripheral zone of the prostate, and its incidence increases dramatically with patient age.[3] Unless detected at an early stage, prostate cancer tends to spread to the pelvic lymph nodes and ultimately localizes at distant lymph nodes or bone. With these sites being hardly amenable to surgery or radiation therapy, treatment of metastatic disease commonly relies on the hormone-responsive nature of the cancer. Like normal prostate, prostate tumors rely on male sex hormones for development and growth. To induce apoptosis in cancer cells, the hypothalamus–pituitary–sex gland signaling axis, which supplies cancer cells with testicular androgen, is inhibited. [15-17] Such androgen deprivation initially results in a reduction in tumor mass as androgen-dependent cancer cells undergo apoptotic death.[15] Response to hormone therapy is, however, only temporary, and development of androgen-independent disease with tumor progression occurs in virtually all patients 12 to 18 months after the onset of treatment. At this stage, prostate carcinoma is incurable by current treatment strategies, with five-year survival rates as low as 15%. [6,15-17]

7.3 PROSTATE EPITHELIAL STEM CELLS

Evidence for the existence of a stem cell subpopulation in the prostate has been accumulating since the 1980s. Initial experiments in the murine prostate demonstrated the presence of a small number of androgen-independent cells which were able to reconstitute a functional prostate gland. [18-20] More recent experiments provided indirect evidence that basal and luminal cells were derived from a common precursor [21-24] before prostate epithelial stem cells were finally isolated and their stem cell characteristics demonstrated.[25,26]

7.3.1 Evidence for Prostate Epithelial Stem Cells

Androgen Cycling Studies   The experiment that most vividly demonstrates that cells exhibiting stem cell characteristics must exist in the prostate are the castration experiments conducted by English et al.[19,27] The authors showed that androgen withdrawal in rats led to an involution of the prostate gland caused by apoptosis of androgen-dependent epithelial, endothelial, and periacinar stromal cells. Interestingly, these cell compartments could be restored by readministration of androgen. This cycle of gland involution–restoration can be repeated several times, which argues for the existence of a long-lived, androgen-independent population of prostate stem cells which can give rise to an androgen-dependent luminal population. In support of these findings, Montpetit et al. [20] were able to isolate clones of androgen-independent cells from primary cultures of androgen-dependent epithelial cells from the rat ventral prostate. Based on these seminal papers, Isaacs and Coffey[28] proposed a stem cell theory that linked basal and luminal cells in the prostate in a precursor–progeny relationship. According to their hypothesis, a population of androgen-independent stem cells in the basal epithelial compartment gives rise to an androgen-responsive transit-amplifying population, which in turn produces the fully differentiated androgen-dependent luminal population (Fig. 7.2).

Around the same time, however, cell kinetic and morphological investigations raised the opposing view that basal and luminal cells in the prostate form part of independent differentiation lineages. Cell labeling studies demonstrated that

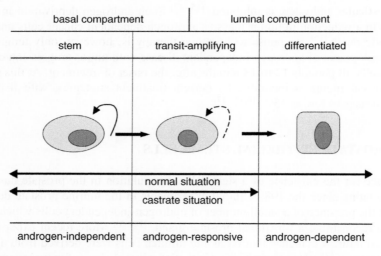

FIGURE 7.2 Stem cell model for the organization of the prostate epithelium. A stem cell located in the basal epithelial compartment gives rise to a population of transit-amplifying cells, which in turn differentiate to form the secretory luminal cells. Stem as well as transit-amplifying cells are androgen independent and therefore persist under castrate conditions. Luminal cells in contrast are dependent on androgens and undergo apoptotic death upon androgen withdrawal. (Adapted from A. T. Collins and N. J. Maitland, Prostate cancer stem cells, *Eur. J. Cancer*, 2006; 42:1213–1218.)

both basal and luminal epithelial prostate cells were residual after castration and that proliferation could be observed in both compartments after restoration of androgen levels.[18,29,30] However, the persistence of luminal-like cells after androgen cycling experimentation does not contradict the stem cell theory proposed by Isaacs and Coffey[28] if we assume that the residual luminal-like cells mentioned above constitute the androgen-insensitive but responsive transit-amplifying population (Fig. 7.2).

Phenotypic Relationship Between Epithelial Cell Types in the Prostate   The theory that basal and luminal cells in the prostate are linked in a hierarchical manner was strengthened by the identification of morphologically and immunophenotypically intermediate cell types. As early as 1978, Dermer[31] was able to distinguish between basal and luminal cells in the human prostate on the basis of their differential affinity for toluidine blue. He identified cells with a basal phenotype exhibiting a high affinity for toluidine blue not only in contact with the basement membrane but also close to the gland lumen.[31] Following these fundamental studies, sophisticated methods of multiple immunostaining for differentiation stage-specific cytokeratins identified an intermediate cell population in the prostate. [21–24] Thus, a gradual shift in keratin expression can be observed, as basal cells acquire a luminal phenotype via a defined transitional cell population. Bonkhoff et al.[32] demonstrated co-localization of basal cytokeratins with the proliferation-associated antigens Ki67, PCNA, and MIB 1 in normal and hyperplastic prostate, indicating that the proliferative compartment in the prostate resides within the basal cell layer. The same group reported co-localization of basal cell–specific cytokeratins and prostate-specific antigen, a luminal marker indicating terminal differentiation, as well as coexpression of basal cytokeratins and the neuroendocrine marker chromogranin A, in the basal cell compartment.[33] Similarly, Robinson et al. observed a gradual shift from expression of basal to luminal cytokeratins via a transit-amplifying population.[34] These findings indicate considerable phenotypic plasticity in the basal cell compartment of the prostate and point to a common cell differentiation lineage comprising basal, luminal, and neuroendocrine cells. Keratin-based differentiation pathways for the prostate have been hypothesized based on simultaneous immunostaining for up to three different cytokeratins. [21–23] According to these studies, basal stem cells expressing cytokeratins 5 and 14 enter the differentiation pathway by transiently expressing cytokeratin 19 while gradually down-regulating expression of keratins 5 and 14. Further down the lineage, cells transiently coexpress cytokeratins 19, 8, and 18 before acquiring the keratin 8/18-positive luminal phenotype.[21] Transient expression of cytokeratin 19 in intermediate cells complies with the findings of Stasiak et al., who identified it as a "neutral" keratin characteristic of regions of labile or variable cellular differentiation.[35] However, Robinson et al. demonstrated stable expression of keratin 19, and its co-localization with the luminal markers keratin 18, androgen receptor, and prostate-specific antigen (PSA).[34]

7.3.2 Isolation of Human Prostate Epithelial Stem Cells and Demonstration of Their Stem Cell Character

Its current definition requires a tissue stem cell to be capable of giving rise to all cell types that are characteristic of the particular tissue or organ. Addressing this issue of multipotency requires isolation of the respective tissue stem cells and differentiation of the pure population in vitro or in vivo. Isolation of the stem cell subpopulation, however, has to be performed on the basis of a marker that is exclusive to this population. Hudson et al. undertook clonal analysis of human prostate epithelial cells in order to identify the stem cell subpopulation.[36] Upon cloning of cells from biopsies of benign prostate hyperplasia, two types of colony could be distinguished. Type II colonies exhibited a smooth periphery and densely packed cells, whereas cells in type I colonies were less densely packed and colony borders were irregular; which is indicative of the presence of cell types with differing proliferative activity. Type II colonies were capable of generating three-dimensional duct like structures in Matrigel, [36−38] with evidence of luminal-specific cytokeratin and androgen receptor expression in a subset of the cells. These results implied that type II colonies were founded by stem cells, whereas type I colonies were probably founded by transit-amplifying cells. Due to a lack of known prostate stem cell markers at this time, these experiments provided indirect evidence of the lineage of prostate epithelia. Collins et al.[25] took advantage of the observed association of stem cells with basement membranes in skin[39,40] and testis[41] to enrich for prostate epithelial stem cells from BPH tissue samples. About 1% of basal cells in the prostate exhibit a high affinity for the extracellular matrix component collagen I, which is attributed to a high-level expression of integrin $\alpha_2\beta_1$ in these cells. Cells that adhere rapidly to collagen I are clonogenic in vitro and capable of regenerating prostatelike acini in vivo, with evidence of expression of the luminal cell markers PAP, AR, and PSA.[25] Consistent with the findings reported by Hudson et al.,[36] however, differential adhesion to collagen I does not separate stem and transit-amplifying cells, as the retrieved cell population is capable of giving rise to both type I and type II colonies. Furthermore, rapidly adherent cells contain a subpopulation of approximately 5% of committed basal cells, as indicated by expression of the luminal-specific cytokeratin 18 in these cells. Cells adhering to collagen I after 20 minutes found small, terminal colonies of irregular shape (type III colonies), which indicates the high degree of differentiation of these cells.[25]

Richardson et al.[26] extended the cell surface markers for the isolation of prostate epithelial stem cells to include CD133, a five-transmembrane-domain cell surface glycoprotein which had originally been identified as a marker for hematopoietic stem cells[42] and had previously been used to enrich for endothelial[43] and neuronal[44] stem cells. CD133 cells comprise a subset of the basal epithelial cells in prostate (0.75%; Fig. 7.3) and are restricted to the $\alpha_2\beta_1{}^{hi}$ population of basal cells, which had previously been shown to be enriched for prostate epithelial stem cells.[25] Immunostaining patterns for CD133 and the proliferation antigen Ki-67 did not overlap, which suggests a replicative quiescence of the $\alpha_2\beta_1{}^{hi}/CD133^+$ cell fraction. This fraction, however, exhibits a higher

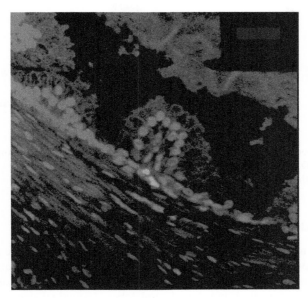

FIGURE 7.3   Expression of CD133 by a rare subset of basal cells in prostate epithelium. A paraffin section of prostatic acini is labeled with the nuclear stain DAPI (shown in blue) and anti-CD133 directly conjugated to PE (shown in red). (From A. T. Collins and N. J. Maitland, Prostate cancer stem cells, *Eur. J. Cancer*, 2006; 42:1213–1218.) (*See insert for color representation of figure.*)

proliferative potential and colony-forming ability than that of the $CD133^-$ cells within the $\alpha_2\beta_1{}^{hi}$ cell fraction. Basal cells selected for $\alpha_2\beta_1{}^{hi}/CD133$ expression are also capable of regenerating a fully differentiated prostatic epithelium in vivo with evidence of expression of the common differentiation markers, whereas $\alpha_2\beta_1{}^{hi}/CD133^-$ cells lack this ability.

In contrast to the subpopulation retrieved from basal prostate epithelial cells on the basis of high-level expression of $\alpha_2\beta_1$ integrin alone, the $\alpha_2\beta_1{}^{hi}/CD133^+$ cell fraction consists solely of epithelial stem cells, whereas transit-amplifying cells are restricted to the $CD133^-$ fraction among $\alpha_2\beta_1{}^{hi}$ cells. Integrin $\alpha_2\beta_1{}^{low}$ cells among the basal population probably represent cells committed to luminal differentiation (Fig. 7.4). A different method for the isolation of prostate stem cells from primary tissues was published by Bhatt et al.[45] The isolation procedure was based on the SP (side population) phenotype, which is commonly exhibited by murine hematopoietic stem cells and is characterized by the ability of the respective cells to efflux Hoechst dye 33342. The dye-effluxing stem cells constitute an unstained side population upon flow cytometric analysis.[46] Using this method for prostate epithelia, the authors estimated the stem cell population to comprise 1.4% of all prostate epithelial cells. However, gland regeneration activity has not been demonstrated in association with this putative stem cell fraction.[45]

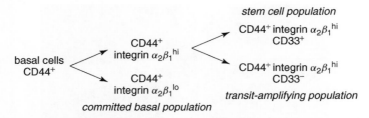

FIGURE 7.4   Fractionation of basal prostate epithelial cells. Basal prostate epithelial cells can be separated into a committed basal (integrin $\alpha_2\beta_1{}^{lo}$), transit-amplifying (integrin $\alpha_2\beta_1{}^{hi}$ CD133$^-$), and stem cell population (integrin $\alpha_2\beta_1{}^{hi}$ CD133$^+$) on the basis of expression of the surface markers integrin $\alpha_2\beta_1$ and CD133.

7.3.3 Epithelial Stem Cells in the Murine Prostate

With respect to the location of epithelial stem cells within the prostate gland, different results have been obtained using the mouse model. Unlike the human prostate, the murine organ can be subdivided into ventral and dorsolateral prostates, which in turn consist of ventral and dorsolateral lobes. Individual ducts are made up of proximal, intermediate, and distal regions.[47] Label-retention studies on mice demonstrated that the majority of quiescent cells are located in the proximal region of the ducts, whereas luminal cells are located at the ductal tips. Label retention is a common method of mapping the location of stem cells, which is based on the assumption that cells exhibiting stem cell properties will enter the cell cycle very infrequently and therefore retain a DNA label for an extended period of time. In contrast, label density will decrease rapidly in actively dividing cells.[48] Interestingly, label-retaining cells in the mouse prostate are located among both the basal and luminal cells in the proximal region, which either suggests that basal and luminal cells in the murine prostate form separate lineages or that label-retaining cells in the luminal compartment constitute an early transit-amplifying population derived from basal stem cells. Proximal, label-retaining cells have a higher proliferative potential than that of distal cells in vitro and are capable of forming large glandular structures in collagen gels,[49] which is in accordance with a stem cell phenotype. Burger et al. enriched for murine prostate epithelial stem cells from the proximal region of prostatic ducts on the basis of expression of Sca-1 (stem cell antigen 1).[50] The Sca-1-purified cells efficiently regenerated prostatic ducts in vivo and expressed α_6-integrin as well as the antiapoptotic marker bcl-2,[7, 8] which is commonly expressed among stem cells. Xin et al. enriched for Sca-1-expressing cells from murine prostate by flow-sorting and demonstrated prostate-regenerating activity for the subpopulation selected.[51] Lawson et al. went on to show that Sca-1 expression in cells in the proximal murine duct region co-localizes with both basal (cytokeratin 5) and luminal (cytokeratin 8) cytokeratins,[52] which implies that Sca-1 expression is not restricted to the stem cell population but might be retained by early progenitor cells, which could be the luminal-like label retaining cells proposed by Tsujimura et al.[49]

7.3.4 Other Markers of Prostate Epithelial Stem Cells

Unlike the bulk of cells in a tissue, stem cells are long-lived and capable of self-renewal. Enhanced life span and proliferative potential must be reflected in the cells' gene expression program, such that gene products involved in self-renewal and antiapoptotic pathways will be exclusive to, or up-regulated in, the stem/progenitor cell fraction. Due to the common mechanisms underlying stem cell function, stem cells from different tissues are also expected to share markers to some extent.

Bcl-2 is an antiapoptotic protein[7,8] which is expressed primarily in long-lived cells.[53] Expression of bcl-2 can also be detected in the basal epithelial compartment in the prostate.[54,55] Salm et al. suggested that high levels of bcl-2 expression in stem cells, in the proximal region of mouse prostatic ducts, prevent the cells from undergoing apoptosis due to high levels of TGF-β signaling, which maintains stem cell quiescence.[56] Association of bcl-2 expression with the stem cell compartment in prostate, however, remains to be demonstrated.

Telomerase is a ribonucleoprotein enzyme that prevents progressive shortening of telomeres upon cell division, thus preventing chromosome end fusion and ultimately cell death. It is therefore expressed in various cancers where sustained self-renewal capacity of cells is required, whereas expression is absent from adult somatic cells.[57] Telomerase is expressed throughout the basal compartment in adult human prostate.[58] Whether it is expressed in the stem cell fraction, however, has yet to be elucidated. Owing to the infrequent cycling of stem cells, telomerase expression in this cell fraction might not be required. Supporting this assumption, Jaras et al. demonstrated that an increase in hTERT expression in murine hematopoietic stem cells was concomitant with decreased self-renewal capacity.[59]

Notch signaling has a role in the maintenance of hematopoietic stem cells.[60] Shou et al. mapped the Notch1 receptor to basal epithelial cells in the murine prostate using a transgenic mouse model that expressed enhanced green fluorescent protein (eGFP) under the control of the Notch1 promoter.[61] To further assess the role of Notch on prostate development, Wang et al. generated a transgenic mouse model expressing the prodrug-converting enzyme nitroreductase, under the control of the Notch1 promoter, which allows controlled elimination of Notch1 expressing cells by administration of nitroreductase substrate.[62] Ablation of Notch1-expressing cells severely affected morphogenesis, growth, and differentiation in murine prostate, suggesting that it is expressed in the stem cell compartment.

A further marker of basal epithelial cells in the prostate is *p63*, a homolog of tumor suppressor *p53*, which acts on the expression of a wide spectrum of target genes via multiple protein isoforms.[63] Mice bearing a germline inactivation of *p63* fail to develop a prostate gland,[64] which suggests that the fraction of *p63*-expressing cells in the prostate comprises the stem cells. Another gene selectively expressed in basal prostate epithelium is *pp32*, which codes for a nuclear phosphoprotein and has been identified in anatomically defined stem cell compartments of normal human tissues such as intestinal crypt epithelial cells.[65] Whether *pp32* expression is restricted to the stem cell compartment in prostate has yet to be elucidated.

As stated previously, prostate epithelial stem cells express high levels of integrin $\alpha_2\beta_1$ and CD133[26] (see Section 7.3.2), the latter being expressed exclusively in the stem cell fraction of the human prostate. A putative prostate stem cell fraction has also been isolated on the basis of a side population phenotype[45] (see Section 7.3.3). On a molecular level, this phenotype is determined by the presence of membrane efflux pumps of the ATP-binding cassette (ABC) transporter superfamily.[46] Prostate stem cell antigen (PSCA), a homolog to thymocyte stem cell antigen 2, maps to a subset of basal cells in sections of benign prostate.[66] Rather than identifying it as a prostate stem cell marker, however, studies in a transgenic mouse model identified PSCA as a marker of transit-amplifying cells.[13,67] Markers shown to typify other types of stem cells include Oct4, nanog (embryonic stem cells[68,69]), and β-catenin (intestinal stem cells[70]) as well as nestin (neural stem cells[711]). Whether these markers are expressed in prostate epithelial stem cells has yet to be determined.

7.4 PROSTATE CANCER STEM CELLS

Stem cells have long been assigned a role in the development of cancer,[72,73] but it is only over the last decade that our methodologies and our understanding of stem cell biology have advanced to such an extent that we are able to prove the involvement of tissue stem cells in the development of a range of malignancies. [74–79] Only very recently, such evidence has been obtained for prostate cancer.[1] With no effective treatment being available for prostate cancer in its advanced stage, research into prostate cancer stem cells holds the potential to yield novel treatment strategies.

7.4.1 Role of Stem Cells in Prostate Cancer

Similar to many other types of cancer, it is widely assumed that prostate tumors rely on stemlike epithelial cells for tumor maintenance and the subsequent emergence of therapy resistance under androgen blockade. Controversy exists, however, as far as the cell of origin of pathological alterations in prostatic epithelium is concerned. Generally, there are two basic mechanisms by which tumorigenic, stemlike cells can arise. The first mechanism is in accordance with the cancer stem cell hypothesis and is the occurrence of transforming mutations in a tissue stem cell. The second mechanism suggests that a more restricted progenitor cell, [80–82] or even fully differentiated cell[83] is the target of mutational hits and thereby reacquires stem cell properties. Despite the requirement for extensive genetic alterations, the latter mechansim of de-differentiation has long been suggested to be the molecular basis of prostate carcinomas. It was postulated that secretory luminal cells are the targets of mutations leading to prostate cancer, as prostate cancer cells predominantly express the luminal cell markers cytokeratin 8 and 18 and prostate-specific antigen (PSA).[83] The predominant phenotype of prostate carcinomas being luminal-like, however, does not preclude the tumor from being derived from a prostate epithelial stem cell, as the cancer stem cell hypothesis states that tumor stem cells comprise a minority population among tumor cells and give rise to the bulk of tumor cells through differentiation.

De Marzo et al. suggested that PIN and prostate carcinomas are derived from transit-amplifying cells in prostate epithelium, which acquire stemlike features of unlimited self-renewal by mutation.[84] The benign condition of BPH, in contrast, was suggested to arise from aberrantly proliferating stem cells in the basal compartment. This model for the development of prostate disease was seeking to explain regional differences in the occurrence of BPH and prostate carcinomas and the rare progression of BPH to prostate cancer by proposing different cells of origin for both pathologies.

Expression of *c-Met* (the gene coding for hepatocyte growth factor receptor) in intermediate cells in normal and malignant prostate epithelium[85] and identification of keratin expression patterns in prostate cancers associated with intermediate stages of differentiation[23,86] pointed similarly to progenitor cells as being the founder cells of prostatic carcinomas. Nevertheless, the cancer stem cell model postulates that the transformed stem cells, although making up only a minority population of the tumor, are still capable of giving rise to transit-amplifying and luminal populations through differentiation. It is therefore not unexpected that transit-amplifying or luminal cells resembling their counterparts in benign tissue are present in a tumor in large numbers, thereby potentially masking the stem cell phenotype.

Stem cells rather than intermediate cells have, however, also been suggested to be the founder cells of prostate malignancies. De Marzo et al. demonstrated different proliferative compartments, in normal as well as neoplastic prostate tissue,[87] with respect to expression of the cell cycle inhibitor $p27^{Kip1}$. Quiescent stem cells, as well as fully differentiated luminal cells, in normal prostate express high levels of $p27^{Kip1}$, whereas the cell layer sandwiched between basal and luminal cells does not express $p27^{Kip1}$, which is indicative of active cell cycling. It was therefore suggested that the $p27^{Kip1}$-negative population represented the transit-amplifying cells. $p27^{Kip1}$ is down-regulated in high-grade PIN as well as invasive prostate carcinoma, which is in accordance with the expansion of the transit-amplifying cell population in prostate tumorigenesis, possibly through a slight increase in the self-renewal rate of prostate cancer stem cells. These results also imply that down-regulated or absent $p27^{Kip1}$ might be part of the prostate cancer stem cell phenotype. Liu et al. performed expression analysis on a variety of prostate tumors and observed a trend toward expression of basal genes with disease progression.[88] Thus, most primary prostate tumors exhibit a luminal phenotype overall, whereas metastases are basal-like.

Further studies that point to prostate epithelial stem cells as the targets of transforming mutations show overexpression of genes involved in stem cell maintenance which are overexpressed in prostate cancer. Bcl-2 is expressed at high levels in androgen-independent prostate cancers,[89] and similarly, telomerase is expressed in basal epithelial cells in the prostate as well as PIN and prostate carcinoma.[58] Wang et al. generated a *Pten^{loxP/loxP};PB-Cre4* mouse model bearing a conditional knockout for tumor suppressor gene *PTEN*, which is frequently mutated in human prostate cancers.[90] The mouse model recapitulates the disease course seen in humans, progressing sequentially from hyperplasia to PIN, invasive adenocarcinoma, and ultimately, metastasis. Only very recently, the same group managed to demonstrate the direct involvement of basal prostate stem cells in this disease process.[91] Using the same *PTEN* knockout model, the researchers revealed that *PTEN* suppresses proliferation of basal cells, directly while still allowing them to differentiate. *PTEN^-* basal cells undergo extensive proliferation, which translates into an increase in the transit-amplifying population and thus tumor formation. Xin et al. perturbed *PTEN/Akt* signaling in murine prostate cells by transduction with a lentivirus expressing constitutively active *Akt1*[51]. The researchers showed that a cell population selected for the murine prostate epithelial stem cell marker Sca-1[50–52] and infected with *Akt1*-expressing lentivirus was much more effective at giving rise to PIN lesions after grafting than a lentivirus-infected Sca−1^- cell population.[51] This clearly demonstrates that epithelial stem cells can be a target for prostate tumorigenesis and that aberrant activation of the *PTEN/Akt* signaling axis may be an initiating event.

The implication of a stem cell as the initiating cell for prostate cancer requires explanation not just of the phenotype of the primary tumor but also the commonly observed treatment resistance following androgen ablation therapy. Conventional therapy eliminates the androgen-dependent bulk of tumor cells but will allow androgen-independent cancer stem cells and transit-amplifying cells to survive, which leads to an undifferentiated tumor phenotype. With continued therapy, more differentiated progeny become adapted to an androgen-depleted environment and eventually, androgen-receptor signaling may be reestablished by amplification or mutation of the androgen-receptor gene and/or crosstalk with other signaling pathways, as has been shown to be the case in prostate cancer cell lines and patients suffering from hormone-refractory prostate cancer [92–94] (Fig. 7.5).

7.4.2 Prospective Isolation of Prostate Cancer Stem Cells from Human Tissue Samples

Proof that prostate tumors contain a subpopulation of highly tumorigenic prostate cancer stem cells has been obtained from a study by Collins et al. The prospective isolation of the tumor stem cell fraction was based on a CD44$^+$/$\alpha_2\beta_1$^{hi}/CD133$^+$ phenotype. This combination of cell surface markers had previously been used to isolate the stem cell population from normal prostate,[26] which demonstrates that tissue stem cells and their malignant counterparts may share antigenic properties.

Irrespective of tumor grade, the cancer stem cell population makes up about 0.1% of prostate tumor cells. CD44$^+$/$\alpha_2\beta_1$^{hi}/CD133$^+$ cells exhibit an enhanced

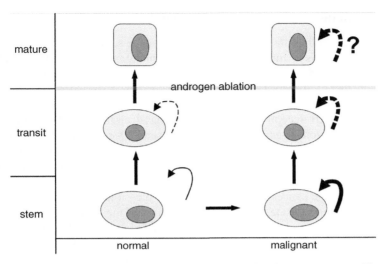

FIGURE 7.5   Stem cell model of normal tissue renewal and prostate cancer. The malignant stem cell arises from transformation of the normal stem cell and gives rise to more differentiated progeny. Androgen ablation induces apoptosis of the androgen-dependent mature cells, and the resulting overproduction of malignant stem cells and their progenitors imparts an undifferentiated appearance to the tumor unless androgen-receptor signaling in these cells is reactivated via alternative pathways. (From refs. 92 to 94.)

colony-forming efficiency compared to an unselected population, demonstrating the increased proliferative capacity of this fraction. There is no significant difference between the primary and secondary colony-forming efficiency of the stem cell fraction, which emphasizes its high self-renewal ability. $CD44^{+}/\alpha_2\beta_1^{hi}/CD133^{-}$ cells, the putative transit-amplifying population, in contrast, exhibit a significantly lower capacity for self-renewal. $CD44^{+}/\alpha_2\beta_1^{hi}/CD133^{+}$ cells selected from prostate tumors are considerably more invasive in vitro than stem cells selected from benign prostate, which complies with a cancer stem cell phenotype. Similar to stem cells from normal prostate, they express HMW keratins 5 and 14 as well as cytokeratin 19. Expression of the differentiation markers' androgen receptor and prostatic acid phosphatase can, however, be induced by incubation in the presence of serum and androgen in vitro, indicating that prostate cancer stem cells can differentiate to a luminal phenotype and therefore recapitulate the differentiation program that is seen in the original tumor.

$CD44^{+}/\alpha_2\beta_1^{hi}/CD133^{+}$ cells have proven highly tumorigenic in preliminary experiments with NOD/SCID mice (A. Collins; unpublished observations). Serial transplantation studies are currently under way to demonstrate the ability of the putative prostate cancer stem cells to give rise repeatedly to phenocopies of the original tumor.

7.4.3 Role of the Stem Cell Niche in Prostate Cancer

In prostate, epithelial–stromal interactions influence survival and proliferation of epithelial cells. This is true for both normal and malignant prostate. In prostate disease, epithelial–stromal interactions are dynamic, allowing for disease progression and establishment of metastases.[95,96]

The Notch signaling pathway may be important in maintaining the stem cell niche. The Notch1 receptor is expressed in basal epithelial cells in the mouse prostate,[61] and Notch1-expressing cells are essential for prostate development.[62] Expression of Notch ligand, however, is undetectable in cultured prostate cancer cell lines and malignant prostate cells in transgenic adenocarcinoma of the mouse prostate (TRAMP)[97] mice, which suggests that the Notch signaling pathway is inactive in prostate cancer.[61] Moreover, overexpression of a constitutively active form of Notch1 in prostate cancer cell lines inhibited their proliferation,[61] which suggests that Notch1 is acting as a tumor suppressor in prostate. Notch also appears to be acting as a tumor suppressor in skin, as conditional knockout of Notch1 in a mouse model led to the development of hyperplasia and skin tumors.[98]

Recent findings suggest that TGF-β signaling is involved in prostate stem cell fate. Using a mouse model, Bhowmick et al. demonstrated the profound influence of prostate stromal cells on tumorigenesis in adjacent epithelium.[99] Inactivation of TGF-β type II receptor in fibroblasts and concomitant loss of TGF-β responsiveness in these cells leads to the development of PIN lesions in prostate epithelium and proliferation in the stromal compartment. Salm et al. demonstrated the direct involvement of TGF-β signaling in the maintenance of stem cell quiescence in murine prostate, whereupon high levels of TGF-β signaling in the proximal duct region inhibit stem cell self-renewal.[56] The level of TGF-β signaling declines along the proximal–distal duct axis, allowing proliferation of transit-amplifying cells, which are located more distally. Temporary cycling of stem cells can be induced by a transient increase in levels of mitotic cytokines. It seems obvious that disturbance of the delicate balance between inhibitory and proliferation-promoting signals supplied by the niche could derail tissue homeostasis and lead to cancer.

7.4.4 Putative Markers of Prostate Cancer Stem Cells

Just like normal stem cells from different tissues, cancer stem cells from different types of tumors are expected to share certain molecular properties. Moreover they are likely to share molecular characteristics of the corresponding tissue stem cells. So far, unique markers have not been identified that distinguish prostate cancer stem cells from benign or normal prostate epithelial stem cells. However, there is evidence that expression of certain prostate stem cell markers is retained in prostate cancer stem cells. The prospective isolation of prostate cancer stem cells by Collins et al.[1] was based on the same surface markers that had been used to isolate prostate epithelial stem cells.[26] Consequently, prostate cancer stem cells continue to express high levels of integrin $\alpha_2\beta_1$ and CD133. The putative prostate stem cell marker pp32[87, 100] is expressed in PIN and prostate carcinoma, which suggests that it also forms part of the prostate cancer stem cell phenotype. Telomerase expression can be detected in normal prostate, PIN, and prostate carcinoma.[58] Whether it is expressed by prostate stem cells and cancer stem cells, however, remains to be elucidated. Expression of telomerase in normal and malignant prostate tissue could also be due to activation of telomerase expression in a more differentiated progenitor cell.[59] The basal cell marker bcl-2 is expressed at high levels in androgen-independent prostate cancers,[89] which indicates that it may also be expressed in prostate cancer stem cells. Similarly, expression levels of basal cell marker Notch1 are elevated in primary and metastatic tumors formed in TRAMP mice,[61,97] which indicates that Notch1 expression may be retained in prostate cancer stem cells founding these tumors. Due to the supposed involvement of the Notch pathway in tumor suppression in the prostate[61] (see Section 4.4), it is unlikely that the pathway is active in these cells. Interestingly, the prostate basal cell marker p63 is absent from PIN and invasive prostate cancer, suggesting that it functions as a tumor suppressor that is lost upon stem cell-to-cancer stem cell transition in prostate.[64] It is possible, however, that continued expression of p63 in prostate cancer stem cells is masked by the bulk of tumor cells, which will be luminal in phenotype.

Although there is evidence that the markers noted above form part of the cancer stem cell phenotype, ultimate proof for this has not yet been obtained and will require gene expression profiling of these cells. Comparative analysis of gene expression programs in normal prostate stem cells and prostate cancer stem cells has the potential to reveal the molecular mechanisms underlying the prostate stem cell-to-cancer stem cell transition.

7.5 STEM CELL TRACKING IN THE PROSTATE

There is little doubt regarding the existence of epithelial stem cells in the prostate and their involvement in tumorigenesis. Evidence suggests that basal stem cells in the prostate can give rise to differentiated epithelial phenotypes[26] and that malignant prostate stem cells can give rise to the complex and heterogeneous mixture of cells that make up a prostate tumor.[1] However, ultimate proof of this has not yet been obtained, as the ability to recapitulate tumorigenesis in vivo has not been demonstrated to date. To investigate the lineage capacity of both normal prostate stem cells and prostate cancer stem cells, the progeny and differentiation of a marked or isolated stem cell must be tracked. Using a mouse model, benign prostate stem cells could potentially be followed through several cycles of androgen depletion–repletion in order to investigate gland-regenerating activity. Similarly, labeled prostate cancer stem cells could be tracked through multiple rounds of xenotransplantation or experimental androgen ablation to assess their tumorigenicity and therapy refractiveness.

Current approaches in cell tracking include the application of fluorescent semi-conductor quantum dots,[101,102] transient transfection of reporter constructs,[103] ultra small iron particles for magnetic resonance imaging,[104] fluorescent cytoplasmic dyes,[105] and stable genome modifications using retroviral vectors.[105,106] Trackable particles and dyes as well as transfected reporter constructs will be diluted below detection level after a limited number of cell divisions, which makes them unsuitable for long-term studies. Retroviral vectors, in contrast, integrate into the host cell genome and thus allow for permanent reporter gene expression in transduced cells.[107] Among retroviruses, lentiviruses, especially, have gained widespread application, due to their unique ability to infect nondividing and terminally differentiated cells[108,109] and due to easy pseudotyping with vesicular stomatitis virus G glycoprotein, which greatly broadens tfhe virus host range.[110] Lentivirus cloning systems are readily available, and the vectors have been used successfully to deliver transgenes to stem cell populations. [106,111–113] Potential drawbacks of the lentivirus strategy are insertional inactivation or deregulation of host cell genes by means of virus integration,[114] or inactivation of the transgene by methylation and/or chromatin modification.[115–117] Both problems can, however, potentially be avoided by performing transductions at low levels of multiplicity of infection and by selecting several clones of transduced cells, as transgene inactivation may be dependent on the integration site.

With a view to lineage tracking in stem cells, the lentiviral approach allows reporter genes to be placed under the control of differentiation-stage-specific pro-moters, thereby translating lineage transitions into a visible signal. Using this strategy, Suter et al. monitored neuron-specific transgene expression during the differentiation of lentivirus-transduced embryonic stem cells by combining two lentivectors expressing different fluorescent marker genes under the control of pro-moters specific for the neuronal lineage.[106] This approach could prove valuable in establishing detailed differentiation lineages for tissue stem cells in that it allows investigation of the temporal relationship of promoter activation as well as assess-ment of similarities and differences in the differentiation pathways of benign and malignant tissue stem cells. Moreover, such a tracking system could be used to evaluate the ability of novel drugs to eliminate or differentiate cancer stem cells. In the prostate, this approach could involve keratin promoters[21] as well as promoters of genes whose temporal expression patterns in prostate epithelium have not been elucidated in as much detail (e.g., PSCA[67]).

In our laboratory, a range of lentiviruses have been generated, expressing different fluorescent marker genes under the control of the constitutive CMV (cytomegalovirus) promoter. Prostate cancer stem cells isolated from the P4E6 cell line[118] have been infected with lentiviruses expressing the eGFP or DsRed fluorescent marker genes (Fig. 7.6) and are now available for tracking experiments in vitro and in vivo. Driving marker gene expression in prostate (cancer) stem cells from differentiation stage-specific promoters would allow us real-time monitoring of prostate tumorigenesis. The generation of suitable lentivectors is currently under way in our laboratory.

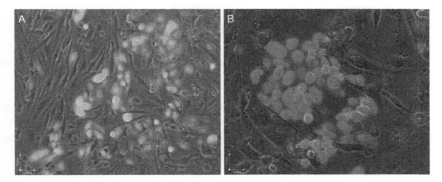

FIGURE 7.6   Transgenic prostate cancer stem cells cultured on feeder cells. Prostate cancer stem cells were isolated from clonal populations of P4E6 prostate cancer cells infected with a lentivirus delivering an (A) CMV-eGFP or (B) CMV-DsRed expression cassette, and were cultured on STO feeder cells. Pictures were taken after (A) 15 or (B) 10 days of culture. (A): 200 × magnification, (B): 400 × magnification. (From S. Hager, unpublished observations.) (*See insert for color representation of figure.*)

7.6 CONCLUSIONS AND FUTURE PERSPECTIVES

The prospective isolation of prostate cancer stem cells from human tissue samples[1] has paved the way for their extensive molecular characterization. There is little consensus among researchers regarding the cell of origin for prostate cancer. However, a fraction of highly tumorigenic cells has been identified in a range of prostate tumor samples. Whatever the differentiation stage of the originating cell, the target for future therapies against prostate cancer has been defined.

Progress will ultimately require tracking of prostate cancer stem cells in vitro and in vivo to prove their tumorigenicity and lineage capacity. Gene expression analysis of purified cancer stem cell populations will be required with a view to identifying markers that can be targeted in therapy. Due to the anticipated similarity between tissue stem cells and the corresponding cancer stem cells, markers of prostate cancer stem cells would be expected to be shared by the normal epithelial stem cells to at least some extent. Use of one or more of these overlapping markers as a target for therapy would be unacceptable in the majority of tissues, but would be acceptable in the prostate. Therapies against prostate cancer, modified to eradicate the prostate cancer stem cells, are highly likely to result in more durable responses than current approaches and possibly even cures for metastatic disease.

REFERENCES

1. Collins AT, Berry PA, Hyde C, Stower MJ, Maitland NJ. Prospective identification of tumorigenic prostate cancer stem cells. *Cancer Res.* 2005; 65:10946–10951.

2. McNeal JE. The zonal anatomy of the prostate. *Prostate.* 1981; 2:35–49.

3. Abate-Shen C, Shen MM. Molecular genetics of prostate cancer. *Genes Dev.* 2000; 14:2410–2434.

4. Wernert N, Seitz G, Achtstatter T. Immunohistochemical investigation of different cytokeratins and vimentin in the prostate from the fetal period up to adulthood and in prostate carcinoma. *Pathol Res Pract.* 1987; 182:617–626.

5. Bonkhoff H, Stein U, Remberger K. Endocrine–paracrine cell types in the prostate and prostatic adenocarcinoma are postmitotic cells. *Hum Pathol.* 1995; 26:167–170.

6. Chodak G. Prostate cancer: epidemiology, screening and biomarkers. *Rev Urol.* 2006; Suppl 8: S3–S8.

7. Adams JM, Cory S. The Bcl-2 protein family: arbiters of cell survival. *Science.* 1998; 281:1322–1326.

8. Hockenbery D, Nunez G, Milliman C, Schreiber RD, Korsmeyer SJ. Bcl-2 is an inner mitochondrial membrane protein that blocks programmed cell death. *Nature.* 1990; 348:334–336.

9. Bonkhoff H. Role of the basal cells in premalignant changes of the human prostate: a stem cell concept for the development of prostate cancer. *Eur Urol.* 1996; 30:201–205.

10. Magi-Galluzzi C, Loda M. Molecular events in the early phases of prostate carcinogenesis. *Eur Urol.* 1996; 30:167–176.

11. Moll R, Franke WW, Schiller DL, Geiger B, Krepler R. The catalog of human cytokeratins: patterns of expression in normal epithelia, tumors, and cultured cells. *Cell.* 1982; 31:11–24.

12. Sawicki JA, Rothman CJ. Evidence for stem cells in cultures of mouse prostate epithelial cells. *Prostate.* 2002; 50:46–53.

13. Tran CP, Lin C, Yamashiro J, Reiter RE. Prostate stem cell antigen is a marker of late intermediate prostate epithelial cells. *Mol Cancer Res.* 2002; 1: 113–121.

14. Kyprianou N, Isaacs JT. Activation of programmed cell death in the rat ventral prostate after castration. *Endocrinology.* 1988; 122:552–562.

15. Foster CS, Cornford P, Forsyth L, Djamgoz MB, Ke Y. The cellular and molecular basis of prostate cancer. *BJU Int.* 1999; 83:171–194.

16. Culig Z, Steiner H, Bartsch G, Hobisch A. Mechanisms of endocrine therapy-responsive and -unresponsive prostate tumors. *Endocr Relat Cancer.* 2005; 12: 229–244.

17. Rau KM, Kang HY, Cha TL, Miller SA, Hung MC. The mechanisms and managements of hormone-therapy resistance in breast and prostate cancers. *Endocr Relat Cancer.* 2005; 12:511–532.

18. English HF, Drago JR, Santen RJ. Cellular response to androgen depletion and repletion in the rat ventral prostate: autoradiography and morphometric analysis. *Prostate.* 1985; 7:41–51.

19. English HF, Santen RJ, Isaacs JT. Response of glandular versus basal rat ventral prostatic epithelial cells to androgen withdrawal and replacement. *Prostate.* 1987; 11:229–242.

20. Montpetit M, Abrahams P, Clark AF, Tenniswood M. Androgen-independent epithelial cells of the rat ventral prostate. *Prostate.* 1988; 12:13–28.

21. Hudson DL, Guy AT, Fry P, O'are MJ, Watt FM, Masters JR. Epithelial cell differentiation pathways in the human prostate: Identification of intermediate phenotypes by keratin expression. *J Histochem Cytochem.* 2001; 49:271–278.

22. van Leenders G, Dijkman H, Hulsbergen-van de Kaa C, Ruiter D, Schalken J. Demonstration of intermediate cells during human prostate epithelial differentiation in situ and invitro using triple-staining confocal scanning microscopy. *Lab Invest.* 2000; 80:1251–1258.

23. Verhagen AP, Ramaekers FC, Aalders TW, Schaafsma HE, Debruyne FM, Schalken JA. Colocalization of basal and luminal cell-type cytokeratins in human prostate cancer. *Cancer Res.* 1992; 52:6182–6187.

24. Xue Y, Smedts F, Debruyne FM, de la Rosette JJ, Schalken JA. Identification of intermediate cell types by keratin expression in the developing human prostate. *Prostate.* 1998; 34:292–301.

25. Collins AT, Habib FK, Maitland NJ, Neal DE. Identification and isolation of human prostate epithelial stem cells based on $\alpha_2\beta_1$-integrin expression. *J Cell Sci.* 2001; 114:3865–3872.

26. Richardson GD, Robson CN, Lang SH, Neal DE, Maitland NJ, Collins AT. CD133, a novel marker for human prostatic epithelial stem cells. *J Cell Sci.* 2004; 117:3539–3545.

27. English HF, Kyprianou N, Isaacs JT. Relationship between DNA fragmentation and apoptosis in the programmed cell death in the rat prostate following castration. *Prostate.* 1989; 15:233–250.

28. Isaacs JT, Coffey DS. Etiology and disease process of benign prostatic hyperplasia. *Prostate Suppl.* 1989; 2:33–50.

29. van dersKwast TH, Tetu B, Suburu ER, Gomez J, Lemay M, Labrie F. Cycling activity of benign prostatic epithelial cells during long-term androgen-blockade: evidence for self-renewal of luminal cells. *J Pathol.* 1998; 186:406–409.

30. Evans GS, Chandler JA. Cell proliferation studies in the rat prostate: II. The effects of castration and androgen-induced regeneration upon basal and secretory cell proliferation. *Prostate.* 1987; 11:339–351.

31. Dermer GB. Basal cell proliferation in benign prostatic hyperplasia. *Cancer.* 1978; 41:1857–1862.

32. Bonkhoff H, Stein U, Remberger K. The proliferative function of basal cells in the normal and hyperplastic human prostate. *Prostate.* 1994; 24:114–118.

33. Bonkhoff H, Stein U, Remberger K. Multidirectional differentiation in the normal, hyperplastic, and neoplastic human prostate: simultaneous demonstration of cell-specific epithelial markers. *Hum Pathol.* 1994; 25:42–46.

34. Robinson EJ, Neal DE, Collins AT. Basal cells are progenitors of luminal cells in primary cultures of differentiating human prostatic epithelium. *Prostate.* 1998; 37:149–160.

35. Stasiak PC, Purkis PE, Leigh IM, Lane EB. Keratin 19: predicted amino acid sequence and broad tissue distribution suggest it evolved from keratinocyte keratins. *J Invest Dermatol.* 1989; 92:707–716.

36. Hudson DL, O'Hare M, Watt FM, Masters Jr. Proliferative heterogeneity in the human prostate: evidence for epithelial stem cells. *Lab Invest.* 2000; 80:1243–1250.

37. Lang SH, Sharrard RM, Stark M, Villette JM, Maitland NJ. Prostate epithelial cell lines form spheroids with evidence of glandular differentiation in three-dimensional Matrigel cultures. *Br J Cancer.* 2001; 85:590–599.

38. Lang SH, Stark M, Collins A, Paul AB, Stower MJ, Maitland NJ. Experimental prostate epithelial morphogenesis in response to stroma and three-dimensional Matrigel culture. *Cell Growth Differ.* 2001; 12:631–640.

39. Jones PH, Harper S, Watt FM. Stem cell patterning and fate in human epidermis. *Cell.* 1995; 80:83–93.

40. Li A, Simmons PJ, Kaur P. Identification and isolation of candidate human keratinocyte stem cells based on cell surface phenotype. *Proc Natl Acad Sci U S A.* 1998; 95:3902–3907.

41. Shinohara T, Avarbock MR, Brinster RL. β_1- and α_6-integrin are surface markers on mouse spermatogonial stem cells. *Proc Natl Acad Sci U S A.* 1999; 96:5504–5509.

42. Yin AH, Miraglia S, Zanjani ED, et al. AC133, a novel marker for human hematopoietic stem and progenitor cells. *Blood.* 1997; 90:5002–5012.

43. Peichev M, Naiyer AJ, Pereira D, et al. Expression of VEGFR-2 and AC133 by circulating human CD34(+) cells identifies a population of functional endothelial precursors. *Blood.* 2000; 95:952–958.

44. Uchida N, Buck DW, He D, et al. Direct isolation of human central nervous system stem cells. *Proc Natl Acad Sci U S A.* 2000; 97:14720–14725.

45. Bhatt RI, Brown MD, Hart CA, et al. Novel method for the isolation and characterization of the putative prostatic stem cell. *Cytometry A.* 2003; 54:89–99.

46. Challen GA, Little MH. A side order of stem cells: the SP phenotype. *Stem Cells.* 2006; 24:3–12.

47. Lee C, Sensibar JA, Dudek SM, Hiipakka RA, Liao ST. Prostatic ductal system in rats: regional variation in morphological and functional activities. *Biol Reprod.* 1990; 43:1079–1086.

48. Braun KM, Watt FM. Epidermal label-retaining cells: background and recent applications. *J Invest Dermatol Symp Proc.* 2004; 9:196–201.

49. Tsujimura A, Koikawa Y, Salm S, et al. Proximal location of mouse prostate epithelial stem cells: a model of prostatic homeostasis. *J Cell Biol.* 2002; 157:1257–1265.

50. Burger PE, Xiong X, Coetzee S, et al. Sca-1 expression identifies stem cells in the proximal region of prostatic ducts with high capacity to reconstitute prostatic tissue. *Proc Natl Acad Sci U S A.* 2005; 102:7180–7185.

51. Xin L, Lawson DA, Witte ON. The Sca-1 cell surface marker enriches for a prostate-regenerating cell subpopulation that can initiate prostate tumorigenesis. *Proc Natl Acad Sci U S A.* 2005; 102:6942–6947.

52. Lawson DA, Xin L, Lukacs R, Xu Q, Cheng D, Witte ON. Prostate stem cells and prostate cancer. *Cold Spring Harb Symp Quant Biol.* 2005; 70:187–196.

53. Hockenbery DM, Zutter M, Hickey W, Nahm M, Korsmeyer SJ. BCL2 protein is topographically restricted in tissues characterized by apoptotic cell death. *Proc Natl Acad Sci U S A.* 1991; 88:6961–6965.

54. Bonkhoff H, Remberger K. Differentiation pathways and histogenetic aspects of normal and abnormal prostatic growth: a stem cell model. *Prostate.* 1996; 28:98–106.

55. Rizzo S, Attard G, Hudson DL. Prostate epithelial stem cells. *Cell Prolif.* 2005; 38:363–374.

56. Salm SN, Burger PE, Coetzee S, Goto K, Moscatelli D, Wilson EL. TGF-β maintains dormancy of prostatic stem cells in the proximal regions of ducts. *J Cell Biol.* 2005; 170:81–90.

57. Armanios M, Greider CW. Telomerase and cancer stem cells. *Cold Spring Harb Symp Quant Biol.* 2005; 70:205–208.

58. Paradis V, Dargere D, Laurendeau I, et al Expression of the RNA component of human telomerase (hTR) in prostate cancer, prostatic intraepithelial neoplasia, and normal prostate tissue. *J Pathol.* 1999; 189:213–218.

59. Jaras M, Edqvist A, Rebetz J, Salford LG, Widegren B, Fan X. Human short-term repopulating cells have enhanced telomerase reverse transcriptase expression. *Blood.* 2006; 108:1084–1091.

60. Suzuki T, Chiba S. Notch signaling in hematopoietic stem cells. *Int J Hematol.* 2005; 82:285–294.

61. Shou J, Ross S, Koeppen H, de Sauvage FJ, Gao WQ. Dynamics of notch expression during murine prostate development and tumorigenesis. *Cancer Res.* 2001; 61:7291–7297.

62. Wang XD, Shou J, Wong P, French DM, Gao WQ. Notch1-expressing cells are indispensable for prostatic branching morphogenesis during development and regrowth following castration and androgen replacement. *J Biol Chem.* 2004; 279:24733–24744.

63. Barbieri CE, Pietenpol JA. p63 and epithelial biology. *Exp Cell Res.* 2006; 312:695–706.

64. Signoretti S, Waltregny D, Dilks J, et al. p63 is a prostate basal cell marker and is required for prostate development. *Am J Pathol.* 2000; 157:1769–1775.

65. Walensky LD, Coffey DS, Chen TH, Wu TC, Pasternack GR. A novel M(r) 32,000 nuclear phosphoprotein is selectively expressed in cells competent for self-renewal. *Cancer Res.* 1993; 53:4720–4726.

66. Reiter RE, Gu Z, Watabe T, et al. Prostate stem cell antigen: a cell surface marker overexpressed in prostate cancer. *Proc Natl Acad Sci U S A.* 1998; 95:1735–1740.

67. Watabe T, Lin M, Ide H, et al. Growth, regeneration, and tumorigenesis of the prostate activates the PSCA promoter. *Proc Natl Acad Sci U S A.* 2002; 99:401–406.

68. Moorthy PP, Kumar AA, Devaraj H. Expression of the *Gas7* gene and *Oct4* in embryonic stem cells of mice. *Stem Cells Dev.* 2005; 14:664–670.

69. Zhang X, Stojkovic P, Przyborski S, et al. Derivation of human embryonic stem cells from developing and arrested embryos. *Stem Cells.* 2006; 24:2669–2676.

70. Kuhnert F, Davis CR, Wang HT, et al. Essential requirement for Wnt signaling in proliferation of adult small intestine and colon revealed by adenoviral expression of Dickkopf-1. *Proc Natl Acad Sci U S A.* 2004; 101:266–271.

71. Kawaguchi A, Miyata T, Sawamoto K, et al. Nestin-EGFP transgenic mice: visualization of the self-renewal and multipotency of CNS stem cells. *Mol Cell Neurosci.* 2001; 17:259–273.

72. Fiala S. The cancer cell as a stem cell unable to differentiate: a theory of carcinogenesis. *Neoplasma.* 1968; 15:607–622.

73. Pierce GB, Johnson LD. Differentiation and cancer. *In Vitro* 1971; 7:140–145.

74. Al-Hajj M, Wicha MS, Benito- Hernandez A, Morrison SJ, Clarke MF. Prospective identification of tumorigenic breast cancer cells. *Proc Natl Acad Sci U S A.* 2003; 100:3983–3988.

75. Bapat SA, Mali AM, Koppikar CB , Kurrey NK. Stem and progenitor-like cells contribute to the aggressive behavior of human epithelial ovarian cancer. *Cancer Res.* 2005; 65:3025–3029.

76. Bonnet D, Dick JE. Human acute myeloid leukemia is organized as a hierarchy that originates from a primitive hematopoietic cell. *Nat Med.* 1997; 3:730–737.

77. Lapidot T, Sirard C, Vormoor J , et al A cell initiating human acute myeloid leukaemia after transplantation into SCID mice. *Nature.* 1994; 367:645–648.

78. Singh SK, Hawkins C, Clarke ID, et al. Identification of human brain tumor initiating cells. *Nature.* 2004; 432:396–401.

79. Singh SK, Clarke ID, Terasaki M, et al. Identification of a cancer stem cell in human brain tumors. *Cancer Res.* 2003; 63:5821–5828.

80. Clarke MF, Fuller M. Stem cells and cancer: two faces of Eve. *Cell.* 2006; 124:1111–1115.

81. Pardal R, Clarke MF, Morrison SJ. Applying the principles of stem-cell biology to cancer. *Nat Rev Cancer.* 2003; 3:895–902.

82. Reya T, Morrison SJ, Clarke MF, Weissmann IL. Stemcells, cancer, and cancer stem cells. *Nature.* 2001; 414:105–111.

83. Nagle RB, Ahmann FR, McDaniel KM, Paquin ML, Clark VA, Celniker A. Cytokeratin characterization of human prostatic carcinoma and its derived cell lines. *Cancer Res.* 1987; 47:281–286.

84. DeMarzo AM, Nelson WG, Meeker AK, Coffey DS. Stem cell features of benign and malignant prostate epithelial cells. *J Urol.* 1998; 160:2381–2392.

85. vanLeenders G, vanBalken B, Aalders T, Hulsbergen-van de Kaa C, Ruiter D, Schalken J. Intermediate cells in normal and malignant prostate epithelium express c-MET: implications for prostate cancer invasion. *Prostate.* 2002; 51:98–107.

86. Schalken JA, van Leenders G. Cellular and molecular biology of the prostate: stem cell biology. *Urology.* 2003; 62:11–20.

87. DeMarzo AM, Meeker AK, Epstein JI, Coffey DS. Prostate stem cell compartments: expression of the cell cycle inhibitor p27Kip1 in normal, hyperplastic, and neoplastic cells. *Am J Pathol.* 1998; 153:911–919.

88. Liu AY, Nelson PS, van den Engh G, Hood L. Human prostate epithelial cell-type cDNA libraries and prostate expression patterns. *Prostate.* 2002; 50:92–103.

89. McDonnell TJ, Troncoso P, Brisbay SM, et al. Expression of the protooncogene bcl-2 in the prostate and its association with emergence of androgen-independent prostate cancer. *Cancer Res.* 1992; 52:6940–6944.

90. Wang S, Gao J, Lei Q, et al. Prostate-specific deletion of the murine Pten tumor suppressor gene leads to metastatic prostate cancer. *Cancer Cell.* 2003; 4:209–221.

91. Wang S, Garcia AJ, Wu M, Lawson DA, Witte ON, Wu H. Pten deletion leads to the expansion of a prostatic stem/progenitor cell subpopulation and tumor initiation. *Proc Natl Acad Sci U S A.* 2006; 103:1480–1485.

92. Hobisch A, Eder IE, Putz T, et al. Interleukin-6 regulates prostate-specific protein expression in prostate carcinoma cells by activation of the androgen receptor. *Cancer Res.* 1998; 58:4640–4645.

93. Tilley WD, Buchanan G, Hickey TE, Bentel JM. Mutations in the androgen receptor gene are associated with progression of human prostate cancer to androgen independence. *Clin Cancer Res.* 1996; 2:277–285.

94. Visakorpi T, Hyytinen E, Koivisto P , et al In vivo amplification of the androgen receptor gene and progression of human prostate cancer. *Nat Genet.* 1995; 9:401–406.

95. Chung LW, Hsieh CL, Law A, et al. New targets for therapy in prostate cancer: modulation of stromal–epithelial interactions. *Urology.* 2003; 62:44–54.

96. Hsieh CL, Gardner TA, Miao L, Balian G, Chung LW. Cotargeting tumor and stroma in a novel chimeric tumor model involving the growth of both human prostate cancer and bone stromal cells. *Cancer Gene Ther.* 2004; 11:148–155.

97. Greenberg NM, DeMayo F, Finegold MJ , Prostate cancer in a transgenic mouse. *Proc Natl Acad Sci U S A.* 1995; 92:3439–3443.

98. Nicolas M, Wolfer A, Raj K, et al. Notch1 functions as a tumor suppressor in mouse skin. *Nat Genet.* 2003; 33:416–421.

99. Bhowmick NA, Neilson EG, Moses HL. Stromal fibroblasts in cancer initiation and progression. *Nature.* 2004; 432:332–337.

100. Kadkol SS, Brody JR, Epstein JI, Kuhajda FP, Pasternack GR. Novel nuclear phosphoprotein pp32 is highly expressed in intermediate- and high-grade prostate cancer. *Prostate.* 1998; 34:231–237.

101. Chan WC, Maxwell DJ, Gao X, Bailey RE, Han M, Nie S. Luminescent quantum dots for multiplexed biological detection and imaging. *Curr Opin Biotechnol.* 2002; 13:40–46.

102. Gao X, Cui Y, Levenson RM, Chung LW, Nie S. In vivo cancer targeting and imaging with semiconductor quantum dots. *Nat Biotechnol.* 2004; 22:969–976.

103. Wolbank S, Peterbauer A, Wassermann E, et al. Labelling of human adipose-derived stem cells for non-invasive in vivo cell tracking. *Cell Tissue Bank.* 2007; 8:163–177.

104. Giesel FL, Stroick M, Griebe M, et al. Gadofluorine uptake in stem cells as a new magnetic resonance imaging tracking method: an in vitro and in vivo study. *Invest Radiol.* 2006; 41:868–873.

105. McMahon SS, McDermott KW. A comparison of cell transplantation and retroviral gene transfection as tools to study lineage and differentiation in the rat spinal cord. *J Neurosci Methods.* 2006; 152:243–249.

106. Suter DM, Cartier L, Bettiol E, et al. Rapid generation of stable transgenic embryonic stem cell lines using modular lentivectors. *Stem Cells.* 2006; 24:615–623.

107. Tang H, Kuhen KL , Wong-Staal F. Lentivirus replication and regulation. *Annu Rev Genet.* 1999; 33:133–170.

108. Lewis PF, Emerman M. Passage through mitosis is required for oncoretroviruses but not for the human immunodeficiency virus. *J Virol.* 1994; 68:510–516.

109. Trobridge G, Russell DW. Cell cycle requirements for transduction by foamy virus vectors compared to those of oncovirus and lentivirus vectors. *J Virol.* 2004; 78:2327–2335.

110. Burns JC, Friedmann T, Driever W, Burrascano M, Yee JK. Vesicular stomatitis virus G glycoprotein pseudotyped retroviral vectors: concentration to very high titre and efficient gene transfer into mammalian and nonmammalian cells. *Proc Natl Acad Sci U S A.* 1993; 90:8033–8037.

111. Consiglio A, Gritti A, Dolcetta D, et al. Robust in vivo gene transfer into adult mammalian neural stem cells by lentiviral vectors. *Proc Natl Acad Sci U S A.* 2004; 101:14835–14840.

112. Pfeifer A, Ikawa M, Dayn Y, Verma IM. Transgenesis by lentiviral vectors: lack of gene silencing in mammalian embryonic stem cells and preimplantation embryos. *Proc Natl Acad Sci U S A*. 2002; 99:2140–2145.

113. Zhang XY, La Russa VF, Reiser J. Transduction of bone-marrow-derived mesenchymal stem cells by using lentivirus vectors pseudotyped with modified RD114 envelope glycoproteins. *J Virol*. 2004; 78:1219–1229.

114. Nienhuis AW, Dunbar CE, Sorrentino BP. Genotoxicity of retroviral integration in hematopoietic cells. *Mol Ther*. 2006; 13:1031–1049.

115. Ellis J. Silencing and variegation of gammaretrovirus and lentivirus vectors. *Hum Gene Ther*. 2005; 16:1241–1246.

116. Ellis J, Yao S. Retrovirus silencing and vector design: relevance to normal and cancer stem cells? *Curr Gene Ther*. 2005; 5:367–373.

117. Yao S, Sukonnik T, Kean T, Bharadwaj RR, Pasceri P, Ellis J. Retrovirus silencing, variegation, extinction, and memory are controlled by a dynamic interplay of multiple epigenetic modifications. *Mol Ther*. 2004; 10:27–36.

118. Maitland NJ, MacIntosh CA, Hall J, Sharrard M, Quinn G, Lang S. In vitro models to study cellular differentiation and function in human prostate cancers. *Radiat Res*. 2001; 155:133–142.

8 Molecular Signatures of Highly Malignant Melanoma Stem Cells

SURAIYA RASHEED

8.1 GENERAL PROPERTIES OF HUMAN MELANOMAS

Melanomas are the most common malignant tumors that arise in various epithelial tissues of the skin, eyes, meninges, digestive tract, and other parts of the body.[1–3] Currently, melanomas account for approximately 77% of all deaths from skin cancers, and the incidence of these tumors is increasing in all ages, especially among persons exposed to excessive ultraviolet radiation from the sun.[1,2,4–7] Although genetic and familial predisposition to the evolution of malignant melanoma has been proposed,[4,8–10] the vast majority of these tumors are sporadic.

Melanomas represent a group of heterogeneous tumors with diverse morphology, histology, and pigmentations that range from black, brown, tan, and gray to blue, pink, or red with areas of white, unpigmented cells. Primary amelanotic melanomas are uncommon but have been reported in various populations.[11,12] Among the major types of melanomas, the superficially spreading melanomas are most common in the United States, and they can develop at any age.[13] Although melanomas are less frequent in people with dark skin, when they do occur, they belong primarily to the acral lentiginous (AL) type. The AL melanomas (ALMs) are also called "hidden" tumors because they develop in plantar skin and acral or mucosal sites such as soles, palms, under the fingernails, tonsils, mucous membranes of the mouth, nose, and genitourinary tracts, and other areas that escape detection during routine physical examinations.[14] ALMs can occur in people of all ages,[15] but the chances of developing nonpigmented tumors increase with age, especially in sun-damaged skin of the elderly.[4,5] The uncommon subtypes of ALMs are the desmoplastic, neutrotrophic, and amelanotic/nonpigmented melanomas, which usually appear as pink or red nodules on the head and neck.[11,12] A more common cancer in the sun-damaged skin of elderly is the lentigo maligna melanoma, which grows slowly for years under the skin in the form of brownish patches. One of the most aggressive

Cancer Stem Cells: Identification and Targets, Edited by Sharmila Bapat
Copyright © 2009 John Wiley & Sons, Inc.

skin tumors is nodular melanoma (NM), which accounts for about 15% of all diagnosed melanomas.[16] NMs can occur at any age but are more common in older people; an important distinguishing characteristic is that NMs do not grow out of a preexisting nevus or mole.[16] Initially, the tumor appears as a cyst, scar tissue, or a brownish to colorless nodule. Since NMs grow deeper (not wider) more rapidly than other melanomas, they do not change color for long periods or show any external signs of growth. NMs are therefore not detected until the lesions bleed or ulcerate and are often diagnosed as a "primary tumor" of that site. Thus, different types of melanomas can present themselves as multiple primary melanomas.[12,14]

Early stages of melanomas are difficult to diagnose[17,18] because most of the transformed melanocytes move downward from the surface and develop into microscopic tumors in deeper parts of the dermis, where they find their niches. Benign tumors remain confined to the tissue of origin, but as soon as the cells acquire metastatic potential, they begin to invade neighboring tissues and metastasize to lymph nodes, liver, lung, brain, bone, and other organs of the body.[13,17,18] Over a period of time, the pigmentation in the lesion may darken, developing increasingly irregular borders and areas of inflammation. Metastatic tumors spread rapidly and are lethal in most cases.[13]

Although *Braf, ras,* and other genes have been associated with the development of various human melanomas, most of the current diagnostic and prognostic markers do not correlate well with their clinical outcome.[19–27] The prognosis of highly malignant melanomas is mostly unfavorable, as they develop resistance to conventional radiation and chemotherapy, while patients who respond to therapy are often at high risk of developing new tumors.[26,28–33] The resistance to therapeutic agents has been associated with sequestration of cytotoxic drugs by melanosomes and other factors associated with pigmentation of melanocytes.[34–37] However, it may be that the more aggressive drug-resistant tumors are derived from stem cells and therefore can survive adverse conditions and continue to spread more rapidly.

8.2 CHARACTERISTICS OF STEM CELL–DERIVED MELANOMAS

The first clue of a stem cell–derived melanoma was provided by our studies on a highly malignant cat melanoma that differentiated into neuronal cells.[38,39] More recently, several clones of melanoma stem cells have also been isolated from established human melanoma cell lines in vitro.[40]

The molecular mechanisms involved in the development of human melanomas are not fully understood, and it is not clear which types of human melanoma may be derived from stem cells. Melanomas develop from melanocytes present in the inner epithelial cell lining and mucous membranes of many organs. In developing embryos, melanocytes arise from the neural crest, which also gives rise to stem cell progenitors of various cell types, including hematopoietic, endothelial, skeletal, connective tissue, bone, cartilage, neurons, glial, and other sensory cells of both the peripheral and central nervous system.[41] Although it is well established that only a small number of stem cells are involved in the development of various tissues and organ systems, how the progenitor cells make fate choices for differentiation into specific tissue types is unclear.[42] Moreover, the mechanisms by which some tissues regenerate and others do not.

To identify genes involved in the development of melanomas, investigators are studying gene expression profiles of human melanoma cell lines.[23,40,43–49] Since proteins are the functional effectors of all cells, the gene expression profiles may not necessarily translate into the respective encoded proteins that are responsible for conferring oncogenic potential to the healthy cells of a "normal" genome. Melanomas and cancer stem cells have also been characterized by quantitative RNA, reverse transcriptase polymerase chain reaction (RT-PCR), single nucleotide polymorphisms (SNPs) of specific genes, and changes in the gene expression profiles by microarray analyses.[50–59] However, although microarray-based studies have enhanced our knowledge of comprehensive RNA expression profiles of different cell types, these databases cannot be translated directly to the respective encoded functional proteins primarily because a single gene could conceivably produce multiple proteins by frame shifting, mutations, mRNA splicing, and variance in translational start or stop codons. In addition, proteins are posttranslationally modified by phosphorylation, glycosylation, acety-lation, and O-linked-N-acetylglucosamine (O-GlcNAc), each of which is uniquely significant in cell signaling and performance of cellular functions.[60–68] As a result, the number of proteins encoded by a genome far exceeds the number of genes present in the DNA. In addition, proteins fold in three-dimensional struc-tures and form specific aggregates or complexes with interacting proteins or other molecules. In an unrelated project, our laboratory had compared mRNA expres-sion results with proteomic protein profiles of the same cells harvested at the same time. The results showed only about 10 to 12% correlation between the RNA and protein expression profiles of cloned cells (S. Rasheed, manuscript in preparation).

8.3 THE CAT MODEL SYSTEM FOR STEM CELL MELANOMAS

To gain an insight into the molecular processes involved in the progression of melanomas, we studied a naturally occurring, highly malignant melanoma of a domestic cat, designated CT1413 (Fig. 8.1), which exhibits similar clinical and pathological properties that are presented by human melanomas. Over the past two decades our laboratory has mined a wealth of information indicating that this cat melanoma is derived from stem cells.[38,39,69] Most significantly, through the appli-cation of proteomics and bioinformatics technologies, we have identified unique stem cell–related proteins that were not described previously for any other human or animal melanoma.[69] This melanoma model system has provided a unique opportunity for experimental in vitro manipulation in order to gain an insight into the relationship between stem cells, differentiation, and cancer (melanoma). This knowledge can be applied further to identify protein markers that can be used to distinguish stem cell–derived human melanomas from those that may not be

FIGURE 8.1   Athymic nude mouse with melanoma induced by in vitro cultured CT-1413 cells. Melanoma developed within 10 days of subcutaneous inoculation cells. (*See insert for color representation of figure.*)

developed from stem cells. In the following sections we focus on the biological, biochemical, and molecular properties of the foregoing system and provide evidence for the involvement of differentially regulated immediate, early, and germ cell proteins in self-renewal, tumorigenesis, and differentiation of tumor cells.

8.3.1 Biological Characteristics of Highly Malignant Stem Cell Melanomas

The highly metastatic cat melanoma cells contain many black/brown pigmented cells that synthesize and secrete melanin in vivo as well as in vitro.[38] These tumor cells were initially grown in the minimum essential medium containing 5 to 10% fetal bovine serum.[38] However, since the discovery of its stem cell properties, all cultures are being maintained in 1 to 2% serum or no serum-containing medium. Sparsely plated melanoma cells appear fibroblastic in morphology, as they attach to the plastic surface and occasionally produce small bipolar extensions[38] (Fig. 8.2). When the same cells are densely plated, they tend to attach loosely and grow as microspheres, especially in a low-serum medium.

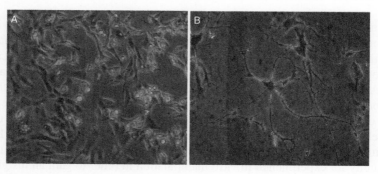

FIGURE 8.2   Cat melanoma cell cultures before and after trans-differentiation in vitro: (A) uninfected (control) melanoma cells 48 hours following culture; (B) trans-differentiated neuronal cells 48 hours after infection with the endogenous cat retrovirus RD114.

The cat melanoma cells maintained in low- or no-serum-containing media exhibit no significant difference in their differentiation capabilities. Cloning of tumor cells in these media enhanced sphere formation, and cells that were isolated 25 years ago differentiated again in vitro.[72] In contrast, melanoma cells cultured for prolonged periods in 10 to 20% serum do not differentiate well. This indicated that these not only self-renew and survive for long periods in low-serum-like stem cells, but also retain early embryonic gene expression, which influences their differentiation into neuronal cells (discussed in the following sections). Subcutaneous inoculation of in vitro cultured melanoma cells in young kittens or nude mice produced pigmented, highly metastatic melanomas within 10 to 15 days[38] (Fig. 8.1). The histopathology and clinical features of these tumors were similar to those observed in metastatic human melanomas.

8.3.2 Trans-differentiation of the Malignant Cat Melanoma into Neuronal Cells

One of the most important properties of all stem cells is their ability to differentiate into different lineage-specific cell types. Early studies indicated that exposure of cat melanoma cells to the nonpathogenic endogenous cat retrovirus RD114 in vitro could induce neuronal cell differentiation within 24 to 48 hours.[38,69] Trans-differentiated melanoma cells show long, multipolar, cytoplasmic processes and giant multinucleated cells with long neuritelike extensions that form a network of connections with smaller neural glial or astrocytelike cells[38,69] (Fig. 8.2).

The melanoma cells do not produce any virus, including the feline leukemia virus (FeLV) or the RD114,[70,71] although low levels of endogenous RD114-related RNA were detected in the cytoplasm on one occasion. [38] Chemicals that induce endogenous retroviruses from cat or mouse cells failed to produce any virus after treatment of these cells. In addition, other feline and murine retroviruses capable of replicating in cat cells were not detected, nor could they induce neuronal cell differentiation or any morphological changes in these cultures.[38] Thus, the susceptibility of melanoma cells to differentiate is not due to viral replication,[38] nor does neural cell differentiation in these cells depend on the primary binding site of the RD114 virus. It is mediated by a unique multifaceted interaction with other cell surface proteins, as suggested by our proteomic studies.[69] The binding of the RD114 viral envelope glycoproteins to the NAA receptors present on the surface of melanoma cells is a highly ordered event that results in novel protein–protein interactions capable not only of activating the external cues but of sending signals that alter differentiation-specific pathways that facilitate neuronal cell differentiation.[69] Conformational changes in the cell surface proteins due to heterodimer formation and possible interactions with other proteins in close proximity to the receptor may activate proteins and trigger a cascade of events that result in altering genome-wide transcriptional and translational programs and induce neuronal cell differentiation.[69]

Production of new cells during neuronal cell differentiation and maintenance of healthy neural phenotypes are highly intricate biological processes that involve concurrent multifaceted mechanisms. Our studies on cat melanoma tumors have indicated that long-term in vitro cultures of cell lines may compromise and reduce stem cell characteristics of certain tumor cells unless grown under specialized conditions (S. Rasheed, personal observations). An important observation during the neuronal cell differentiation of melanoma cells was that the cellular environment played a direct role in cell differentiation. The two factors that reduced the ability of CT1413 cells to trans-differentiate were the continuous maintenance of cultures as monolayers in high serum concentration and frequent trypsinization. Further, since the cat melanoma cells express stem cell proteins, cloning of these tumors does not make any difference during the trans-differentiation process. Although these differentiation studies had predicted a possible stem cell origin of this melanoma over two decades ago,[38] concrete evidence toward the phenomenon was provided through analyses of protein profiles of both cell types.[69]

An interesting phenomenon associated with the neuronal cell differentiation of cat melanoma is that neurogenic transformation is accompanied by concomitant loss of the tumorigenic potential of melanoma cells. No tumors were produced when the differentiated neuronal cells were inoculated in nude mice.[38] Since both melanocytes and neuronal cells arise from the embryonic neural crest, the neuronal cell differentiation of melanoma cells has been considered stem cell specific.[38]

8.3.3 Proteins Associated with Neuronal Cell Differentiation

To identify differentiation- or tumor-specific proteins associated with development of the highly malignant cat melanoma, we performed genome-wide protein profiling of these cells at different stages of growth. Rather than study the entire cellular proteome, we preferred to identify the differentially expressed proteins (up- or down-regulated) by subtractive analyses of protein profiles of the tumor cells from those expressed or suppressed in the counterpart neuronal cells or those present in the normal cat embryo fibroblast cultures.[69] All differentially expressed proteins were confirmed unambiguously by mass spectrometry, and each protein was scrutinized thoroughly for its known functions in the global databases. The unique accession number for each protein has been identified from the SwissProt database (Fig. 8.3).

To relate the protein expression profiles with specific functions, we applied several computational bioinformatics techniques and delineated protein–protein interaction pathways within the cat genome.[72,73] This analysis indicated that most of the proteins expressed in the cat melanoma cells were of immediate early type and were multifunctional. The proteins identified could be grouped into functional families of proto-oncogenes, cytokines, growth-stimulating hormones, neurotrophins, neurosensory proteins, receptors, transcriptional regulators, germ cell–associated proteins, cell cycle–related proteins, peptidases, phosphatases, kinases, enzymes, and other signaling proteins. These proteins could be associated with a variety of stem cell functions including self-renewal, dysregulation of the cell cycle, tumorigensis, and neurogenesis. In addition, the tumor cells also express proteins that are produced by a wide range of specialized organs, indicating potential capabilities of these cells to differentiate into multiple cell types.

Comparison of the comprehensive protein profiles of neuronal cells with those present in the highly malignant cat melanoma cells indicated that although several proteins were induced de novo (i.e., were expressed exclusively in neuronal cells and were absent in melanoma cells), over 90% of the proteins expressed or suppressed during neurogenesis were shared by the tumorigenic melanoma cells (Fig. 8.3). These results indicate that it is a critical balance between various differentially expressed proteins rather than simply the presence or absence of a protein that affects cellular phenotypes.

Trans-differentiated neuronal cells expressed several multifunctional proteins, including neurogenic enzymes, neurotrophic factors, cytokines, hormones, and the cell cycle regulatory gene complex. The polycomb group protein PCGF4 was expressed early during differentiation and appears to be necessary for neurogenesis (Fig. 8.3). The presence of PCGF4 is critical for self-renewal of the cat melanoma stem cells, and some of the same signaling proteins are also involved in the early stages of embryonic development. Our results suggest that the functions of these differentially expressed proteins may be interconnected, and the up-regulation of PCGF4 in neuronal cells points to its significant role in neurogenesis. This is indeed the case, as the PCGF4 complex has been shown recently to be absolutely essential for embryonic brain development.[74,75] The proteomics data on the cat stem cell melanoma has now added a new dimension to the PCGF4 functions, as this protein complex is not only important for neuronal cell differentiation of the cat melanoma cells, but its expression is also critical for maintaining a transcriptional balance between many of the coordinately expressed early proteins involved in cell differentiation. These findings are also consistent with the role of PCGF4 in preserving the stemness of melanoma cells over a long period of time.

The expression of neuronal cell phenotypes depends on neurotrophic factors such as the brain-derived neurotrophic factor (BDNF) and neurotrophic factor 3 (NT3). These factors are critical for the development of both the peripheral and central nervous system cells.[76–78] Both BDNF and NT3 factors were up-regulated in neuronal cells compared to their levels in the melanoma. These neurotrophins are essential for the survival of neuronal stem cells.[79] In adult humans, ischemia or stroke leads to increased production of BDNF in cortical and hippocampal neurons, indicating its role in repair or regeneration of neuronal cells.[80–82] Patients treated with recombinant BDNF show reduced damage when infused intracerebroventrically after hypoxia-induced ischemia.[90] The BDNF activity in neuronal cells may therefore be directly related to neurogenesis in the differentiated neuronal cells.[83] During neuronal cell differentiation, the neurotrophins work in concert with other neurogenic growth factors, cytokines, neuron-specific enzymes, kinases, and membrane receptors that are already present in melanoma cells.[69] For example, BDNF interacts with PDGFB to increase the length of the neurites.[84] β1-adrenergic receptor (ADRβ1), peripherin (RDS), parathyroid hormone (PTH), and Myc proto-oncoprotein were also up-regulated during neuronal cell differentiation, and each of these proteins has been shown to be directly or indirectly associated with neuronal cell activities.[85–87]

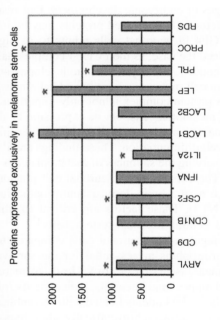

Proteins shared by melanoma stem cells and differentiated neuronal counterpart cells

Proteins expressed exclusively in melanoma stem cells

FIGURE 8.3 Protein expression profiles of stem cell melanoma: quantitative protein expression profiles of 77 proteins identified by matrix-assisted laser desorption ionization–time-of-flight mass spectrometry from multiple gels derived both from melanoma and neuronal cells. X-axis: protein name abbreviations (full names, abbreviations, and accession numbers are provided below). Protein names, abbreviations, and accession numbers are from the SwissProt database. Y-axis: normalized quantities for each protein as detected and confirmed by mass spectrometry from multiple gels. One asterisk indicates that a protein has also been identified as being involved in self-renewal, growth, and proliferation. Two asterisks indicate proteins expressed in neuronal cells after trans-differentiation (i.e., not detected in melanoma). From left to right: ACES, acetylcholinesterase [precursor] (O62763); ADRB2, β-2 adrenergic receptor (Q9TST5); ALBU, serum albumin [precursor] (P49064); AMPN, aminopeptidase N (P79171); ARSB, arylsulfatase B [precursor] (P33727); AT2A2, sarcoplasmic/endoplasmic reticulum calcium ATPase 2 (Q00779); BDNF, brain-derived neurotrophic factor [precursor] (Q9TST3); BGAL, β-galactosidase [precursor] (O19015); BGLR, β-glucuronidase [precursor] (O97524); CASP1, caspase 1 [precursor] (Q9MZV6); CASP3, caspase 3 [precursor] (Q8MJU1); CATW, cathepsin W [precursor] (Q9TST1); CD8B, T-cell surface glycoprotein CD8 β chain [precursor] (P79336); CDN1A, cyclin-dependent kinase inhibitor 1 (O19002); CHLE, cholinesterase [precursor] (O62760); CNR1, cannabinoid receptor 1 (O02777); CP17A, cytochrome P450 17A1 (Q9GMC8); CSF1R, macrophage colony-stimulating factor 1 receptor [precursor] (P13369); DCE1, glutamate decarboxylase 1 (P14748); DPP4, dipeptidyl peptidase 4 (Q9N2I7); FEL1B, major allergen I polypeptide chain 2 [precursor] (P30440); FES, proto-oncogene tyrosine-protein kinase FES/FPS (P14238); HEXB, β-hexosaminidase β chain [precursor] (P49614); IFNG, interferon gamma [precursor] (P46402); IL12B, interleukin-12 β chain [precursor] (O02744); IL15, interleukin-15 [precursor] (O97687); IL1A, interleukin-1α [precursor] (O46613); IL1B, interleukin-1β [precursor] (P41687); IL4, interleukin-4 [precursor] (P53713); IL6, interleukin-6 [precursor] (Q28889); KPYM, pyruvate kinase, isozyme M1 (P11979); LIPL, lipoprotein lipase [precursor] (P55031); LRCH1, leucine-rich repeats and calponin homology domain containing protein 1 [fragment] (P41737); MA2B1, lysosomal α-mannosidase [precursor] (O46432); MDM2, ubiquitin-protein ligase E3 Mdm2 (Q7YRZ8); MLR5, superfast myosin regulatory light chain 2 (P41691); MMP7, matrilysin [precursor] [fragment] (P55032); NAC1, sodium/calcium exchanger 1 [precursor] (P48767); NT3, neurotrophin-3 [precursor] (Q9TST2); NU2M, NADH-ubiquinone oxidoreductase chain 2 (P48905); NU5M, NADH-ubiquinone oxidoreductase chain 5 (P48921); OPSR, red-sensitive opsin (O18913); P53, cellular tumor antigen p53 (P41685); PCGF4, polycomb group ring finger 4 (Q9TST0); PDGFB, platelet-derived growth factor B chain [precursor] (P12919); PGCB, brevican core protein [precursor] [fragment] (P41725); PHOS, phosducin (P41686); PIM1, proto-oncogene serine/threonine-protein kinase pim-1 (Q95LJ0); PPBT, alkaline phosphatase, tissue-nonspecific isozyme [precursor] (Q29486); PPIA, peptidyl-prolyl *cis-trans* isomerase A (Q8HXS3); PRIO, major prion protein [precursor] (O18754); RS3A, 40S ribosomal protein S3a [fragment] (P61246); RS4X, 40S ribosomal protein S4, X isoform (P62705); SCF, kit ligand [precursor] (P79169); SOMA, somatotropin [precursor] (P46404); SPYA, serine-pyruvate aminotransferase, mitochondrial [precursor] (P41689); TFR1, transferrin receptor protein 1 (Q9MYZ3); TLR4, toll-like receptor 4 [precursor] (P58727); TNFL5, tumor necrosis factor ligand superfamily member 5 (O97605); ZP2, zona pellucida sperm-binding protein 2 [precursor] (P47984); ZP3, zona pellucida sperm-binding protein 3 [precursor] (P48832); ZP4, zona pellucida sperm-binding protein 4 [Precursor] (P48834); ARYL, arylamine *N*-acetyltransferase [fragment] (O62696); CD9, CD9 antigen (P40239); CDN1B, cyclin-dependent kinase inhibitor 1B (O19001); CSF2, granulocyte-macrophage colony-stimulating factor [precursor] (O62757); IFNA, interferon α [precursor] (P35849); IL12A interleukin-12α chain [precursor] (O02743); LACB1, β-lactoglobulin-1 (P33687); LACB2, β-lactoglobulin-2 (P21664); LEP, leptin [precursor] (Q9N2C1); PRL, prolactin [precursor] (P46403); PROC, vitamin K-dependent protein C [fragment] (serine protease zymogen) (Q28412); RDS, peripherin (P35906). Proteins expressed exclusively in differentiated neuronal cells include ADRB1, β-1 adrenergic receptor (Q9TST6); IL10, interleukin-10 [precursor] (P55029); IL2, interleukin-2 [precursor] (Q07885); IL2RA, interleukin-2 receptor α-chain [precursor] (P41690); MYC, Myc proto-oncogene protein (c-myc) (P68271); PENK, proenkephalin A [fragment] (Q28409); PTHY, parathyroid hormone [precursor] (Q9GL67).

Many cytokines were also up-regulated in transdifferentiated neuronal cells compared to melanoma.[69] A key cytokine, interferon gamma (IFNγ), was induced as the first line of defense against the RD114 virus infection.[69] The presence of IFNγ is important for the production of several multifunctional cytokines that are activated by phosphatases and peptidases present in melanoma cells.[69] These include interleukin-1α (IL1α), IL1β, IL1, IL2, IL2 receptor α-chain (IL2Rα), IL4, IL10, IL12β-chain (IL12β), and IL15 (S. Rasheed, manuscript in preparation). In Vivo, IL1β is induced in response to stress or hypoxia in the central nervous system (CNS), while IL1β-converting enzyme (ICE) or caspase 1 activates IL1β and many other pro-inflammatory cytokines.[88,89] Caspase 1 and caspase 3 are cysteine proteases that regulate a large number of biological processes, including activation of proteins by cleavage.[90,91] IL1β is an essential factor for neurogenesis, as it induces astrocytes and glial cells to produce many growth factors and other cytokines in the central nervous system.[92–94] In the presence of tumor necrosis factor (TNF-α), IL1β also induces the expression of IL1 receptor (IL1R1) on astrocytes and neurons.[95]

In addition to the neurogenic proteins and factors produced by differentiated neuronal cells, the influx and efflux of Ca^{2+}, and a number of other activities associated with Ca^{2+}, K^+, and Na^+ channels, are deemed essential in regulating signal transductions that influence neuronal cell differentiation.[96,97] Among the various mechanisms that control neural cell homeostasis, the concentration of cellular Ca^{2+} is crucial for cell activation, as it influences growth and neuronal cell differentiation.[98] For example, active communication between astrocytes and other neuronal cells is through the Ca^{2+} pump designated sarcoplasmic/endoplasmic reticulum calcium ATPase 2 (ATA2).[99,100] This protein displayed the highest frequency of detection in the trans-differentiated neuronal cells, and it is reasonable to assume that it is involved in controlling the amount of Ca^{2+} efflux and influx that is essential for the translocation of calcium from the cytosol to the sarcoplasmic reticulum lumen.[69,97]

8.3.4 Cell Cycle Dysregulation and Antitumorigenic Effects During Cell Differentiation

An important characteristic of malignant cat melanoma cells is that after differentiation into neuronal cells, the ability to produce tumors in nude mice or young cats is lost.[38,39] This melanoma–neuronal cell differentiation model system therefore provides a unique resource for studying stem cell–tumorigenesis on the one hand, and antitumorigenic and antineurogenic properties on the other. As far as we could determine, this is the first demonstration of neuronal cell differentiation with concomitant expression of tumor-inhibitory proteins and suppression of tumorigenic potential of melanoma cells. A significant finding of our proteomic analysis is that the cell cycle genes are differentially regulated in melanoma and differentiated neuronal cells.[69] These include the tumor suppressor protein p53 (TP53), cyclin-dependent kinase inhibitor 1 (CDKN1A), and cyclin-dependent kinase inhibitor 1B (CDKN1B/p21) (Fig. 8.3), which control cellular proliferation by inhibiting cyclin-dependent kinases.[101–103] The CDK inhibitors affect the G1 and S phases of the cell cycle, the cyclin-dependent kinases drive the cell cycle, and inhibition of this kinase arrests the cell cycle.[101,104] Whereas the CDKN1A kinase was overexpressed in the rapidly growing cat melanoma cells, this protein was absent or detected at extremely low levels during the neuronal cell differentiation.[69] In contrast, p53 tumor suppressor protein was down-regulated in melanoma cells but was overexpressed in the trans-differentiated neuronal cells.[69] The ubiquitin-protein ligase E3 Mdm2 (MDM2) counteracts the growth arrest and apoptosis of p53, which indicates that a disruption in cell cycle regulation may influence the initiation of tumorigenesis.[65,105,106]

After the binding of RD114 retrovirus to the melanoma cell surface receptors, the cells were arrested in the G1 phase due to the expression of IL6, IL1A, ITGB, PTH, and c-Myc proteins. The expression of c-Myc can be both oncogenic and antitumorigenic, depending on the other proteins expressed at that time. Similarly, the presence of multifunctional cytokines during cell differentiation may contribute to their antitumorigenic effects in the initial phases of growth arrest and neurogenesis. The parathyroid hormone (PTH) is up-regulated during neuronal cell differentiation, and this hormone has been shown to use multiple mechanisms to arrest cell cycle progression of osteoclasts from G1 to S phase and is therefore required for neural cell growth[88,104] (Fig. 8.3). In addition, PTH regulates expression of CDKN1A, Ca^{2+} channels, alkaline phosphatase (a neuronal marker), and other neurogenic proteins, all of which arrest tumor cell growth.[86]

Dipeptidyl peptidase IV (DPP4) is a cell surface peptidase that is normally expressed on healthy skin epithelial cells, including melanocytes.[107] This enzyme is a critical component of K^+ channels, and it enhances growth of astrocytes.[108,109] However, DPP4 also inhibits invasion of melanoma,[110,111] which suggests that it may play a role as an antitumorigenic protein during cell differentiation. Activation of caspases is required for cell differentiation[119]; both caspase 1 and caspase 3 were up-regulated in neuronal cells. These enzymes are involved in the spontaneous regression of neuroblastomas and therefore may confer antitumorigenic properties on melanoma cells.[112]

8.3.5 Molecular Signatures of Self Renewal and Long-Term Proliferation of Tumor Cells

Analyses of differentially expressed protein profiles of rapidly growing melanoma cells in comparison with those present in the normal cat embryo fibroblasts or those expressed in the terminally differentiated, nondividing neuronal cells suggested that most significant indicators of self-renewal included regulation of cell cycle and growth-related genes ($p < 0.001$). An important protein that was detected at various stages of the growth of the cat melanoma cells over a period of two years belonged to the Polycomb group Ring finger 4 (PCGF4), a Bmi1-associated protein.[113–116] Although this gene was initially reported as a proto-oncogene involved in the development of hematopoietic and other cancers,[113,114,117] the PCG proteins Bmi1 and Ring fingers 1 to 4 have now been shown to represent subunits of Polycomb repressive complexes (PRCs), which are ubiquitous in nature and are critical for the regulation of genes expressed in health and disease.[118–122] Expression of PCG promotes growth of both normal hematopoietic (HSC) and leukemic human stem cells; a lack of this gene results in defects in the self-renewal capacity of HSC [75,123–125] and growth retardation and neurological deficits in mice. This indicates that these transcriptional repressors may be essential for preserving growth and survival of both stem cells and stem cell–derived tumors.[117,126,127] The PCG complexes interact with a large number of proteins, including the stem cell regulators OCT4, SOX2, and NANOG,[74] that maintain the transcriptional repressive state of several genes and modify chromatin such that the changes are heritable.[126] The repression is essential for embryogenesis and self-renewal, as a number of cell cycle genes are dysregulated and CDK inhibitor is suppressed.[74,128,129]

Based on the reported functions for each of the proteins expressed in melanoma cells and using different aspects of protein–protein interaction programs (Ingenuity Pathway Analysis; www.ingenuity.com), we have identified an additional set of proteins (marked with an asterisk in Fig. 8.3) that may also cooperate in protecting stem cell characteristics of cat melanoma cells within the confines of their ever-changing environment both in vivo and in vitro. Among the many significant proteins that are coexpressed in cat melanoma cells, the high-affinity stem cell tyrosine kinase growth factor receptor (c-KIT) is a key regulator for the survival of stem cells and tissue repair.[130] c-KIT is considered a germ cell marker and is also expressed in stem cells derived from cardiomyocytes and thymocytes.[130–132] Although the levels of c-KIT and its ligand [i.e., stem cell factor (SCF)] were lower in malignant melanoma than those observed in the counterpart, differentiated neuronal cells,[69] the binding of SCF to KIT activates and autophosphorylates the KIT receptor, and this activation initiates a cascade of interactions with the SCF growth factor. This enhances autocrine mechanisms of self-renewal and cell proliferation.[133] In contrast to cat melanoma, the expression of SCF has been reported to be totally lost in human malignant melanoma cell lines,[134] and coexpression of both KIT and SCF can inhibit metastatic growth of human melanomas.[135,136] These differences could arise from the fact that the human melanomas tested may not be derived from stem cells. Alternatively, the stem cell–derived cat melanoma may be unique. SCF is closely related to receptors for platelet-derived growth factor beta (PDGF-B), macrophage colony-stimulating factor (M-CSF), and the FLT3 ligand.[137] Since all these proteins were expressed in the metastatic cat melanoma, it is possible that these factors may also use c-KIT to promote self-renewal and long-term proliferation of these cells. Our results thus support the regulatory role of c-KIT in preserving the stemness of these tumor cells.

Several cytokines expressed in cat melanoma cells may also contribute to the production of hormones and other growth factors necessary for the self-renewal of these cells.[138] Tumor necrosis factor ligand superfamily member 5 (TNFL5) is expressed in high quantities in cat melanoma cells (Fig. 8.3), and this factor, together with other cytokines, has been shown to sustain regeneration of stem cells in the bone marrow following extensive damage.[139] Expression of IL12A induces production of other cytokines, human leukocyte antigen (HLA) classes I and II.[140]. Expression of both IL1β and TNF increases granulocyte–macrophage colony-stimulating factor 2 (CSF2), and GMCSF stimulates stem cells by interacting with Kit and SCF.[141] The macrophage colony-stimulating factor I receptor (CSF1R) is expressed in early growth and differentiation of embryos, and its expression in multipotential cells has been shown to favor "self-renewal versus differentiation."[142]

The cat melanoma cells also express many immediate early signaling proteins that can be associated with numerous growth-regulating functions and survival of embryonic proteins in these cells. For instance, CDKN1B, arylamine N-acetyltransferase, CD9, IL12β, vitamin K–dependent protein C, prolactin, and granulocyte–macrophage colony-stimulating factor (GM-CSF) were expressed exclusively in melanoma stem cells and were not present in detectable levels in trans-differentiated neuronal cells.[69] It is therefore probable that these proteins may be working in concert to promote an undifferentiated state in melanoma cells and are essential for self-renewal.[141,143–145] These proteins have been shown to promote stem cell–related functions in embryos, placenta, or other cell types in the body. Although most of the proteins expressed in melanoma cells are essential for the growth of the embryos, these immediate early proteins have not been reported previously in relation to self-renewal or growth of melanocytes. For example, two hormone precursors (leptin and prolactin) were up-regulated in melanoma cells, and each has been shown to influence proliferation of different cell types by regulating their growth through amino acid uptake.[146–148]

Our studies also indicate that both up- and down-regulated proteins are important because several proteins act synergistically, and suppression of one protein may lead to overexpression of another protein. In both cases, protein–protein interactions that are necessary for the performance of specific cellular functions, including self-renewal, will be affected. To validate the functional significance of several proteins implicated in self-renewal, growth, and maintenance of undifferentiated or a quasi-differentiated state of stem cell melanoma, further studies to compare cellular proteomes before and after treatment with small interfering RNAs (RNAi) for silencing multiple genes will be undertaken toward a better understanding of specific roles of critical proteins in the self-renewal, growth, and differentiation-related characteristics of melanoma stem cells.

8.3.6 Proteins Involved in Tumorigenesis and Metastasis

Tumor development is a multistep process that involves de novo expression of signaling proteins that are essential for tumorigenesis. Most of the proteins expressed in cat melanoma cells are involved in signal transduction and cell–cell signaling. Although paracrine and autocrine signals are transient, the extracellular cues are central to the induction of new signals that generate specific biological responses required for transformation. These proteins include several multifunctional proto-oncogenes expressed in cat melanoma cells, and they perform a variety of stepwise functions, from initiation and promotion of tumorigenesis to metastasis. The c-KIT receptor may interact with platelet-derived growth factor B chain (PDGFB) and maintain melanoma cells in an activation mode. The FES/FPS tyrosine kinase and the serine/threonin–protein kinase pim-1 (PIM1) play vital roles in growth, mitosis, and cell survival, respectively.[155–157]

Expression of PIM1 kinase in hematopoietic cells has been reported to result in the development of lymphoma and other tumors.[149–151] Further, PDGFB is expressed at the highest levels compared to other proto-oncogenes in cat melanoma cells (Fig. 8.3). This growth factor cooperates with the proto-oncogenes c-KIT, FES, FOS, and PIM1, which are also expressed in melanoma cells, albeit at low levels, and helps in regulating oncogenesis, stimulating cytokine production, and controlling proliferation of stem cells[152,153] (Fig. 8.3). Studies with Bmi-1 deletion mutants have also suggested that PCG4–Bmi-1 complex can mediate cellular transformation by translocating itself from the nucleus to cytoplasm but not by its transcriptional suppression activity.[154,155]

The immediate early genes expressed in cat melanoma cells represent both embryonic and developmental proteins that are regulated by many overlapping multifunctional cytokines present in these cells [e.g., granulocyte–macrophage colony-stimulating factor (GMCSF) and IL4], which regulate the proliferation of multipotential stem cells while interacting with IL6, granulocyte colony-stimulating factor, interleukin-12, and other factors during tumorigenesis.[156] The expression of CSF1R (receptor for CSF) stimulates tumor necrosis factor (TNF), which is overexpressed in cat melanoma cells. This in turn activates caspases, which are indirectly involved in tumorigenesis, as they activate a number of proinflammatory cytokines: IFNG, IL1B, IL2, and IL.10[157] Expression of these factors can be mitogenic, particularly in the presence of Kit receptors and their ligand SCF.[134,158,159] Matrilysin (MMP7) was overexpressed in melanoma cells; this metalloprotease is essential for metastasis,[69] and secretion of this enzyme from tumor cells facilitates cell–cell communication, modulates cell membranes, and facilitates tumor cell invasion.[160,161] A unique combination of the intrinsic and extrinsic factors present in these cells thus creates a milieu conducive to expression of embryonic proteins while promoting the malignant potential of the tumor cells (Fig. 8.3). This signaling further creates novel protein–protein interactions along numerous pathways that enhance the malignant potential of transformed cells and invade various tissues and organs of the body.

Thus, we have identified distinct sets of cellular proteins that are involved in the maintenance of stem cell characteristics of this melanoma as well as those that play major roles in the cell differentiation and malignant transformation of these cells.[69] Establishing molecular definitions of stem cells, their differentiated cellular partners, and different stages in the development of stem cell–derived tumors would lead to a reconceptualization of mechanisms by which resident stem cells of certain organs in particular individuals develop tumors more aggressively than in other organs.

8.3.7 Expression of Germline and Embryonic Proteins in Cat Melanoma

In addition to the tumorigenic and neurogenic proteins, the stem cell cat melanoma cells express a number of early proteins that are necessary for the growth and development of many differentiated cell types, including mammary gland, retinal cells, and oocyte membranes: for example, the proteins leptin and lipoprotein lipase are associated with the development of adipose tissue,[162–164] the superfast myosin light chain is involved in the development of muscle cells,[165] and the CD8, the Toll-like receptor 4 and cathepsin W, are expressed in hematopoietic cells.[166–168] The mammary gland–related proteins β-lactoglobulins and prolactin, which are usually expressed in placenta, pituitary, and endometrium,[169] were also expressed in cat melanoma cells. CD9 is a multifunctional protein that is expressed exclusively in melanoma but not in neuronal cells (Fig. 8.3). CD9 sends signals for induction of membrane metalloproteases,[173] and therefore this protein may be involved in promoting the metastatic potential of cat melanoma cells. This membrane protein is also required for gamete fusion and cell motility.[170–172] Peripherin was up-regulated exclusively in the stem cell cat melanoma (i.e., not detected in neuronal cells or in cat embryo cells). This is an intermediate filament protein produced by neurons and is involved in the growth and development of peripheral nervous system cells.[85] In addition, stem cell melanoma expresses the retinal photo-transducing proteins phosducin and the red-sensitive opsin receptor, and the germ cell–associated extracellular membrane protein zona pellucida (Fig. 8.3). Although the roles of these proteins in the development of cat melanoma are not known, their presence in this tumor may correlate with the stemlike characteristics of melanoma cells.

The presence of differentiation-specific proteins in melanoma cells suggests that these cells may be capable of differentiating into multiple cell types. However, signals that induce these specific proteins have yet to be identified, and factors that influence progression toward certain cell lineages are also not known. Currently, we are developing in vitro techniques for directing melanoma cells to differentiate into other cell types by manipulating culture conditions and by using different ligand-binding proteins, peptides, and cues that may generate novel signals and trigger lineage-specific pathways. It would also be important to identify naturally occurring factors, peptides, and other small molecules that are present in the microenvironments of different types of resident stem cells in vivo. These triggers could then influence growth or differentiation into cell types of interest. This information would provide better tools for controlling type-specific spatial and temporal environments for the production of specialized cell types in vitro or in vivo.

8.3.8 Naturally Occurring Protein–Protein Interaction Complexes in Melanomas

All living cells have dynamic intracellular molecular programs that facilitate protein–protein interactions in order to perform specialized functions in various cell types. These interaction complexes are constantly renewed to maintain cellular homeostasis. When the normal physiological pathways are altered, different complexes are formed that alter the cellular decisions of differentiation, apoptosis, metabolic breakdown, and disease susceptibility.

Conventionally, protein interaction complexes are separated by protein tagging, affinity binding of specific proteins to monoclonal antibodies, protein chips painted with numerous ligands, or by the use of baits in the yeast two-hybrid system.[174,175] The separated complexes are identified by mass spectrometry and protein–protein interactions validated by computational techniques.[72] Although these methods are excellent for purification of specific proteins of interest, they may not identify functional differences between cells of the same phenotype since their specific affinity ligands are unknown. To probe the specificity of a particular protein within these complexes and to validate the interacting proteins, Martin et al. have applied bioinformatics and computational technologies to analyze protein–protein interactions in these complexes.[176] A previously established computational/bioinformatics methodology was used to analyze the 31 protein pairs that occurred in multiple complexes in cat melanoma cells.[72] The significance of interacting amino acid sequences in the protein pairs and occurrence of these proteins within a specific complex was defined at below the 0.05 confidence level, and this analysis was able to predict, with about 96% accuracy, when a protein pair would occur in a complex.[176] Using the sequence data alone, these computational analyses could predict the protein pairs in the complexes independently, regardless of whether the interaction pairs occurred in melanoma cells or in the trans-differentiated counterpart neuronal cells. This suggests that the protein pairs isolated by our proteomics technology do interact and have biological relevance.[176]

Examination of the protein network in cat melanoma cells has also revealed that the most significant protein–protein interactions identified in the tumor cells or the counterpart trans-differentiated neuronal cells involve signal-transducing proteins that influence activation or suppression of cellular functions.[69] The highest frequencies of interactions in both (melanoma and neuronal) cell types were detected with integrin β1 (ITGB1), mass/stem cell growth factor receptor (c-KIT), proto-oncogene tyrosine-protein kinase (FES), and sarcoplasmic endoplasmic reticulum (ATA2) (Table 8.1). Integrins, transmenbrane proteins present on the surfaces of all cell types, are essential for focal adhesion, vesicle transport, regulation of actin cytoskeleton, and targeting to other membrane proteins. Interactions of integrins with extracellular matrix proteins are critical for cell proliferation, migration, and differentiation.[177–179] Our computational analyses of the experimental proteomics data indicates that integrin β1 interacts most significantly with 13 signal-transducing proteins, including membrane-bound kinases, enzymes, phosphatases, growth factors, and receptors (Table 8.1). These interactions may result in the assembly of distinct protein complexes that send signals through membrane proteins and regulate expression and/or suppression of other regulatory proteins that may be necessary for altering cellular functions.

TABLE 8.1   Protein–Protein Interactions of Naturally Occurring Multiprotein Complexes[a]

Protein	Interacting Protein	p-Value	Protein	Interacting Protein	p-Value
Mast/stem cell growth factor receptor (KIT)	Integrin beta-1 precursor	<0.001	Integrin beta-1 precursor (ITGB1)	Sarcoplasmic/ endoplasmic reticulum	<0.001
	Sarcoplasmic/ endoplasmic reticulum	<0.001		Transferrin receptor protein 1	<0.001
	Serum albumin precursor	<0.001		Zona pellucida sperm-binding protein	<0.001
	Proto-oncogene tyrosine-protein kinase	<0.001		Serum albumin precursor	<0.001
	Pyruvate kinase, M1 isozyme	0.002		Pyruvate kinase, M1 isozyme	<0.001
	Sodium/calcium exchanger 1 precursor	0.002		Aminopeptidase N	<0.001
				Alkaline phosphatase	0.007
	Zona pellucida sperm-binding protein	0.007		Toll-like receptor 4 precursor	0.038
	Aminopeptidase N	0.016		Sodium/calcium exchanger 1 precursor	<0.001
	Transferrin receptor protein 1	<0.001		Beta-glucuronidase precursor	0.016
	Interleukin-1 beta convertase	0.016		Cathepsin W precursor	0.016
Proto-oncogene tyrosine-protein kinase (FES)	Integrin beta-1 precursor	<0.001		Lysosomal α-mannosidase	0.038
	Sarcoplasmic/ endoplasmic reticulum	0.007		Glutamate decarboxylase	0.016
Sarcoplasmic/ endoplasmic reticulum (ATA)	Transferrin receptor protein 1	<0.001			
	Serum albumin precursor	<0.001			
	Zona pellucida sperm-binding protein	0.007			

[a]Computational analyses of protein–protein interactions and evidence that the experimentally identified protein complexes isolated by two-dimensional gel electrophoresis and identified by mass spectrometry have biological relevance (see the text).

Despite the discovery of many signal-transducing proteins, simple networks of interactions involved in cellular processes such as cell growth and differentiation have not been delineated. In melanoma cells the c-KIT protein-tyrosine kinases interact with 10 distinct proteins, and FES exhibits significant complexes with two proteins (Table 8.1). These proteins are essential for phosphorylation, dephosphorylation, and activation and deactivation of a wide range of molecular processes that are critical for cell differentiation and neurogenesis. Our probability calculations and predictions indicate that the protein pairs identified in multiple complexes from cat melanoma cells represent naturally interactive proteins (S. Martin and S. Rasheed, unpublished data). Further, computational and bioinformatics analyses have been helpful in authenticating the experimentally identified protein–protein interactions in naturally occurring cat melanoma (Table 8.1).

8.3.9 Networks of Protein Interaction Pathways

To further define the biological relevance of our experimental data, we have utilized the software program Ingenuity Pathway Analysis (www.ingenuity.com) to analyze protein–protein interactions pathways and network connections of proteins identified in the stem cell melanoma. As shown in Fig. 8.4, most of the proteins identified in melanoma cells do indeed interact with each other and with other proteins in the cell. Although the biological significance of each of these interactions is not clear at present, the discovery of novel protein interactions and pathway networks that may lead to cell transformation would be crucial in designing strategies to alter specific pathways and correct biological defects by interfering with the formation of functional complexes. This knowledge would ultimately lead to the development of new tools for the diagnosis, prognosis, and therapeutic interventions of stem cell cancers.

8.4 CHALLENGES OF RESEARCH IN CANCER STEM CELLS AND THERAPEUTICS

Analyses of differentially expressed protein profiles of melanoma cells at different stages of growth and differentiation have provided compelling evidence that multifunctional embryonic stem cell genes are involved directly in tumor development and cell differentiation. Proteins expressed in stem cells can be distinguished from those present in fully differentiated tissues, which produce distinct sets of proteins that are necessary for the maintenance of a healthy cellular state. Such studies have just begun to provide new insights into molecular processes involved in neural crest–derived tumors and cell differentiation.

Our knowledge of the self-renewal properties of progenitors of various differentiated tissue types is extremely limited, and cues or pathways that direct these cells toward type-specific differentiation are not known. The cat melanoma model system offers a unique opportunity to understand how some stem cells that retain embryonic properties within the confines of an adult organ or tissue may be driven

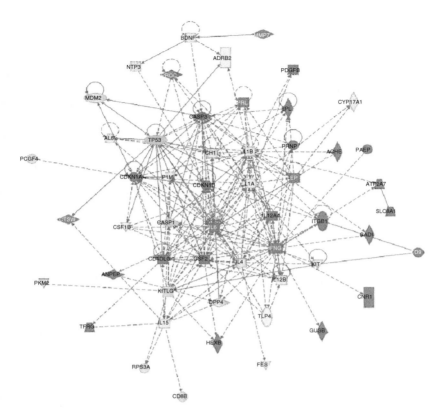

FIGURE 8.4   Graphical representation of the molecular relationship between proteins expressed in stem cell cat melanoma and network connections generated through the use of Ingenuity Pathway Analysis systems (www.ingenuity.com). Each protein is represented as a node, and the biological relationship between two nodes is represented by a line. All lines are supported by at least one reference from the literature, a textbook, or from canonical information stored in the Ingenuity Pathway knowledge base. The intensity of the node color indicates normalized quantities of expressed proteins [i.e., degree of expression, up- (red) or down- (green) regulation in melanoma cells] compared to those present in counterpart neuronal cells. Nodes are displayed using various shapes that represent the functional class of the protein. Fischer's exact test was used to calculate a p-value that determines the probability that each biological function assigned to that network is due to chance alone. (*See insert for color representation of figure.*)

to change direction by external cues from oncogenic to nononcogenic pathways. Understanding the biology of stem cell cancers and molecular cues that trigger differentiation or maintain undifferentiated state of cells would lead to designing better strategies for identifying novel therapeutic targets. These studies may also guide us in learning how different cell types can be manipulated to change the directions of their pathways toward functional regeneration or health.

8.5 CONCLUSIONS AND FUTURE PERSPECTIVES

For a number of years, scientists have been discussing the relationship between the presence of embryonic genes and the development of cancer, but no conclusive evidence had been available until genome-wide proteins were analyzed by proteomic studies. Although many crucial questions regarding embryonic or adult stem cell growth and differentiation have yet to be answered, in this chapter some of the most fundamental biological and molecular properties of a stem cell–derived melanoma have been presented and a distinctive relationship between differentiation and cancer has been demonstrated. This model system has provided insights into the immediate early or embryonic gene products that are involved in regulating self-renewal, growth, proliferation, and neural cell differentiation of a stem cell melanoma.

At present, clinically relevant proteins that may be involved in the expression of heterogeneous tumors such as melanomas are not known. Although many techniques are currently available to analyze the gene expression profiles and proteins involved in various cellular functions, we have taken a direct approach to identifying some naturally occurring events specific to a stem cell melanoma compared with those of normal healthy and trans-differentiated counterpart cells. Since proteins are the ultimate functional effectors, we have compared comprehensive protein profiles of different types of cellular phenotypes under different conditions and analyzed proteins systematically by "reductionism." This differential analysis of protein expression profiles of melanoma and trans-differentiated neuronal cells has demonstrated how the expression of stem cell proteins maintains plasticity for long periods of time. This strategy has also facilitated multivariate analyses of naturally occurring molecular processes inside different cell types and tissues.

Analyses of naturally occurring protein complexes by mass spectrometry followed by validation of interacting proteins by computational analyses has indicated that specific protein–protein interactions are responsible for changing the direction of a cell from tumorigenic pathways to lineage-specific networks of interactions leading to neuronal cell differentiation. These studies may eventually lead to a better understanding of how various molecular interactions guide the larger structures inside living cells to perform certain functions. This knowledge would be conducive in applying novel approaches for studying human melanomas and identifying disease-specific proteins in relation to tumor development.

We have identified distinct protein profiles that can be used to distinguish differences between neuronal cells and tumor cells. Thus, the protein profiles of pigmented tumorigenic cells can be distinguished from those of normal embryonic cells or neuronal, nontumorigenic, unpigmented cells. Establishment of molecular definitions of different stem cell progenitors would lead to a reconceptualization of mechanisms by which stem cells of certain organs in particular people develop tumors more aggressively than in others.

Our experimental data also show that the tumorigenic potential of melanoma cells is lost or inhibited during neurogensis, due to the expression of many early embryonic and cell cycle–related genes whose products have antitumorigenic effects. Correlation of cellular phenotypes with specific signature patterns of expressed proteins in response to epigenetic or environmental changes and validation of test results by computational and bioinformatics analyses would lead to establishing molecular classification of stem cell cancers and a better understanding of stem cell differentiation.

The finding that infection of melanoma stem cells by RD114 virus can exert neurogenic potential raises the possibility that endogenous retroviruses may be involved in differentiation of cells in early embryos and the repair of various organs and tissues by the resident stem cells present in adults.[38] Since both melanocytes and neuronal cells are derived from the embryonic neural crest, our experimental findings have provided evidence that the precursor of this melanoma is a stem cell rather than the primitive melanoblast committed to melanocytic differentiation only.

Genetic or epigenetic factors that may confer susceptibility or resistance to transformation in different resident stem cell populations within a tissue may vary from person to person and from organ to organ in the same person. Identification of stem cell–specific proteins is important not only for establishing molecular signatures of stem cell–derived tumors from various tissues but also for understanding proteins involved in maintaining normal functions of stem cells that are resident in fully differentiated adult organs. Characterization of proteins that induce cell differentiation is crucial because they can be utilized as targets for developing inhibitors or promoters of differentiation-specific pathways. These data would ultimately provide innovative translational research tools for clinical diagnosis and/or for developing interventional therapeutics for a variety of cancers and related disorders.

The relationship between factors that confer oncogenic potential to specific cell types and those that are involved in embryonic cell differentiation or development are not known. Analyses of the intracellular interaction pathways and signaling proteins involved in controlling normal cellular differentiation and those that are responsible for the breakdown of normal regulatory processes that guide toward tumorigenic pathway may lead to a better understanding of molecular mechanisms for the development of melanomas. Finally, this study is just the beginning of a road that may lead to an ocean of knowledge and a better understanding of the complex inner workings of a cell's life.

Acknowledgments

Sincere thanks are due to the summer students of 2005–2006: Jasper Yan for his help in literature searches and functional analyses of proteins expressed in melanoma cells; Ronica Patel for her enthusiasm in performing protein–protein interaction pathways using the Ingenuity Pathway Analysis program; Adil Hussain, Joshua Khalili, and Sara Hussain for their help in scanning figures and reading the manuscript; Shawn Martin for Table 8.1 data analysis; Zisu Mao for technical assistance; and support scientists Jaya Ramkumar, Jyothi Paniyadi, Jinah Kim, and Amy Mendenhall at Ingenuity Systems for their assistance with the use of the program (www.ingenuity.com). The Rasheed Research and Endowment Funds at USC supported this research.

REFERENCES

1. Jemal A, Tiwari RC, Murray T, et al.; American Cancer Society. Cancer statistics. *CA Cancer J Clin.* 2004; 54:8–29.

2. Geller AC, Miller DR, Annas GD, Demierre MF, Gilchrest BA, Koh HK. Melanoma incidence and mortality among US whites, 1969–1999. *JAMA.* 2002; 288:1719–1720.

3. Seddon JM, Polivogianis L, Hsieh CC, Albert DM, Gamel JW, Gragoudas ES. Death from uveal melanoma: number of epithelioid cells and inverse SD of nucleolar area as prognostic factors. *Arch Ophthalmol.* 1987; 105:801–806.

4. Lazovich D, Sweeney C, Weinstock MA, Berwick M. Re: A prospective study of pigmentation, sun exposure, and risk of cutaneous malignant melanoma in women. *J Natl Cancer Inst.* 2004; 96:335–338.

5. Naldi L, Altieri A, Imberti GL, Gallus S, Bosetti C, La VC. Sun exposure, phenotypic characteristics, and cutaneous malignant melanoma: an analysis according to different clinico-pathological variants and anatomic locations (Italy). *Cancer Causes Control.* 2005; 16:893–899.

6. Weir HK, Thun MJ, Hankey BF, et al. Annual report to the nation on the status of cancer, 1975–2000, featuring the uses of surveillance data for cancer prevention and control. *J Natl Cancer Inst.* 2003; 95:1276–1299.

7. Edwards BK, Howe HL, Ries LA, et al. Annual report to the nation on the status of cancer, 1973–1999, featuring implications of age and aging on U.S. cancer burden. *Cancer.* 2002; 94:2766–2792.

8. Azizi E, Friedman J, Pavlotsky F, et al. Familial cutaneous malignant melanoma and tumors of the nervous system: a hereditary cancer syndrome. *Cancer.* 1995; 76:1571–1578.

9. Albert LS, Rhodes AR, Sober AJ. Dysplastic melanocytic nevi and cutaneous melanoma: markers of increased melanoma risk for affected persons and blood relatives. *J Am Acad Dermatol.* 1990; 22:69–75.

10. Marsh D, Zori R. Genetic insights into familial cancers: update and recent discoveries. *Cancer Lett.* 2002; 181:125–164.

11. Hauser J, Homann HH, Drucke D, Kuhnen C, Esenwein SA, Steinau HU. [Desmoplastic neurotropic melanoma: diagnosis and therapeutic procedures for a rare clinical entity.] *Chirurg.* 2006; 77:939–942.

12. Stringa O, Valdez R, Beguerie JR, et al. Primary amelanotic melanoma of the esophagus. *Int J Dermatol.* 2006; 45:1207–1210.

13. Weyers W, Euler M, az-Cascajo C, Schill WB, Bonczkowitz M. Classification of cutaneous malignant melanoma: a reassessment of histopathologic criteria for the distinction of different types. *Cancer.* 1999; 86:288–299.

14. Hutcheson AC, McGowan JW, Maize JC Jr, Cook J. Multiple primary acral melanomas in African-Americans: a case series and review of the literature. *Dermatol Surg.* 2007; 33:1–10.

15. Kageshita T, Hamby CV, Hirai S, Kimura T, Ono T, Ferrone S. Differential clinical significance of alpha(v)beta(3) expression in primary lesions of acral lentiginous melanoma and of other melanoma histotypes. *Int J Cancer.* 2000; 89:153–159.

16. Kelly JW, Chamberlain AJ, Staples MP, McAvoy B. Nodular melanoma: no longer as simple as ABC. *Aust Fam Physician.* 2003; 32:706–709.

17. Abbasi NR, Shaw HM, Rigel DS, et al. Early diagnosis of cutaneous melanoma: revisiting the ABCD criteria. *JAMA.* 2004; 292:2771–2776.

18. Augsburger JJ. Diagnostic biopsy of selected intraocular tumors. *Am J Ophthalmol.* 2005; 140:1094–1095.

19. Brose MS, Volpe P, Feldman M, et al. BRAF and RAS mutations in human lung cancer and melanoma. *Cancer Res.* 2002; 62:6997–7000.

20. Menrad A, Speicher D, Wacker J, Herlyn M. Biochemical and functional characterization of aminopeptidase N expressed by human melanoma cells. *Cancer Res.* 1993; 53:1450–1455.

21. Easty DJ, Herlyn M, Bennett DC. Abnormal protein tyrosine kinase gene expression during melanoma progression and metastasis. *Int J Cancer.* 1995; 60:129–136.

22. Gilhar A, Ullmann Y, Kalish RS, Berkutski T, Azizi E, Bank I. Favourable melanoma prognosis associated with the expression of the tumor necrosis factor receptor and the alpha1beta1 integrin: a preliminary report. *Melanoma Res.* 1997; 7:486–495.

23. Goto Y, Matsuzaki Y, Kurihara S, et al. A new melanoma antigen fatty acid–binding protein 7, involved in proliferation and invasion, is a potential target for immunotherapy and molecular target therapy. *Cancer Res.* 2006; 66:4443–4449.

24. Fang D, Kute T, Setaluri V. Regulation of tyrosinase-related protein-2 (TYRP2) in human melanocytes: relationship to growth and morphology. *Pigment Cell Res.* 2001; 14:132–139.

25. Ghosh S, Rosenthal R, Zajac P, et al. Culture of melanoma cells in 3-dimensional architectures results in impaired immunorecognition by cytotoxic T lymphocytes specific for Melan-A/MART-1 tumor-associated antigen. *Ann Surg.* 2005; 242:851–857.

26. Elias EG, Zapas JL, Beam SL, Brown SD. GM-CSF and IL-2 combination as adjuvant therapy in cutaneous melanoma: early results of a phase II clinical trial. *Oncology.* 2005; 19:15–18.

27. Kitaura H, Shinshi M, Uchikoshi Y, Ono T, Iguchi-Ariga SM, Ariga H. Reciprocal regulation via protein–protein interaction between c-Myc and p21(cip1/waf1/sdi1) in DNA replication and transcription. *J Biol Chem.* 2000; 275:10477–10483.

28. Daponte A, Ascierto PA, Gravina A, et al. Temozolomide and cisplatin in advanced malignant melanoma. *Anticancer Res.* 2005; 25:1441–1447.

29. Alatrash G, Hutson TE, Molto L, et al. Clinical and immunologic effects of subcutaneously administered interleukin-12 and interferon alfa-2b: phase I trial of patients with metastatic renal cell carcinoma or malignant melanoma. *J Clin Oncol.* 2004; 22:2891–2900.

30. Yoo YC, Watanabe S, Watanabe R, Hata K, Shimazaki K, Azuma I. Bovine lactoferrin and lactoferricin inhibit tumor metastasis in mice. *Adv Exp Med Biol.* 1998; 443:285–291.

31. Vuoristo MS, Hahka-Kemppinen M, Parvinen LM, et al. Randomized trial of dacarbazine versus bleomycin, vincristine, lomustine and dacarbazine (BOLD) chemotherapy combined with natural or recombinant interferon-alpha in patients with advanced melanoma. *Melanoma Res.* 2005; 15:291–296.

32. Shellman YG, Ribble D, Miller L, et al. Lovastatin-induced apoptosis in human melanoma cell lines. *Melanoma Res.* 2005; 15:83–89.

33. Di Lauro V, Scalone S, La MN, et al. Combined chemoimmunotherapy of metastatic melanoma: a single institution experience. *Melanoma Res.* 2005; 15:209–212.

34. Chen KG, Valencia JC, Lai B, et al. Melanosomal sequestration of cytotoxic drugs contributes to the intractability of malignant melanomas. *Proc Natl Acad Sci U S A.* 2006; 103:9903–9907.

35. Chi A, Valencia JC, Hu ZZ, et al. Proteomic and bioinformatic characterization of the biogenesis and function of melanosomes. *J Proteome Res.* 2006; 5:3135–3144.

36. Stein U, Jurchott K, Schlafke M, Hohenberger P. Expression of multidrug resistance genes MVP, MDR1, and MRP1 determined sequentially before, during, and after hyperthermic isolated limb perfusion of soft tissue sarcoma and melanoma patients. *J Clin Oncol.* 2002; 20:3282–3292.

37. Helmbach H, Kern MA, Rossmann E, et al. Drug resistance towards etoposide and cisplatin in human melanoma cells is associated with drug-dependent apoptosis deficiency. *J Invest Dermatol.* 2002; 118:923–932.

38. Rasheed S. Role of endogenous cat retrovirus in cell differentiation. *Proc Natl Acad Sci U S A.* 1982; 79:7371–7375.

39. Rasheed S. Characterization of a differentiated cat melanoma cell line. *Cancer Res.* 1983; 43:3379–3384.

40. Fang D, Nguyen TK, Leishear K, et al. A tumorigenic subpopulation with stem cell properties in melanomas. *Cancer Res.* 2005; 65:9328–9337.

41. Cramer SF. The origin of epidermal melanocytes: implications for the histogenesis of nevi and melanomas. *Arch Pathol Lab Med.* 1991; 115:115–119.

42. Fuchs E, Segre JA. Stem cells: a new lease on life. *Cell.* 2000; 100:143–155.

43. Herlyn M. Molecular targets in melanoma: strategies and challenges for diagnosis and therapy. *Int J Cancer.* 2006; 118:523–526.

44. Gyorffy B, Lage H. A Web-based data warehouse on gene expression in human malignant melanoma. *J Invest Dermatol.* 2007; 127:394–399.

45. Smith AP, Hoek K, Becker D. Whole-genome expression profiling of the melanoma progression pathway reveals marked molecular differences between nevi/melanoma in situ and advanced-stage melanomas. *Cancer Biol Ther.* 2005; 4:1018–1029.

46. Ghosh S, Spagnoli GC, Martin I, et al. Three-dimensional culture of melanoma cells profoundly affects gene expression profile: a high density oligonucleotide array study. *J Cell Physiol.* 2005; 204:522–531.

47. Matsuzaki Y, Hashimoto S, Fujita T, et al. Systematic identification of human melanoma antigens using serial analysis of gene expression (SAGE). *J Immunother.* 2005; 28:10–19.

48. Wilson LL, Tran L, Morton DL, Hoon DS. Detection of differentially expressed proteins in early-stage melanoma patients using SELDI-TOF mass spectrometry. *Ann NY Acad Sci.* 2004; 1022:317–322.

49. Vogl A, Sartorius U, Vogt T, et al. Gene expression profile changes between melanoma metastases and their daughter cell lines: implication for vaccination protocols. *J Invest Dermatol.* 2005; 124:401–404.

50. Bachmann IM, Straume O, Puntervoll HE, Kalvenes MB, Akslen LA. Importance of P-cadherin, beta-catenin, and Wnt5a/frizzled for progression of melanocytic tumors and prognosis in cutaneous melanoma. *Clin Cancer Res.* 2005; 11:8606–8614.

51. Berger AJ, Davis DW, Tellez C, et al. Automated quantitative analysis of activator protein-2alpha subcellular expression in melanoma tissue microarrays correlates with survival prediction. *Cancer Res.* 2005; 65:11185–11192.

52. Haqq C, Nosrati M, Sudilovsky D, et al. The gene expression signatures of melanoma progression. *Proc Natl Acad Sci U S A.* 2005; 102:6092–6097.

53. Nambiar S, Mirmohammadsadegh A, Doroudi R, et al. Signaling networks in cutaneous melanoma metastasis identified by complementary DNA microarrays. *Arch Dermatol.* 2005; 141:165–173.

54. Bittner M, Meltzer P, Chen Y, et al. Molecular classification of cutaneous malignant melanoma by gene expression profiling. *Nature.* 2000; 406:536–540.

55. Nishizuka S. Profiling cancer stem cells using protein array technology. *Eur J Cancer.* 2006; 42:1273–1282.

56. Carr KM, Bittner M, Trent JM. Gene-expression profiling in human cutaneous melanoma. *Oncogene.* 2003; 22:3076–3080.

57. Dooley TP, Curto EV, Davis RL, Grammatico P, Robinson ES, Wilborn TW. DNA microarrays and likelihood ratio bioinformatic methods: discovery of human melanocyte biomarkers. *Pigment Cell Res.* 2003; 16:245–253.

58. McArdle L, Rafferty MM, Satyamoorthy K, et al. Microarray analysis of phosphatase gene expression in human melanoma. *Br J Dermatol.* 2005; 152:925–930.

59. Sherlock G, Hernandez-Boussard T, Kasarskis A, et al. The Stanford Microarray Database. *Nucleic Acids Res.* 2001; 29:152–155.

60. Areces LBF, Dello SPF, Jucker MF, Stanley ERF, Feldman RA. Functional specificity of cytoplasmic and transmembrane tyrosine kinases: identification of 130- and 75-kilodalton substrates of c-fps/fes tyrosine kinase in macrophages. *Mol Cell Biol.* 1994; 14:4606–4615.

61. Bernard K, Litman E, Fitzpatrick JL, et al. Functional proteomic analysis of melanoma progression. *Cancer Res.* 2003; 63:6716–6725.

62. Hatch WC, Ganju RK, Hiregowdara D, Avraham S, Groopman JE. The related adhesion focal tyrosine kinase (RAFTK) is tyrosine phosphorylated and participates in colony-stimulating factor-1/macrophage colony-stimulating factor signaling in monocyte-macrophages. *Blood.* 1998; 91:3967–3973.

63. Irish JM, Hovland R, Krutzik PO, et al. Single cell profiling of potentiated phospho-protein networks in cancer cells. *Cell.* 2004; 118:217–228.

64. Zisch AH, D'Alessandri L, Amrein K, Ranscht B, Winterhalter KH, Vaughan L. The glypiated neuronal cell adhesion molecule contactin/F11 complexes with src-family protein tyrosine kinase Fyn. *Mol Cell Neurosci.* 1995; 6:263–279.

65. Woods DB, Vousden KH. Regulation of p53 function. *Exp Cell Res.* 2001; 264:56–66.

66. Smelt VA, Upton A, Adjaye J, et al. Expression of arylamine *N*-acetyltransferases in pre-term placentas and in human pre-implantation embryos. *Hum Mol Genet.* 2000; 9:1101–1107.

67. Nandi A, Sprung R, Barma DK, et al. Global identification of O-GlcNAc-modified proteins. *Anal Chem.* 2006; 78:452–458.

68. Sprung R, Nandi A, Chen Y, et al. Tagging-via-substrate strategy for probing O-GlcNAc modified proteins. *J Proteome Res.* 2005; 4:950–957.

69. Rasheed S, Mao Z, Chan JM, Chan LS. Is melanoma a stem cell tumor? Identification of neurogenic proteins in trans-differentiated cells. *J Transl Med.* 2005; 3:14.

70. McAllister RM, Nicolson M, Gardner MB, et al. C-type virus released from cultured human rhabdomyosarcoma cells. *Nat New Biol.* 1972; 235:3–6.

71. Rasheed S, McAllister RM, Henderson BE, Gardner MB. Brief communication: In Vitro host range and serologic studies on RD-114 virus. *J Natl Cancer Inst.* 1973; 51:1383–1385.

72. Martin S, Roe D, Faulon JL. Predicting protein–protein interactions using signature products. *Bioinformatics.* 2005; 21:218–226.

73. Akasaka T, Takahashi N, Suzuki M, Koseki H, Bodmer R, Koga H. MBLR, a new Ring finger protein resembling mammalian Polycomb gene products, is regulated by cell cycle–dependent phosphorylation. *Genes Cells.* 2002; 7:835–850.

74. Lee TI, Jenner RG, Boyer LA, et al. Control of developmental regulators by Polycomb in human embryonic stem cells. *Cell.* 2006; 125:301–313.

75. Vogel T, Stoykova A, Gruss P. Differential expression of Polycomb repression complex 1 (PRC1) members in the developing mouse brain reveals multiple complexes. *Dev Dyn.* 2006; 235:2574–2585.

76. Lachyankar MB, Condon PJ, Quesenberry PJ, Litofsky NS, Recht LD, Ross AH. Embryonic precursor cells that express Trk receptors: induction of different cell fates by NGF, BDNF, NT-3, and CNTF. *Exp Neurol.* 1997; 144:350–360.

77. Zhang J, Geula C, Lu C, Koziel H, Hatcher LM, Roisen FJ. Neurotrophins regulate proliferation and survival of two microglial cell lines in vitro. *Exp Neurol.* 2003; 183:469–481.

78. Tessarollo L, Coppola V, Fritzsch B. NT-3 replacement with brain-derived neurotrophic factor redirects vestibular nerve fibers to the cochlea. *J Neurosci.* 2004; 24:2575–2584.

79. Pyle AD, Lock LF, Donovan PJ. Neurotrophins mediate human embryonic stem cell survival. *Nat Biotechnol.* 2006; 24:344–350.

80. Kokaia Z, Lindvall O. Neurogenesis after ischaemic brain insults. *Curr Opin Neurobiol.* 2003; 13:127–132.

81. Kokaia Z, Andsberg G, Yan Q, Lindvall O. Rapid alterations of BDNF protein levels in the rat brain after focal ischemia: evidence for increased synthesis and anterograde axonal transport. *Exp Neurol.* 1998; 154:289–301.

82. Kokaia Z, Nawa H, Uchino H, et al. Regional brain-derived neurotrophic factor mRNA and protein levels following transient forebrain ischemia in the rat. *Brain Res Mol Brain Res.* 1996; 38:139–144.

83. Lindvall O, Kokaia Z, Bengzon J, Elmer E, Kokaia M. Neurotrophins and brain insults. *Trends Neurosci.* 1994; 17:490–496.

84. Lee SH, Han JH, Choi JH, Huh EY, Kwon YK, Kaang BK. The effect of brain-derived neurotrophic factor on neuritogenesis and synaptic plasticity in aplysia neurons and the hippocampal cell line HiB5. *Mol Cells.* 2003; 15:233–239.

85. Prieto VG, McNutt NS, Lugo J, Reed JA. Differential expression of the intermediate filament peripherin in cutaneous neural lesions and neurotized melanocytic nevi. *Am J Surg Pathol.* 1997; 21:1450–1454.

86. Friedman PA, Goodman WG. PTH(1–84)/PTH(7–84): a balance of power. *Am J Physiol Renal Physiol.* 2006; 290:975–984.

87. Snyder EY, Kim SU. Hormonal requirements for neuronal survival in culture. *Neurosci Lett.* 1979; 13:225–230.

88. Kim NG, Lee H, Son E, et al. Hypoxic induction of caspase-11/caspase-1/interleukin-1beta in brain microglia. *Brain Res Mol Brain Res.* 2003; 114:107–114.

89. Gu Y, Kuida K, Tsutsui H, et al. Activation of interferon-gamma inducing factor mediated by interleukin-1beta converting enzyme. *Science.* 1997; 275:206–209.

90. Yamauchi T, Sawa Y, Sakurai M, et al. ONO-5046 attenuation of delayed motor neuron death and effect on the induction of brain-derived neurotrophic factor, phosphorylated extracellular signal-regulated kinase, and caspase 3 after spinal cord ischemia in rabbits. *J Thorac Cardiovasc Surg.* 2006; 131:644–650.

91. Kothakota S, Azuma T, Reinhard C, et al. Caspase-3-generated fragment of gelsolin: effector of morphological change in apoptosis. *Science.* 1997; 278:294–298.

92. Dame JB, Chegini N, Christensen RD, Juul SE. The effect of interleukin-1beta (IL-1beta) and tumor necrosis factor-alpha (TNF-alpha) on granulocyte–macrophage colony-stimulating factor (GM-CSF) production by neuronal precursor cells. *Eur Cytokine Netw.* 2002; 13:128–133.

93. Docagne F, Campbell SJ, Bristow AF, et al. Differential regulation of type I and type II interleukin-1 receptors in focal brain inflammation. *Eur J Neurosci.* 2005; 21:1205–1214.

94. Hailer NP, Vogt C, Korf HW, Dehghani F. Interleukin-1beta exacerbates and interleukin-1 receptor antagonist attenuates neuronal injury and microglial activation after excitotoxic damage in organotypic hippocampal slice cultures. *Eur J Neurosci.* 2005; 21:2347–2360.

95. Friedman WJ. Cytokines regulate expression of the type 1 interleukin-1 receptor in rat hippocampal neurons and glia. *Exp Neurol.* 2001; 168:23–31.

96. Gloss B, Villegas S, Villarreal FJ, Moriscot A, Dillmann WH. Thyroid hormone–induced stimulation of the sarcoplasmic reticulum $Ca(2+)$ ATPase gene is inhibited by LIF and IL-6. *Am J Physiol Endocrinol Metab.* 2000; 278:738–743.

97. Scemes E. Components of astrocytic intercellular calcium signaling. *Mol Neurobiol.* 2000; 22:167–179.

98. Racay P, Kaplan P, Lehotsky J. Control of Ca^{2+} homeostasis in neuronal cells. *Gen Physiol Biophys.* 1996; 15:193–210.

99. Magnier-Gaubil C, Herbert JM, Quarck R, et al. Smooth muscle cell cycle and proliferation: relationship between calcium influx and sarco-endoplasmic reticulum Ca^{2+} ATPase regulation. *J Biol Chem.* 1996; 271:27788–27794.

100. Jiang D, Xiao B, Yang D, et al. RyR2 mutations linked to ventricular tachycardia and sudden death reduce the threshold for store-overload-induced Ca^{2+} release (SOICR). *Proc Natl Acad Sci U S A.* 2004; 101:13062–13067.

101. Sherr CJ, Roberts JM. CDK inhibitors: positive and negative regulators of G1-phase progression. *Genes Dev.* 1999; 13:1501–1512.

102. Cheng M, Olivier P, Diehl JA, et al. The p21(Cip1) and p27(Kip1) CDK 'inhibitors' are essential activators of cyclin D–dependent kinases in murine fibroblasts. *EMBO J.* 1999; 18:1571–1583.

103. Murai T, Nakagawa Y, Maeda H, Terada K. Altered regulation of cell cycle machinery involved in interleukin-1-induced G(1) and G(2) phase growth arrest of A375S2 human melanoma cells. *J Biol Chem.* 2001; 276:6797–6806.

104. Qin L, Li X, Ko JK, Partridge NC. Parathyroid hormone uses multiple mechanisms to arrest the cell cycle progression of osteoblastic cells from G1 to S phase. *J Biol Chem.* 2005; 280:3104–3111.

105. Burch L, Shimizu H, Smith A, Patterson C, Hupp TR. Expansion of protein interaction maps by phage peptide display using MDM2 as a prototypical conformationally flexible target protein. *J Mol Biol.* 2004; 337:129–145.

106. Vogelstein B, Lane D, Levine AJ. Surfing the p53 network. *Nature.* 2000; 408:307–310.

107. Sorrell JM, Brinon L, Baber MA, Caplan AI. Cytokines and glucocorticoids differentially regulate APN/CD13 and DPPIV/CD26 enzyme activities in cultured human dermal fibroblasts. *Arch Dermatol Res.* 2003; 295:160–168.

108. Nadal MS, Ozaita A, Amarillo Y, et al. The CD26-related dipeptidyl aminopeptidase-like protein DPPX is a critical component of neuronal A-type K^+ channels. *Neuron.* 2003; 37:449–461.

109. Struckhoff G. Dipeptidyl peptidase II in astrocytes of the rat brain: meningeal cells increase enzymic activity in cultivated astrocytes. *Brain Res.* 1993; 620:49–57.

110. Pethiyagoda CL, Welch DR, Fleming TP. Dipeptidyl peptidase IV (DPPIV) inhibits cellular invasion of melanoma cells. *Clin Exp Metastasis.* 2000; 18:391–400.

111. Wesley UV, Albino AP, Tiwari S, Houghton AN. A role for dipeptidyl peptidase IV in suppressing the malignant phenotype of melanocytic cells. *J Exp Med.* 1999; 190:311–322.

112. Nakagawara A, Nakamura Y, Ikeda H, et al. High levels of expression and nuclear localization of interleukin-1 beta converting enzyme (ICE) and CPP32 in favorable human neuroblastomas. *Cancer Res.* 1997; 57:4578–4584.

113. Park IK, Qian D, Kiel M, et al. Bmi-1 is required for maintenance of adult self-renewing haematopoietic stem cells. *Nature.* 2003; 423:302–305.

114. Lessard J, Sauvageau G. Bmi-1 determines the proliferative capacity of normal and leukaemic stem cells. *Nature.* 2003; 423:255–260.

115. Trentin A, Glavieux-Pardanaud C, Le Douarin NM, Dupin E. Self-renewal capacity is a widespread property of various types of neural crest precursor cells. *Proc Natl Acad Sci U S A.* 2004; 101:4495–4500.

116. Molofsky AV, Pardal R, Iwashita T, Park IK, Clarke MF, Morrison SJ. Bmi-1 dependence distinguishes neural stem cell self-renewal from progenitor proliferation. *Nature.* 2003; 425:962–967.

117. Pardal R, Molofsky AV, He S, Morrison SJ. Stem cell self-renewal and cancer cell proliferation are regulated by common networks that balance the activation of proto-oncogenes and tumor suppressors. *Cold Spring Harb Symp Quant Biol.* 2005; 70:177–185.

118. Haupt Y, Alexander WS, Barri G, Klinken SP, Adams JM. Novel zinc finger gene implicated as myc collaborator by retrovirally accelerated lymphomagenesis in E mu-myc transgenic mice. *Cell.* 1991; 65:753–763.

119. Jorgensen HF, Giadrossi S, Casanova M, et al. Stem cells primed for action: Polycomb repressive complexes restrain the expression of lineage-specific regulators in embryonic stem cells. *Cell Cycle.* 2006; 5:1411–1414.

120. Boyer LA, Plath K, Zeitlinger J, et al. Polycomb complexes repress developmental regulators in murine embryonic stem cells. *Nature.* 2006; 441:349–353.

121. Bracken AP, Dietrich N, Pasini D, Hansen KH, Helin K. Genome-wide mapping of Polycomb target genes unravels their roles in cell fate transitions. *Genes Dev.* 2006; 20:1123–1136.

122. Guitton AE, Berger F. Control of reproduction by Polycomb group complexes in animals and plants. *Int J Dev Biol.* 2005; 49:707–716.

123. Nakauchi H, Oguro H, Negishi M, Iwama A. Polycomb gene product Bmi-1 regulates stem cell self-renewal. *Ernst Schering Res Found Workshop.* 2005:85–100.

124. Li Z, Cao R, Wang M, Myers MP, Zhang Y, Xu RM. Structure of a Bmi-1-Ring1B polycomb group ubiquitin ligase complex. *J Biol Chem.* 2006; 281:20643–20649.

125. Valk-Lingbeek ME, Bruggeman SW, van Lohuizen M. Stem cells and cancer: the Polycomb connection. *Cell.* 2004; 118:409–418.

126. Iwama A, Oguro H, Negishi M, et al. Enhanced self-renewal of hematopoietic stem cells mediated by the Polycomb gene product Bmi-1. *Immunity.* 2004; 21:843–851.

127. Iwama A, Oguro H, Negishi M, Kato Y, Nakauchia H. Epigenetic regulation of hematopoietic stem cell self-renewal by Polycomb group genes. *Int J Hematol.* 2005; 81:294–300.

128. Squazzo SL, O'Geen H, Komashko VM, et al. Suz12 binds to silenced regions of the genome in a cell-type-specific manner. *Genome Res.* 2006; 16:890–900.

129. Domen J, Weissman IL. Self-renewal, differentiation or death: regulation and manipulation of hematopoietic stem cell fate. *Mol Med Today.* 1999; 5:201–208.

130. Fazel S, Cimini M, Chen L, et al. Cardioprotective c-kit$^+$ cells are from the bone marrow and regulate the myocardial balance of angiogenic cytokines. *J Clin Invest.* 2006; 116:1865–1877.

131. Nayernia K, Lee JH, Drusenheimer N, et al. Derivation of male germ cells from bone marrow stem cells. *Lab Invest.* 2006; 86:654–663.

132. Massa S, Balciunaite G, Ceredig R, Rolink AG. Critical role for c-kit (CD117) in T cell lineage commitment and early thymocyte development in vitro. *Eur J Immunol.* 2006; 36:526–532.

133. Turner AM, Zsebo KM, Martin F, Jacobsen FW, Bennett LG, Broudy VC. Non-hematopoietic tumor cell lines express stem cell factor and display c-kit receptors. *Blood.* 1992; 80:374–381.

134. Welker P, Schadendorf D, Artuc M, Grabbe J, Henz BM. Expression of SCF splice variants in human melanocytes and melanoma cell lines: potential prognostic implications. *Br J Cancer.* 2000; 82:1453–1458.

135. Huang S, Luca M, Gutman M, et al. Enforced c-KIT expression renders highly metastatic human melanoma cells susceptible to stem cell factor–induced apoptosis and inhibits their tumorigenic and metastatic potential. *Oncogene.* 1996; 13:2339–2347.

136. Zakut R, Perlis R, Eliyahu S, Yarden Y, Givol D, Lyman SD, Halaban R. KIT ligand (mast cell growth factor) inhibits the growth of KIT-expressing melanoma cells. *Oncogene.* 1993; 8:2221–2229.

137. Dirnhofer S, Zimpfer A, Went P. [The diagnostic and predictive role of kit (CD117).] *Ther Umsch.* 2006; 63:273–278.

138. Eisengart CA, Mestre JR, Naama HA, et al. Prostaglandins regulate melanoma-induced cytokine production in macrophages. *Cell Immunol.* 2000; 204:143–149.

139. Barak V, Ben-Ishay Z. Cytokine and growth factor gene expression by bone marrow stroma of mice with damaged hematopoiesis and during regeneration. *Leuk Res.* 1994; 18:733–739.

140. Yue FY, Geertsen R, Hemmi S, Burg G, Pavlovic J, Laine E, Dummer R. IL-12 directly up-regulates the expression of HLA class I, HLA class II and ICAM-1 on human melanoma cells: a mechanism for its antitumor activity? *Eur J Immunol.* 1999; 29:1762–1773.

141. Chen J, Carcamo JM, Golde DW. The {alpha} subunit of the granulocyte–macrophage colony-stimulating factor receptor interacts with c-Kit and inhibits c-Kit signaling. *J Biol Chem.* 2006; 281:22421–22426.

142. Bourette RP, Mouchiroud G, Ouazana R, Morle F, Godet J, Blanchet JP. Expression of human colony-stimulating factor-1 (CSF-1) receptor in murine pluripotent hematopoietic NFS-60 cells induces long-term proliferation in response to CSF-1 without loss of erythroid differentiation potential. *Blood.* 1993; 81:2511–2520.

143. Oka M, Tagoku K, Russell TL. CD9 is associated with leukemia inhibitory factor-mediated maintenance of embryonic stem cells. *Mol Biol Cell.* 2002; 13: 1274–1281.

144. Sugiura N, Adams SM, Corriveau RA. An evolutionarily conserved N-terminal acetyltransferase complex associated with neuronal development. *J Biol Chem.* 2003; 278:40113–40120.

145. Jean WC, Spellman SR, Wallenfriedman MA, et al. Effects of combined granulocyte–macrophage colony-stimulating factor (GM-CSF), interleukin-2, and interleukin-12 based immunotherapy against intracranial glioma in the rat. *J Neurooncol.* 2004; 66:39–49.

146. Huttenbach Y, Prieto VG, Reed JA. Desmoplastic and spindle cell melanomas express protein markers of the neural crest but not of later committed stages of Schwann cell differentiation. *J Cutan Pathol.* 2002; 29:562–568.

147. Bertrand G. [Melanotic adenocarcinoma of the uterus: neuroendocrine tumor of the uterus.] *Ann Pathol.* 1988; 8:295–304.

148. Torner L, Maloumby R, Nava G, Aranda J, Clapp C, Neumann ID. In Vivo release and gene upregulation of brain prolactin in response to physiological stimuli. *Eur J Neurosci.* 2004; 19:1601–1608.

149. Chen AP, Essex M, de Noronha F. Detection and localization of a phosphotyrosine-containing onc gene product in feline tumor cells. *Haematol Blood Transfus.* 1983; 28:227–235.

150. Easty DJ, Bennett DC. Protein tyrosine kinases in malignant melanoma. *Melanoma Res.* 2000; 10:401–411.

151. Bhattacharya N, Wang Z, Davitt C, McKenzie IF, Xing PX, Magnuson NS. Pim-1 associates with protein complexes necessary for mitosis. *Chromosoma.* 2002; 111:80–95.

152. Ataliotis P, Symes K, Chou MM, Ho L, Mercola M. PDGF signalling is required for gastrulation of. *Xenopus laevis Development.* 1995; 121:3099–3110.

153. Yan XQ, Brady G, Iscove NN. Platelet-derived growth factor (PDGF) activates primitive hematopoietic precursors (pre-CFCmulti) by up-regulating IL-1 in PDGF receptor–expressing macrophages. *J Immunol.* 1993; 150:2440–2448.

154. Cohen KJ, Hanna JS, Prescott JE, Dang CV. Transformation by the Bmi-1 oncoprotein correlates with its subnuclear localization but not its transcriptional suppression activity. *Mol Cell Biol.* 1996; 16:5527–5535.

155. van der Lugt NM, Domen J, Linders K, et al. Posterior transformation, neurological abnormalities, and severe hematopoietic defects in mice with a targeted deletion of the bmi-1 proto-oncogene. *Genes Dev.* 1994; 8:757–769.

156. Ogawa M. Hematopoiesis. *J Allergy Clin Immunol.* 1994; 94(3 Pt 2): 645–650.

157. Mogi M, Togari A. Activation of caspases is required for osteoblastic differentiation. *J Biol Chem.* 2003; 278:47477–47482.

158. Di Noto R, Lo PC, Schiavone EM, et al. Stem cell factor receptor (c-kit, CD117) is expressed on blast cells from most immature types of acute myeloid malignancies but is also a characteristic of a subset of acute promyelocytic leukaemia. *Br J Haematol.* 1996; 92:562–564.

159. London CA, Kisseberth WC, Galli SJ, Geissler EN, Helfand SC. Expression of stem cell factor receptor (c-kit) by the malignant mast cells from spontaneous canine mast cell tumors. *J Comp Pathol.* 1996; 115:399–414.

160. Tam EM, Morrison CJ, Wu YI, Stack MS, Overall CM. Membrane protease proteomics: isotope-coded affinity tag MS identification of undescribed MT1-matrix metalloproteinase substrates. *Proc Natl Acad Sci U S A.* 2004; 101:6917–6922.

161. Egeblad M, Werb Z. New functions for the matrix metalloproteinases in cancer progression. *Nat Rev Cancer.* 2002; 2:161–174.

162. Huey PU, Marcell T, Owens GC, Etienne J, Eckel RH. Lipoprotein lipase is expressed in cultured Schwann cells and functions in lipid synthesis and utilization. *J Lipid Res.* 1998; 39:2135–2142.

163. Tschematsch MM, Mlecnik B, Trajanoski Z, Zechner R, Zimmermann R. LPL-mediated lipolysis of VLDL induces an upregulation of AU-rich mRNAs and an activation of HuR in endothelial cells. *Atherosclerosis.* 2006; 189:310–317.

164. Friedman JM, Halaas JL. Leptin and the regulation of body weight in mammals. *Nature.* 1998; 395:763–770.

165. Collins C, Schappert K, Hayden MR. The genomic organization of a novel regulatory myosin light chain gene (MYL5) that maps to chromosome 4p16.3 and shows different patterns of expression between primates. *Hum Mol Genet.* 1992; 1:727–733.

166. Wex T, Buhling F, Wex H, et al. Human cathepsin W, a cysteine protease predominantly expressed in NK cells, is mainly localized in the endoplasmic reticulum. *J Immunol.* 2001; 167:2172–2178.

167. Linnevers C, Smeekens SP, Brömme D. Human cathepsin W, a putative cysteine protease predominantly expressed in CD8[+] T-lymphocytes. *FEBS Lett.* 1997; 405:253–259.

168. Akira S, Takeda K, Kaisho T. Toll-like receptors: critical proteins linking innate and acquired immunity. *Nat Immunol.* 2001; 2:675–680.

169. Kontopidis G, Holt C, Sawyer L. Beta-lactoglobulin: binding properties, structure, and function. *J Dairy Sci.* 2004; 87:785–796.

170. Deissler H, Blass-Kampmann S, Kindler-Rohrborn A, Meyer HE, Rajewsky MF. Characterization of rat NCA/CD9 cell surface antigen and its expression by normal and malignant neural cells. *J Neurosci Res.* 1996; 43:664–674.

171. Longo N, Yanez-Mo M, Mittelbrunn M, et al. Regulatory role of tetraspanin CD9 in tumor-endothelial cell interaction during transendothelial invasion of melanoma cells. *Blood.* 2001; 98:3717–3726.

172. Ikeyama S, Koyama M, Yamaoko M, Sasada R, Miyake M. Suppression of cell motility and metastasis by transfection with human motility–related protein (MRP-1/CD9) DNA. *J Exp Med.* 1993; 177:1231–1237.

173. Hong IK, Kim YM, Jeoung DI, Kim KC, Lee H. Tetraspanin CD9 induces MMP-2 expression by activating p38 MAPK, JNK and c-Jun pathways in human melanoma cells. *Exp Mol Med.* 2005; 37:230–239.

174. Deshaies RJ, Seol JH, McDonald WH, et al. Charting the protein complexome in yeast by mass spectrometry. *Mol Cell Proteomics.* 2002; 1:3–10.

175. Zhu H, Bilgin M, Snyder M. Proteomics. *Annu Rev Biochem.* 2003; 72:783–812.

176. Martin S, Mao Z, Chan LS, Rasheed, S. Inferring protein–protein interaction networks from protein complex data. *Int J Bioinf Res Appl.* 2007; 3:480–492.

177. Radford KJ, Thorne RF, Hersey P. CD63 associates with transmembrane 4 superfamily members, CD9 and CD81, and with beta 1 integrins in human melanoma. *Biochem Biophys Res Commun.* 1996; 222:13–18.

178. Staquicini FI, Moreira CR, Nascimento FD, et al. Enzyme and integrin expression by high and low metastatic melanoma cell lines. *Melanoma Res.* 2003; 13:11–18.

179. Clark EA, Brugge JS. Integrins and signal transduction pathways: the road taken. *Science.* 1995; 268:233–239.

9 Invasion Program of Normal and Cancer Stem Cells

DAVID OLMEDA, GEMA MORENO-BUENO, DAVID SARRIÓ, JOSÉ PALACIOS, and AMPARO CANO

9.1 INTRODUCTION

Metastasis represents the most life-threatening event for cancer patients. Unfortunately, although considerable progress has been made in the last two decades in new therapeutic strategies to stop tumor growth, fighting the metastatic spread of tumors is still an unresolved matter for most, if not all, cancer types. This important drawback in effective cancer eradication is due primarily to the complexity of the metastatic process and to the relatively scarce knowledge we have of the process at the molecular level that could provide effective new drug targets. Despite this, it is presently considered that metastases originate from individual cells or reduced groups of cells inside the primary tumor that acquire the capacity to colonize into distant organs. The invasion of tumor cells from the original primary tumor to adjacent tissue, presently recognized as the initiating step of the metastatic process for most solid tumors, is of particular relevance in carcinomas. The increased knowledge generated in recent years on the invasive processes has provided new insights into cellular and molecular cues of metastases.

Epithelial to mesenchymal transition (EMT), a meaningful developmental process, is frequently associated with the acquisition of invasive behavior in carcinoma and melanoma cell lines and in several animal models. The recent recognition of the importance of cancer stem cells (CSCs) in initiation and maintenance of tumor growth can also be extended to malignant conversion. Indeed, it has been speculated that metastasis can originate from certain subpopulations of CSCs that acquire invasive and migratory properties. Furthermore, it has been suggested that EMT could occur specifically in that subpopulation of migratory cancer stem cells. Nevertheless, the implication of EMT in invasion/tumor progression of human carcinomas has still to be proved definitively. Other important open questions in the field relate to the actual occurrence of CSC in human tumors and their relationship to EMT/invasive processes.

In this chapter we discuss current knowledge regarding the tumor invasion program in the context of EMT and CSC biology, focusing on (1) the basic mechanisms and programs that drive invasion and EMT in development and tumor progression, and (2) the existence of EMT in human tumors and its relation to CSC biology.

9.2 BASICS OF TUMOR PROGRESSION: INVASION AND METASTASIS

Observations in human cancers and animal models of cancer have sustained the hypothesis that tumor development is governed by the rules of Darwinian evolution in which a succession of genetic changes, each one conferring an advantage in growth or survival, leads to a progressive conversion of normal human cells into tumor cells. Accumulated evidence has also shown that cancer is a multistep process in which different steps are associated with the accumulation of both genetic and phenotypic alterations in the cells that drive the progressive transformation of normal human cells into malignant cells. Despite the great diversity of human cancers, it is broadly accepted that there are six main steps or acquired properties, also known as *hallmarks of cancer*, required for any normal cell to become a highly tumorigenic and metastatic tumor cell[1]: self-sufficiency to growth signals, insensitivity to antigrowth signals, evasion of the programmed cell death, limitless replicative potential, sustained angiogenesis, and tissue invasion and metastasis. Among them, tissue invasion and metastasis, the final step of most tumors, is responsible for an estimated 90% of the deaths caused by cancer tumors in humans.[2] Acquisition of the invasive and metastatic phenotype confers on tumor cells the capacity to colonize new areas in the body where, at least in principle, nutrients and space are not a limiting factor for tumor growth.[1,3,4] Similar to the primary tumor, the formation of secondary metastatic foci is highly dependent on the other five hallmark acquired capabilities, but with additional capacities.

Metastasis is itself a multistep process that in the case of carcinomas involves basement membrane destruction and local invasion into adjacent tissues, intravasation and survival in the bloodstream, extravasation into distant organs, and finally, their colonization, which requires the capacities of survival plus proliferation at the metastatic site (Fig. 9.1). Invasion is the first step of the metastatic cascade and is also a highly regulated multistep process. Invasion and metastasis are closely allied and extremely complex processes at the mechanistic level. Local invasion of tumor cells involves profound changes in cell adhesion properties, affecting both cell-to-cell and cell-to-extracellular matrix adhesion, in addition to the expression of extracellular proteases. Working in an orchestrated fashion, those changes lead to the remodeling of cell–cell contacts, de-attachment of the cells from their original location in the tissue, degradation of the extracellular matrix, and migration of the tumor cells into adjacent tissues. Thus, during invasion a complete new cellular program needs to be set in motion. Several classes of proteins are involved in the invasion process. Among them, cell adhesion proteins[5–8] and

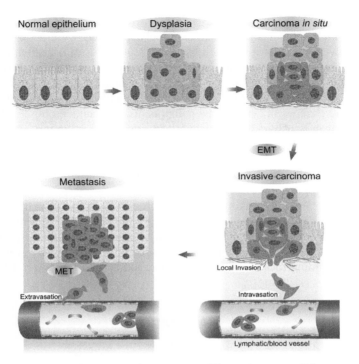

FIGURE 9.1   Metastasis is a multistep process. The major steps of the metastatic process are indicated: local invasion, involving basement membrane destruction and invasion into adjacent tissues; intravasation and survival in the bloodstream; and extravasation into distant organs and proliferation/survival in the new host organ. The specific steps where EMT (epithelial–mesenchymal transition) and the reverse MET (mesenchymal–epithelial transition) process are thought to occur are indicated. (*See insert for color representation of figure.*)

extracellular matrix proteases[9, 10] play key roles. The most relevant molecules involved in cell–cell adhesion and their remodeling during tumor invasion belong to the calcium-dependent adhesion proteins of the cadherin superfamily, with a prominent role for the classic epithelial E-cadherin in carcinomas.[5,6] Integrins, which mediate cell–extracellular matrix interactions, also play an important, although still incompletely understood role in local invasion.[7,8]

9.3 EPITHELIAL-TO-MESENCHYMAL TRANSITION IN DEVELOPMENT AND ITS RELATION TO THE INVASIVE PROCESS

In epithelial tumors (carcinomas), which represent about 90% of all human tumors, the changes in cell behavior accompanying local invasion can be associated with a process known as *epithelial–mesenchymal transition* (EMT). The EMT process was initially characterized as an extremely regulated process during specific stages of development characterized by the loss of cell–cell adhesion, repression of E-cadherin expression, and increased cell mobility[11] (Fig. 9.2). During EMT, epithelial cell layers lose polarity and cell–cell contacts and undergo a dramatic remodeling of the cytoskeleton.[11,12] Concurrent with the loss of epithelial cell adhesion and changes in cytoskeletal components, cells undergoing EMT acquire expression of mesenchymal components and manifest a migratory and, in some cases, invasive phenotype (Fig. 9.2). Epithelial–mesenchymal transition is a highly conserved and fundamental process that governs meaningful events for morphogenesis in multicellular organisms (reviewed in refs. 11 to 13). Cells embedded in an epithelial layer acquire mesenchymal characteristics, become motile, and leave the epithelium. This happens multiple times in the embryo under the control of a range of signaling molecules. EMT in the primitive streak (of amniotes) is the most ancient (in terms of metazoan evolution) of these events and the prerequisite for embryonic body formation. EMT is also essential for numerous developmental processes, including mesoderm formation and neural tube formation. EMTs thus

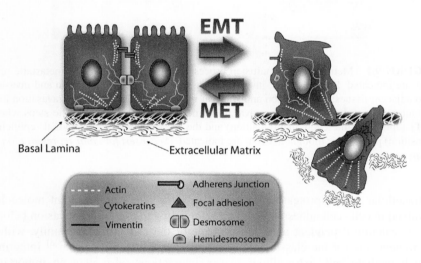

FIGURE 9.2   Epithelial-to-mesenchymal transition and cell plasticity in carcinoma progression. After EMT induction, carcinoma cells decrease epithelial characteristics, lose apical–basal polarity, decrease cell–cell cohesiveness, down-regulate epithelial markers, express mesenchymal markers, and reorganize the actin cytoskeleton. Malignant cells acquire fibroblastic (spindle) morphology and the ability to degrade the basal lamina, invade through the surrounding tissues, and metastasize. Nevertheless, intermediate situations with only partial manifestation of EMT can also be found. EMT is thought to be a transient process; thus, under certain circumstances (i.e., at the site of metastasis) carcinoma cells can reacquire the epithelial phenotype through a reversal mechanism of mesenchymal–epithelial transition (MET).

occur during critical phases of embryonic development in most multicellular animal species. During this transition, mesenchymal cells acquire a morphology that is appropriate for migration in an extracellular environment and for their posterior settlement in areas that are involved in further organ formation. Importantly, developmental EMTs are highly dynamic processes that occur transiently at specific embryonic stages and in a highly regulated spatiotemporal model. In fact, mesenchymal cells participate in organogenesis of the embryo in the formation of specific epithelial organs, such as the kidney, through the reverse process of mesenchymal–epithelial transition (MET).[12] The term *epithelial cell plasticity* has recently been introduced to describe more faithfully the cellular and molecular events governing the remodeling of epithelia and their reversible or irreversible conversion into mesenchymelike cells.[14]

During metazoan development, EMTs are required for morphogenetic movements underlying parietal endoderm formation and gastrulation, as well as during the formation of a range of organs and tissues, such as the neural crest, heart, musculoskeletal system, craniofacial structures, and peripheral nervous system. [11–13] Gastrulation is perhaps the most fundamental developmental process because it leads to establishment of the body plan. In *Drosophila*, genetic studies have identified several genes involved in epithelial cell plasticity during the formation of a furrow in a narrow strip of ventral cells of the blastoderm. The ventral furrow is the site of the invaginating mesodermal and endodermal cells, which first elongate, then shorten, and finally, lose their epithelial morphology, acquiring a fibroblast-like shape. The transcription factors Snail and Twist were identified as key genes in the formation of the ventral furrow.[15] Epithelial cell layer reorganization is even more sophisticated in the amphibian *Xenopus laevis* and involves particular cell movements. Gastrulation begins with radial intercalation in multilayered epithelial sheets, resulting in extension of the layer and its involution, followed by mediolateral intercalation in the invaginated layer. This produces a convergent extension movement, thereby transforming a spherical blastula into an elongated tadpole.[16] Another example of epithelial cell plasticity is provided by the neural crest. In vertebrates, this transient structure arises at the dorsal border of the neural tube. Neural crest cells lose their neuroepithelial morphology, acquiring a migratory phenotype as individual fibroblastlike cells. This EMT is accompanied by a rapid change in the adhesion status,[17] including changes in the expression of cadherin members.[18] Interestingly, the precursor neural crest cells have recently been suggested to share properties with stem cells, supporting the attractive hypothesis that link EMT processes with stemness.[19]

From the tumor progression point of view, clear evidence has emerged in the last years supporting the involvement of EMT in the dissemination of carcinoma cells from the sites of the primary tumors. More generally, EMT might be involved in a de-differentiation program that leads to some malignant carcinoma. In fact, most of the genes that control developmental EMT are expressed during tumor progression (reviewed in refs. 12 and 13). Much of the strong evidence for the association of tumor invasion with EMT comes from studies in cancer cell lines and in animal models,[12,20,21] but the evidence for the occurrence of EMT in human tumors is still scarce, and therefore its actual implication and relevance in human cancer is a matter of debate. [22–24]. One of the main reasons for this is that EMT might not be easily detected in time and space in human tumors, considering that EMT may occur only transiently and in reduced groups of cells or even in isolated cells of tumor-invasive areas.[12,21,25] In fact, a common observation of pathologists is that established metastasis of most carcinomas reacquire the differentiation and morphological appearance of the primary tumor, supporting the transient and dynamic nature of EMT (Figs. 9.1 and 9.2). However, in vitro, various carcinoma cell lines indeed undergo stable EMT in response to different external stimuli or expression of specific genes.[26,27] Moreover, carcinomas that show a diversity of phenotypes and malignant potential can lose most of their epithelial characteristics during tumor progression. A clear example of that situation is represented by carcinosarcomas, mixed tumors with clear sarcomatous and carcinomatous regions that, although generally infrequent, can be highly aggressive and metastatic. We discuss this point further in the last part of the chapter.

In conclusion, the present cumulative knowledge supports the interpretation that the accumulation of genetic mutations in tumor cells, so far considered as the driving force in tumorigenesis, does not necessarily lead to a unidirectional path in tumor progression, but this might rather be considered as a highly regulated process involving dynamic changes in the tumor tissue architecture, more like the morphogenetic process in embryos.[12,28] In this new scenario, EMT and epithelial plasticity might well represent clear manifestations of tumor tissue remodeling at specific stages of tumor progression.

9.4 REGULATION OF EMT: FROM SIGNALS TO MOLECULAR PATHWAYS

In general terms, the EMT process involves setting in motion a complex genetic program that in its simple version will lead to the repression of epithelial markers and to the induction of mesenchymal and migration-related markers. Although some important cellular and molecular aspects of EMT have emerged in recent years, several central questions remain to be resolved: What signals govern EMT in different physiological and pathological contexts? Do the various signals converge on similar or distinct regulatory factors and pathways? Which genetic programs underly EMT? Which programs and signals, if any, are involved in local tumor invasion?

It has become clear that EMT can be induced by different agents/signals or combinations thereof, often dependent on particular cellular models (reviewed in refs. 26 and 27). It is also becoming evident that depending on particular cellular or tissue contexts, EMT can adopt a reversible or stable fate, thus enhancing or repressing to different levels progression of the epithelial cell's gene expression program toward a mesenchymal phenotype. These evolutionarily conserved mechanisms are present in both development and cancer progression, modulating similarly conserved molecular machines that govern both cellular polarity[29] and cellular motility.[30]

The original discovery by Michael Stoker and Michael Perryman in 1985[31] that embryonic fibroblast culture supernatant contained a scatter activity for epithelial Madin–Darby canine kidney (MDCK) cells pointed to the characterization of hepatocyte growth factor [HGF; also known as scatter factor (SF)], the ligand of the c-met receptor, as an inducer of EMT. Evidence for the participation of HGF/c-met in invasion and metastasis has accumulated over recent years and recently been reviewed.[32,33] At present, several other growth factors recognizing tyrosine–kinase surface receptors (RTKs) are known to induce EMT in epithelial cell lines[26,27,34] (Fig. 9.3A). Interestingly, many of the growth factors that have the ability to induce EMT in cell cultures have also been found to be essential for EMT in embryos, calling for the conservation of basic EMT signal inducers. Strong genetic evidence for the direct involvement of some factors, such as FGFs or TGF-β/BMPs, as EMT inducers in invertebrate and mammalian systems has been provided in recent years (reviewed in refs. 12, 13 and 35).

In the majority of epithelial cell types and in transgenic mouse tumor models, TGF-β signaling, one of the main EMT inducers, cooperates with oncogenic Ras or RTKs to cause EMT and metastasis.[26] Besides TGF-β, EGF family members, FGFs, HGF, and IGF1/2 also contribute to EMT via autocrine production, at least in some contexts.[12] In many circumstances, however, physiological and pathological EMT is induced primarily by paracrine production of RTK ligands, secreted by surrounding fibroblast or tumor stroma components rather than by normal or tumoral epithelial cells.[26,27,33] The latter observations highlight the importance of the microenvironment in the modulation of EMT and epithelial cell plasticity in both normal and pathological situations.[36] Importantly, signaling pathways essential for stem cell function and early development, such as Wnt/β-catenin, Notch and Hedgehog signaling, have a major impact on EMT during development and cancer progression.[26] The essential role of the canonical Wnt pathway is to control the cytoplasmic levels of β-catenin through a signaling cascade that involves inhibition of degradation of non-membrane-associated β-catenin. Free cytoplamatic β-catenin can thus translocate to the nucleus, where it interacts with Tcf (T-cell factors), transcription factors activating gene transcription.[37] A plethora of Wnt/β-catenin target genes has been described in various systems in recent years, some of them involved in stemness, others apparently related to EMT and tumor progression.[28] Interestingly, Wnt/β-catenin and TGF-β/Smad signaling can also cooperate via several mechanisms in the induction of EMT.[26] Like TGF-β signaling, the Notch pathway can act in a tumor-suppressive or tumor-promoting fashion, depending on the cellular context and whether cooperating (proto)-oncogenes are present. Recently, the Notch pathway has been shown to collaborate with TGF-β in the induction of Snail,1[38] a known repressor of E-cadherin and inductor of EMT (see below).

Activation of the various pathways described above induces the loss of cell polarity and down-regulation of E-cadherin. Loss of functional expression of

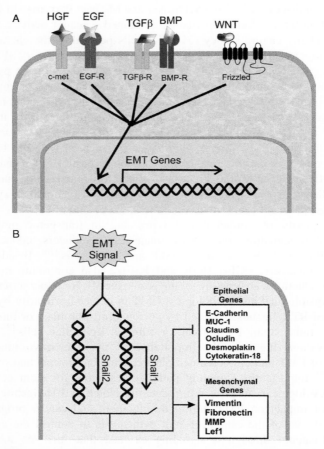

FIGURE 9.3 Different signaling pathways can induce EMT and their mediators. A. Representation of several growth factors and their receptors involved in the regulation of EMT through the direct or indirect induction of EMT-related genes. Some of the signaling pathways indicated can cooperate in the induction of EMT in specific cell contexts (e.g., TGF-β and Wnt or TGF-β and RTK mediated signals[14,26,27]). B. EMT signals lead to the expression of transcription factors implicated in regulation of the genetic program underlying EMT, such as Snail family factors Snail1 and Snail2. The basic genetic program regulated by those factors leads to the down-regulation of epithelial genes and up-regulation of mesenchymal and motility related genes. Whereas some of EMT-related genes are regulated in a similar fashion by Snail1 and Snail2 factors, others are specific to each factor,[86] indicating a distinct role for different EMT regulators and supporting the concept of epithelial cell plasticity.

E-cadherin protein and/or transcriptional repression of its mRNA is presently considered as the hallmark of EMT, both in embryonic development and in cancer progression.[12,21,25] E-cadherin is a central component of cell–cell adherens junctions and is required for the formation of epithelia in the embryo and to maintain epithelial homeostasis in the adult.[18,39] Down-regulation of E-cadherin is a frequent event in most carcinomas, and the evidence accumulated in the last 15 years since its description as a tumor suppressor gene [40–43] has firmly supported its present consideration as a major anti-invasive molecule in carcinomas.[5,20,44] The molecular mechanisms underlying E-cadherin regulation, particularly during tumorigenesis, are reasonably well known at present. During tumor progression, E-cadherin gene expression can be functionally inactivated or silenced by different mechanisms. These mechanisms include somatic mutations (although only in specific tumor types, such as lobular breast and diffuse gastric carcinomas), epigenetic down-regulation through promoter hypermethylation and/or histone deacetylation, and transcriptional repression (reviewed in refs. 25, 45, and 46). In relation to transcriptional regulation, several E-cadherin repressors have been characterized in the last years, acting via interaction with specific E-boxes of the proximal E-cadherin promoter. Importantly, most, if not all of the presently characterized E-cadherin repressors were known as EMT-inducing developmental regulators (reviewed in refs. 13, 25 and 27). The best characterized E-cadherin repressors/EMT inducers are two members of the Snail zinc-finger transcription factors, Snail and Slug, [47–50] presently known as Snail1 and Snail2, respectively.[35] Genetic evidence for the participation of Snail1 in EMT has been obtained. Snail1 mutant mice die at gastrulation as a result of defective EMT and sustained E-cadherin expression.[51] Furthermore, Snail1 has been detected in an increasing number of invasive carcinoma cell lines and, importantly, at invasive areas of some carcinomas (reviewed in refs. 25 and 35; see also below). Other E-cadherin repressors containing zinc-finger domains are SIP1/ZEB-2[52] and δEF-1/ZEB1.[53] SIP1 and Snail1 bind to partly overlapping E boxes in the E-cadherin promoter, with similar repressor effects, and SIP1 is highly expressed in several E-cadherin-deficient human carcinoma cell lines,[52] while δEF-1 is expressed in breast carcinoma cells.[53] In addition, factors of the helix–loop–helix (bHLH) family, E12/E47 (the E2A or TCF3 gene products),[54] and E2-2A/E2-2B (the E2-2 or TCF4 gene products) (V. R. Sobrado and A. Cano, submitted) also show an ability to repress E-cadherin and induce EMT. Interestingly, the E2A protein suppression of the E-cadherin promoter is antagonized by its interaction with the inhibitor of differentiation (Id) proteins Id2 and Id3, at least in cellular context, where they are down-regulated by the EMT inducer TGF-β.[55] Finally, and no less important, the bHLH transcription factor Twist, a protein known to be essential for initiating mesoderm development during gastrulation,[15,56] was also recently characterized as an E-cadherin repressor (by a still unknown mechanism) and inductor of EMT.[57] Importantly, knockdown of Twist expression by RNAi in a metastatic mammary tumor cell line prevented lung metastasis, apparently by specifically inhibiting intravasation,[57] a metastatic step also associated with EMT[12] (see also Fig. 9.1).

9.5 EMT AND CANCER STEM CELLS

The recognition that tumors can be sustained by a rare population of cells with self-renewal properties, thus called *cancer stem cells* (CSCs), has provoked a substantial change in the way scientists are considering tumor biology and its potential implications in clinics. The existence of CSCs has so far been demonstrated in hematological cancers and in an increasing number of solid tumors, such as breast and brain tumors. [58–61] A highly debated and still unresolved issue is whether CSCs are always derived from normal stem cells or whether they can arise from progenitors or more differentiated cells by acquiring specific mutations that endow them with stem cell–like properties, such as self-renewal.[62,63] This specific and fascinating aspect of tumor biology is discussed widely in other chapters in this book. Here, however, we confine our discussion to the relevance of CSCs to tumor progression and the origin of potentially invasive or metastatic cells.

It has long been known that metastasis as a whole is a highly inefficient process, with estimates that only about 1 in 10^7 cells in the primary tumor are able to accomplish successfully all the steps of the metastatic process.[3,4] This inefficiency was assumed to be due, at least in part, to the requirement of acquisition of additional mutations and/or acquired capacities to develop the full metastatic phenotype. However, present genetic evidence has challenged the classical view of metastasis as a late acquired phenotype. Several genetic profiling studies of different tumor series have identified a *metastatic gene signature* or *poor prognosis signature* inside primary tumors, supporting the view that metastatic potential can be an early acquired character in tumor development[64,65] (reviewed in ref. 66). It has also been shown that expression of additional specific genes is required for tumor cells to exhibit full and site-specific metastatic potential, at least for breast tumors.[67,68] Recently, in respect to the current thinking on CSCs, the inefficiency of the metastatic process has been reviewed; it has been hypothesized that metastasis can originate from certain subpopulations of CSCs that acquire a poor prognosis signature.[66] Other authors have speculated that acquisition of migratory properties is an essential requirement for CSCs to become metastatic and proposed the term *migrating cancer stem cells* (MCSs) to define this specific subpopulation of CSCs.[69] MCSCs are indeed proposed as an integrated concept for malignant tumor progression.[69] The MCSC concept has been derived mainly from observations in colorectal carcinomas and is closely linked to the presence of transient EMT in the invasive tumor/stroma interphase areas.[28] Implicit in the existence of MCSCs is the distinction between two types of CSCs: stationary CSCs (SCSCs), located inside the tumor mass, and migratory CSCs (MCSCs), located at the tumor–stroma interphase. An important implication of this model is that MCSCs are derived from CSCs by acquisition of transient EMT.[69] The MCS model is certainly attractive and well supported by data from colorectal tumors. Undoubtedly, it will be interesting to test/confirm its existence in other human tumor types. In this context, it is important to mention that other authors have recently hypothesized that reversible EMT/MET processes might be a functional manifestation of cell plasticity underlying fundamental transitions between "normal" stem cells (embryonic or adult) and cancer stem cells.[19] One of the most interesting aspects of the MCSC concept is its link with EMT/epithelial plasticity and thus with tumor invasion.

9.6 CAN STEM CELL PROPERTIES BE EXTENSIVE TO INVASIVE TUMOR CELLS?

In our opinion, one essential question in the field is: What is the link, if any, between EMT and stem cell biology? Some recent studies tend to support a link. As mentioned above, important signaling pathways involved in stemness also act as potent EMT inducers in different cellular or biological contexts. Importantly, this applies to Wnt/β-catenin and Notch pathways. In colorectal tumors, active Wnt/β-catenin signaling appears to be closely associated with cells at the invasive front with apparent EMT, as indicated by specific nuclear localization of β-catenin in those de-differentiated cells.[70] A similar situation has been observed in endometrioid carcinomas, but its extension to other tumor types has to be established. Essential properties of stem cells are self-renewal and survival capacity against different external stresses, coupled with their low proliferation potential.[63] Therefore, it is pertinent to ask if any of those properties are present in invasive or metastatic cells. Survival is indeed an essential character for tumor cells in the context of metastasis, a fact recognized many years ago. Moreover, in relation to the invasive process, resistance to apoptosis induced by de-attachment from the extracellular matrix (anoikis) is one of the requirements for the success of the process.[71] Indeed, recent data support that survival, considered as a manifestation of the inhibition of cell death, might be a crucial aspect for the entire metastatic process.[72]

Some new data regarding EMT regulators are also raising interesting ideas on the potential link between invasion and stemness. The Snail members Snail1 and Snail2 have recently been described to confer additional properties to EMT, related to proliferation and survival.[35] Thus, Snail1 induces a G1/S cell cycle arrest due, at least in part, to its ability to repress cyclin D2 expression; importantly, Snail1 appears to perform this function in cell culture and in vivo during mouse development.[73] This property could, in fact, confer on cells a low proliferative potential, while permitting their migration in response to specific external cues.[73] Another important function recently recognized for Snail1 and Snail2 is their active participation in protection against cell death induced by external stimuli, including absence of survival factors, presence of TNF-α, and resistance to genotoxic agents, including γ-irradiation and chemotherapeutic agents. [73–76] Interestingly enough, the cell survival properties appear to be shared by all Snail superfamily members from different phyla, supporting the hypothesis that protection against cell death might be a general property of Snail family (reviewed in ref. 35). Importantly, Snail1 has also recently been implicated in the recurrence of primary breast carcinomas.[77] In agreement with this, stable silencing of Snail1 in highly aggressive mouse epidermal carcinoma cell lines induces a dramatic reduction of tumor growth potential.[78] Pending further confirmation in additional human tumors, the present results support the speculation of a role for Snail1 in self-maintenance of the pool of recurrent cells,[79] thus linking Snail1 to CSC biology.

Acquisition of survival properties (e.g., survivin expression) and reduced proliferation potential has been observed in invasive areas of colorectal carcinomas where EMT is thought to occur.[28,70,80] Interestingly, survivin is a Wnt/β-catenin target,[81,82] directly linking at least some aspects of stemness and invasion. In this context it is also worthwhile to mention that Snail2 has been reported as a Wnt/β-catenin target in colorectal carcinoma cells, where it induces transient EMT associated with functional down-regulation of E-cadherin-mediated contacts.[83]

Taken together, all these data seem to support a link between EMT and CSC in tumor progression.

9.7 IS THERE A UNIQUE EMT PROGRAM LINKED TO INVASION?

The present evidence derived from different model systems supports the notion that invasion is a genetic program regulated by transcription factors (reviewed in ref. 84). As already discussed, EMT is presently considered a manifestation of epithelial cell plasticity that can adopt particular or specific aspects, depending on cell/tissue context and external signals. The minimal characteristics defining the EMT processes are loss of stable epithelial cell–cell adhesions and polarity and acquisition of mesenchymal/migratory properties (Fig. 9.2). These properties should be manifested regardless of EMT occurrence as a transient or stable process. As also discussed earlier, we and other authors favor the notion that a transient EMT process is one of the main driving forces for carcinoma invasion.[12,14,21,25,69] But, as outlined above, many questions remain to be answered before a comprehensive understanding of tumor progression can be obtained. One of the open questions we have tried to address recently is whether different E-cadherin repressors drive EMT and the invasive process by inducing similar or distinct genetic programs. In addition, whether the different factors play a specific or redundant role in tumor progression also underlies that question. To approach this issue we have based our studies on a cellular model, the prototypical epithelial MDCK cells that are able to undergo a stable EMT process after expression of various E-cadherin repressors: Snail1, Snail2, or the bHLH E47 factor.[48,49,54] The morphological appearance of the transfected MDCK cells with either factor is very similar, as are most analyzed cellular parameters: complete loss of E-cadherin and other epithelial markers, acquisition of mesenchymal markers (vimentin, fibronectin), and motility and migratory properties (reviewed in ref. 25). However, a closer inspection of both biochemical parameters and more important, in vivo invasion assays, using transplantation chambers of three-dimensional cultures into nude mice indicated important differences in the biological behavior of MDCK cells expressing Snail1 or E47 factors.[85] Whereas Snail1-expressing cells showed highly aggressive invasiveness in vivo, E47-expressing cells invaded much more slowly, but induced a faster and stronger angiogenic response in the host than Snail1-expressing cells.[85] These observations clearly suggested that different genetic programs might be driven by each factor, despite their ability to induce apparently similar EMT processes in the same cellular context. This suggestion has recently been confirmed by genomic studies. Genetic profiling analysis of MDCK cells expressing Snail1, Snail2, or E47 has indicated that of 250 differentially expressed genes, compared to control MDCK cells, only 20% are regulated in a common fashion by the three factors; the majority of genes are regulated by different binary combinations or by only one of the three factors.[86] EMT-related genes are detected among the group of common as well as differentially regulated genes (Fig. 9.3B), suggesting that specific or particular aspects of the EMT program can be undertaken by each specific factor.[86] Studies by other groups related to the gene expression pattern conferred by transient Snail1 expression in colorectal carcinoma cells also support such alteration of epithelial phenoptype as being a main trait.[87] Regarding Snail2, apart from its participation in EMT, it has recently been reported to be required in hematopoietic cells for the functional activity of c-kit signaling, an important contributor to hematological stemness.[88] Interestingly, a recent genomic analysis in fibroblasts has identified additional Snail2 target genes, potentially related to stemness.[89] Whether the common or specific gene expression patterns defined in our recent study, induced by different EMT mediators, can confer a CSC or MCSC "character" is being analyzed presently. Regardless of this important point, our recent genomic study using MDCK cells provides evidence for a differential role of distinct transcription factors acting as EMT mediators that might be important for tumor progression. Of further interest in this genetic analysis were important unsuspected differences in the gene expression patterns driven by highly related members of the Snail family, Snail1 and Snail2.[86] This strongly supports similar yet distinct roles for these transcription factors in tumor progression.

Based on our own results and data accumulated by many groups, we recently proposed a hierarchical model for the action of EMT factors in tumor progression.[25,45] In that model, Snail1 was proposed as the main initiating actor of invasion by inducing local EMT, while Snail2 and E47 were proposed to play a role in maintainance of the migratory phenotype farther away from the invasive front. Recent observations of limited Snail1 expression at the tumor–stroma interface[90] strongly support our hierarchical model of action. As mentioned above, the relation of the different factors to CSC/MCSC remains to be established. In other words, it should be important to determine in different carcinomas whether tumor cells undergoing EMT correspond to CSC/MCSC or if EMT is a paracrine response of epithelial tumor cells to specific factors that can be liberated by CSCs or the tumor microenvironment. In this context it is interesting to mention that Snail1 expression is induced under certain situations that lead to fibroblast activation under normal or pathological situations.[27,35,90]

There is an additional crucial question in relation to EMT, invasion, and tumor progression that perhaps matters more to both basic oncologists and pathologists:

does EMT really occur and to what extent in human tumors? We discussed above the evidence for EMT in colorectal carcinomas; in the next section we expand on this issue in other tumor types.

9.8 EVIDENCE OF EMT IN HUMAN CLINICAL TUMORS

As mentioned previously, the role of EMT in human carcinoma progression is a matter of intense debate.[22–24] One of the major criticisms in accepting a role for EMT in tumor progression and metastasis is that the events defining a full EMT process in vitro (loss of epithelial characteristics, expression of mesenchymal markers, acquisition of spindle/mesenchymal morphology, and the pattern of invasion as single cells) are rarely observed together in human carcinomas. One exception is human carcinosarcoma, which has been suggested to be a true example of complete EMT.

Carcinosarcomas are uncommon but aggressive neoplasias with biphasic histology of carcinomatous and sarcomatous elements. They develop in different organs, preferentially in the endometrium, breast, lung, kidney and upper aerodigestive tract, but rarely in other locations, such as the colon.[91,92] In most organs, the carcinomatous element consists of adenocarcinoma, but in others (e.g., in the esophagus) the epithelial malignant element corresponds to squamous cell carcinoma.[93] The sarcomatous component may be any type of sarcoma, such as fibrosarcoma, leyomiosarcoma, and rhadomyosarcoma. Because the tumors have an intimate mixture of histologically diverse malignant cells, the histogenesis of carcinosarcomas has long been a matter of debate. Thus, three main theories have been proposed: collision theory, composition theory, and combination tumor theory.[94] The first considers these tumors synchronous biclonal tumors that blend together to form carcinosarcomas. The second hypothesis considers the stromal elements reactive and nonneoplastic. However, this theory has been abandoned since metastasis can contain both malignant epithelial and malignant mesenchymal components. Finally, the third and most favored theory postulates that carcinomatous and sarcomatous elements both originate from a single stem cell clone. According to this theory, the sarcomatous component arises within a carcinoma through evolution of specific subclones.

A number of recent cell culture, immunohistochemical, and molecular genetics studies support the monoclonal nature of these neoplasms. Thus, cell lines established from carcinosarcomas have been shown to differentiate into epithelial, mesenchymal, or both components.[95] Furthermore, immunohistochemical studies have documented the expression of epithelial markers in the sarcomatous component of a large proportion of cases.[96] More recently, p53 mutational analyses and loss of heterozygosity (LOH) studies have all shown that carcinomatous and sarcomatous elements share common genetic alterations.[94,97–99] Interestingly,

Gorai et al. (1997) suggested that uterine carcinosarcomas may originate from a precursor (stem) cell.[100]

Unfortunately, at present, we do not know the molecular events that trigger stable EMT in human carcinosarcomas, since the role of E-cadherin repressors and other previously discussed pathways involved in EMT has not yet been analyzed. Although inactivation of p53 (by mutation and LOH of 17p) appear to be frequent in the pathogenesis of human carcinosarcomas in different organs,[98] the relevance of this alteration in the genesis of EMT has not yet been explored.

Apart from carcinosarcomas, another suggested example of full EMT in human cancer is the possibility that stromal components surrounding carcinoma cells (such as myofibroblasts) might originate directly from tumoral cells. In agreement with this, peritumoral stromal cells frequently exhibit the same genetic alterations (e.g., LOH and mutations in the *p53* gene) as those exhibited by the adjacent carcinoma cells, suggesting a common origin for both cell types.[101,102]

However, these extreme examples showing the conversion of carcinoma into mesenchymal cells might not completely reflect the occurrence of EMT in human tumors. Evidence derived from in vitro cell culture studies indicates that the ability to undergo complete EMT varies among different cell lines even when they are exposed to the same stimuli, such as treatment with TGF-β.[14,103] This suggests that complete EMT might not be easy to achieve, but partial EMT may be sufficient for certain cell lines to acquire an aggressive migratory phenotype. Therefore, it has been suggested that expression of mesenchymal markers, loss of epithelial markers (E-cadherin), and/or cadherin switching may be considered independently as signs of partial EMT.[14]

9.9 EXPRESSION OF MESENCHYMAL MARKERS AND CADHERIN SWITCHING IN CARCINOMAS

Although loss of E-cadherin function is a critical event for EMT, the expression of mesenchymal markers is probably required to enable invasive cells to interact with interstitial matrices and to sustain local invasion. In this sense it is interesting to note that a variable number of carcinomas that do not develop a full morphological EMT (and are not considered carcinosarcomas) express specific mesenchymal markers, such as vimentin or smooth muscle actin (SMA). One typical example of that situation is breast carcinoma, in which focal or diffuse vimentin and SMA expression, reported in about 10 to 15% of cases, is generally associated with poor prognosis.[104,105]

Cadherin switching from the prototypical epithelial type (E-cadherin) to mesenchymal types (e.g., N-cadherin) is also a character frequently associated with the EMT processes.[6,12] Different studies have demonstrated the inappropriate expression of nonepithelial cadherins by epithelial cells as a putative mechanism for promoting the interaction with the stroma, thereby facilitating invasion and metastasis. Several cancer cell lines (e.g., human bladder, melanoma, and breast cancer cell lines) that lack E-cadherin-mediated cell–cell adhesion show expression of mesenchymal cadherins, such as N-cadherin and cadherin-11 (reviewed in ref. 106). This phenomenon of cadherin switching has also been observed in human tumors, such as melanoma, prostate, and breast cancer.[107]

In normal human skin, E-cadherin is expressed on the cell surface of keratinocytes and melanocytes, being the major adhesion molecule between both epidermal cell types, while N-cadherin is expressed by dermal fibroblasts and endothelial cells.[108] During melanoma development, a progressive loss of E-cadherin expression has been observed. Disruption of E-cadherin-mediated cell adhesion frees the melanocytic cells from microenvironmental control, whereas restoration of E-cadherin expression in melanoma cells results in keratinocyte-mediated growth control and down-regulated expression of invasion-related adhesion receptors.[109] N-Cadherin seems to play a dual role in cell–cell adhesion in melanoma since it mediates homotypic aggregation between melanoma cells as well as heterotypic adhesion of melanoma cells to dermal fibroblasts and vascular endothelial cells. Together, these results support a role for E- to N-cadherin switching during melanoma development, participating in growth control and possibly providing metastatic advantages to melanoma cells.[110]

The role of N-cadherin expression in tumor progression has also been demonstrated in prostate carcinoma, where it is correlated with clinical parameters including the Gleason score (a measure of tumor differentiation). Interestingly, N-cadherin is found to be expressed in a subset of prostate tumors, which showed mainly aberrant or negative E-cadherin staining,[111,112] reinforcing the concept of the cadherin switch. On the other hand, cadherin-11 (normally expressed in the brain) is expressed in the stroma of all prostatic tumors analyzed at the epithelial–stromal interphase. In addition, cadherin-11 is also expressed in a dotted pattern or at the membrane of the epithelial cells of high-grade cancers. In metastatic prostate cancer, N-cadherin and cadherin-11 are expressed homogeneously.[112] Similarly, the expression of N-cadherin and cadherin-11 in breast cancer has been associated with more aggressive phenotypes.[113,114]

In general terms, present evidence tends to support the notion that the cadherin switching observed in many carcinomas can be interpreted as an additional trait of cell plasticity but may not be sufficient to support overt manifestation of EMT in tumors. Interestingly, in many experimental situations that lead to a complete EMT, E-to-N-cadherin switching also occurs. This is the case in MDCK cells overexpressing Snail1 or Snail,2[86] suggesting that these factors can also induce the expression of mesenchymal-type cadherins. Nevertheless, the action of those factors as direct transcriptional activators of mesenchymal genes has not been yet demonstrated. Regardless of the mechanistic aspects of the action of specific EMT-inducing factors, analysis of their expression in human tumors has attracted great interest in recent years.

9.10 EXPRESSION OF EMT INDUCERS IN HUMAN TUMORS

The expression of several EMT inducers, such as Snail1, Snail2, Twist, and SIP1, has been demonstrated in diverse human tumors using different methodologies. Generally, the expression of these transcription factors is correlated with E-cadherin down-regulation [115–119]. As discussed earlier, based on experimental and developmental studies, we proposed a hierarchical model for the action of different repressors during tumor progression.[25,45] However, the exact contribution of these transcription factors in tumorogenesis in vivo is not fully understood.

Expression of Snail1 mRNA has been detected in biopsies from patients with breast,[77,119–124] ovarian,[121] gastric,[116] colon,[125,126] and hepatocellular carcinomas.[117] Of these reports, only one analyzed Snail1 expression directly in epithelial neoplastic cells. Thus, the study by Blanco et al. (2000) using in situ hibridization reported that mRNA expression was detected at the invasive areas and more frequent among poorly differentiated breast carcinomas with lymph node metastasis.[119] As mentioned before, in breast cancer a role for Snail1 in tumor recurrence has also been proposed.[77] Comparative transcriptome analysis of human primary breast cancers suggests that elevated Snail1 expression is correlated with decreased relapse-free survival.[77] In addition, Toyama et al. observed that node-negative infiltrating ductal carcinomas showing low Snail1 mRNA expression do not developed distant metastasis, whereas 12%, showing high expression of Snail1 mRNA, did.[124] Collectively, these data suggest that Snail1 expression is a poor prognostic factor in breast carcinomas and is probably involved in local recurrence and in lymph node and distant metastasis.

Additionally, it has been reported that Snail1 expression in colon cancer was associated with down-regulation of E-cadherin and *VDR* (vitamin D receptor) gene products.[125,126] These findings suggest that Snail1 may be associated with loss of responsiveness to vitamin D analogs and may thus be used as an indicator of patients who are unlikely to respond to this therapy.[127] Another study reported Snail1 mRNA overexpression in 16% of hepatocellular carcinomas compared with adjacent noncancerous liver tissue.[117] E-Cadherin protein expression was found significantly down-regulated in cases with Snail1 mRNA overexpression. In addition, the tumor and nontumor ratio of Snail1 mRNA correlated independently with tumor invasiveness. This indicates that Snail1 both down-regulates E-cadherin expression and promotes invasion in human hepatocelular carcinoma.[117] Finally, up-regulation of Snail1 associated with reduced or negative E-cadherin expression has also been reported in diffuse gastric cancer.[116,128]

Despite these data, the exact role of Snail1 in clinical tumor invasion and progression remains to be established. In most of the aforementioned studies, the precise contribution of stromal and carcinoma cells to the expression of Snail1 was not assessed. In fact, it has been demonstrated that Snail1 is expressed by tumor stromal fibroblasts[90] and endothelial cells.[129] A recent study using immunohistochemical analysis of Snail1 detected this protein only in occasional neoplastic epithelial cells situated at the edge of the tumor mass. Some of them had an undifferentiated phenotype while others retained an epithelial phenotype.[90] Analysis of E-cadherin in these cells demonstrated that although most of the Snail-positive cells were negative for E-cadherin, others coexpressed both proteins and may represent cells at the initial stages of the EMT, where E-cadherin gene transcription has been switched off but the protein remains.[90] These results fit nicely in our hierarchical model of Snail1 action described previously.[25]

Regarding Snail2, initial in vitro studies suggested that Snail2 is a likely in vivo repressor of E-cadherin in breast cancer[50] and participates in the metastatic potential of melanoma.[130] In human clinical samples, although the expression of Snail2 has been explored only occasionally, some data suggest that Snail2 expression might be an indicator of poor prognosis. Thus, a high expression of Snail2 mRNA in lung cancer tissue was significantly associated with postoperative relapse and shorter patient survival.[131] In addition, positive expression of Snail2 was found in 58% of colorectal carcinomas. The positive expression of Snail2 was significantly associated with higher Dukes stage and distant metastasis and had a negative impact on overall patient survival.[132] In breast cancer, increased Snail2 expression was associated with metastatic disease or tumor recurrence.[123]

With regard to the expression of other EMT-mediating transcription factors in human tumors (e.g., SIP1, Twist), little information is currently available. SIP1 expression has been analyzed in gastric and gynecological cancer and in squamous cell carcinomas.[115,116,121] In 20 intestinal-type gastric cancer samples, reduced E-cadherin expression was found in 60% of the cases, which correlated with up-regulation of SIP1 at the mRNA level. However, in this study, SIP1 overexpression could not be linked to down-regulated E-cadherin in diffuse-type gastric cancer.[116] In ovarian cancer, expression of SIP1 correlated with worse outcome.[121] Twist was studied in primary human gastric, breast, and colon cancers. In diffuse gastric cancer, overexpression of Twist correlated with N-cadherin expression, but this association was not found in intestinal gastric cancer or in colon cancer.[116,133] In breast carcinomas, Twist overexpression was associated with poor outcome.[123] Additionally, Twist expression was up-regulated in lobular breast cancer,[57] a tumor characterized by the lack of E-cadherin expression. Since lobular carcinomas invade in a distinctive diffuse pattern (with cells infiltrating Indian-file or as single cells), the authors suggest that in this tumor type, Twist may be involved in silencing E-cadherin and promoting this EMT-like invasion pattern. However, there is evidence contrary to this assumption. First, lobular tumors consistently express cytokeratins (even in metastsic lesions)[91] but rarely mesenchymal markers such as vimentin.[134] Second, E-cadherin gene is irreversibly inactivated by genetic (mutations, allelic loss) and epigenetic (promoter hypermethylation) alterations in lobular breast cancer.[135] Third, a recent report with a larger series of tumors ($n = 144$) failed to find such an association between Twist expression and the lobular histotype.[136] In contrast, studies in prostate cancer appear to support a role for Twist in this tumor type. Thus, Twist was found highly expressed in the majority (90%) of prostate cancer tissues but only in a small percentage (6.7%) of benign prostate hyperplasia. In addition, the Twist expression levels were positively correlated with Gleason grading and metastasis, indicating its role in the development and progression of prostate cancer.[137]

One important drawback on the studies discussed above is that most of them rely on detection of the factors by RT-PCR (or qRT-PCR) on the whole-tumor samples or on antibodies, the specificity of which has not been clearly established. Provided that EMT can be transient and/or spatiotemporally restricted in most carcinomas, additional studies in a larger series of tumors using highly specific reagents are clearly needed to establish the role of the various EMT factors definitively. It is also possible that overt EMT might not always be a general mechanism for a carcinoma to invade and metastasize. In fact, many carcinomas invade as multi-cellular aggregates and sometimes form glandular structures resembling normal epithelium, in which the carcinoma cells retain epithelial characteristics, including adherens junctions and apical–basal polarity.[22,23] Therefore, these cells can invade and metastasize in a process known as *collective migration*.[138] Thus, it is reasonable to assume that EMT may preferentially occur in specific biological contexts.

9.11 OCCURRENCE OF EMT IN A SPECIFIC SUBSET OF BREAST CARCINOMAS

Breast cancer is a heterogeneous disease encompassing a wide variety of pathological entities and a range of clinical behavior. Recent cDNA microarray studies have demonstrated that breast cancers can be classified according to their gene expression profiles into four main groups: basal-like, luminal, HER2$^+$, and normal breastlike breast carcinomas.[139] Most important, these groups have been shown to be of prognostic and predictive significance.[140] Basal-like breast carcinomas are high-grade tumors, lack estrogen receptor (ER) and HER2 expression, and are consistently positive for basal keratins and/or other markers normally expressed by basal or myoepithelial cells (p63, CK5/6, CK14, EGFR, CD10).[141]

Recent studies tend to indicate that in breast carcinoma the expression of mesenchymal markers (e.g., vimentin, SMA, SPARC) occurs in a specific subtype, the basal-like phenotype.[142,143] This finding might not be surprising considering that normal myoepithelial cells also express some of these markers (e.g., vimentin, SPARC). However, we have observed in a series of 478 invasive ductal carcinomas that basal-like tumors also frequently express N-cadherin and cadherin-11,[152] supporting the association of the mesenchymal phenotype with basal-like tumors. It is also interesting to note that tumors with a basal-like phenotype that express mesenchymal markers have a tendency to produce visceral rather than bone metastasis.[144]

Taken together, the observations above suggest the possibility that in breast tumors the acquisition of an invasive and mesenchymal phenotype might occur preferably in a specific genetic context such as that associated with the basal phenotype. In agreement with this hypothesis, the presence of tumoral areas showing spindle morphology is significantly more frequent in tumors with that phenotype.[145] Additionally, recent studies indicate that breast carcinosarcomas also express myoepithelial markers (e.g., CK5/6, P-cadherin, EGFR), and they are therefore considered to have a basal-like phenotype.[146,147,152] (D. Sarrió et al., unpublished results). Importantly, the recent isolation and characterization of breast stem cells has demonstrated that these cells show an immunophenotype very similar to that observed in basal-like breast carcinomas.[148] Moreover, basal-like tumors are extremely frequent among *BRCA1* (breast Cancer associated gene 1) germline mutation carriers. It has therefore been suggested that these tumors may originate from a pluripotent stem cell rather than differentiated epithelial (luminal) cells (reviewed in ref. 149).

In the normal breast, stem cells are localized in a suprabasal position between luminal and myoepithelial layers, have self-renewal capacity, and have the potential to produce both epithelial and myoepithelial lineages.[150,151] Recent experiments demonstrated that the presence of these cells within breast carcinomas increase tumor aggressiveness,[58] thus providing a link between stem cells, EMT processes, and invasive tumor behavior.

9.12 CONCLUSIONS AND FUTURE PERSPECTIVES

The information accumulated in recent years supports the concept that EMT represents a manifestation of epithelial cell plasticity with important implications for tumor progression. One of the major lessons we are learning regarding EMT is its transient and dynamic nature, which may also underlie dynamic changes in tumor tissue architecture at different stages of the metastatic process. Thus, much of the controversy regarding the actual occurrence of EMT in human tumors might derive from the recognition that stable EMT is probably an infrequent event in most carcinomas but that in some specific tumor types can be detected as carcinosarcomas. More important appreciation of the concept of epithelial cell plasticity can have additional implications as to the origin of metastatic cells. The potential link between the concept of migratory cancer stem cells and reversible EMT/MET processes is certainly a very attractive hypothesis that deserves to be explored in different carcinoma types.

Here, we have reviewed the information available regarding the regulatory mechanisms of the invasive process with an emphasis on the known transcriptional factors operating as mediators of EMT, and have analyzed the current data supporting their implication in human tumors and CSC biology. The present information appears to support a potential role for some EMT mediators, such as Snail family factors, in invasion and stemness. Furthermore, in breast carcinoma a picture seems to emerge linking the occurrence of EMT/cell plasticity to a specific tumor phenotype, the basal-like phenotype, further supporting the link between EMT and CSC biology. Nevertheless, more experimental data and rigorous analysis of expression of EMT/cell plasticity regulators in large series of human tumors are needed to fully understand and link these processes. It will also be important to determine in different carcinomas whether tumor cells suffering EMT correspond to CSC/MCS or if EMT is a paracrine response of epithelial tumor cells to specific microenvironmental factors. This information can be extremely useful in identifying potential new drug targets to block the most lethal process in the tumor progression, metastasis.

Acknowledgments

The authors thank M. R. MacPherson for editorial assistance. We also thank Dr. F. Portillo and members of Dr A. Cano's and Dr. J. Palacios's laboratories for their helpful discussions. Work at the authors' laboratories is supported by grants from the Spanish Ministry of Education and Science (SAF2007-63051 to A.C., SAF2004-08258 to J.P., and SAF2007- to G.M.B.) and the Spanish Ministry of Health, Instituto de Salud Carlos III (PI050656 to A.C., PI051890 to J.P., and RTICCC, FIS03, C03/10). G.M.B is a junior investigator of the Ramón y Cajal Program (2004); D.S. is supported by BEFI fellowship (01/9132).

REFERENCES

1. Hanahan D, Weinberg RA. The hallmarks of cancer. *Cell.* 2000; 100:57–70.

2. Sporn MB. The war on cancer. *Lancet.* 1996; 347:1377–1381.

3. Bergers G, Benjamin LE. Tumorigenesis and the angiogenic switch. *Nat Rev Cancer.* 2003; 3:401–410.

4. Fidler IJ. The pathogenesis of cancer metastasis: the 'seed and soil' hypothesis revisited. *Nat Rev Cancer.* 2003; 3:453–458.

5. Birchmeier W, Behrens J. Cadherin expression in carcinomas: role in the formation of cell junctions and the prevention of invasiveness. *Biochim Biophys Acta.* 1994; 1198:11–26.

6. Cavallaro U, Christofori G. Cell adhesion and signalling by cadherins and Ig-CAMs in cancer. *Nat Rev Cancer.* 2004; 4:118–132.

7. Guo W, Giancotti FG. Integrin signalling during tumor progression. *Nat Rev Mol Cell Biol.* 2004; 5:816–826.

8. Hood JD, Cheresh DA. Role of integrins in cell invasion and migration. *Nat Rev Cancer.* 2002; 2:91–100.

9. Folgueras AR, Pendas AM, Sanchez LM, Lopez-Otin C. Matrix metalloproteinases in cancer: from new functions to improved inhibition strategies. *Int J Dev Biol.* 2004; 48:411–424.

10. Lopez-Otin C, Overall CM. Protease degradomics: a new challenge for proteomics. *Nat Rev Mol Cell Biol.* 2002; 3:509–519.

11. Hay ED. An overview of epithelio-mesenchymal transformation. *Acta Anat (Basel).* 1995; 54:8–20.

12. Thiery JP. Epithelial–mesenchymal transitions in tumor progression. *Nat Rev Cancer.* 2002; 2:442–454.

13. Nieto MA. The snail superfamily of zinc-finger transcription factors. *Nat Rev Mol Cell Biol.* 2002; 3:155–166.

14. Grunert S, Jechlinger M, Beug H. Diverse cellular and molecular mechanisms contribute to epithelial plasticity and metastasis. *Nat Rev Mol Cell Biol.* 2003; 4:657–665.

15. Leptin M. Twist and snail as positive and negative regulators during *Drosophila* mesoderm development. *Genes Dev.* 1995; 5:1568–1576.

16. Keller R. Xenopus laevis: practical uses in cell and molecular biology. In: Kay K, Peng B, eds. *Methods in Cell Biology.* San Diego, CA: Academic Press; 1991: 61–113.

17. Duband JL, Delannet M, Monier F, Garret S, Desban N. Modulations of cellular interactions during development of the neural crest: role of growth factors and adhesion molecules. *Curr Top Microbiol Immunol.* 1996; 212:207–227.

18. Takeichi M. Cadherins: a molecular family important in selective cell–cell adhesion. *Annu Rev Biochem.* 1990; 59:237–252.

19. Prindull G. Hypothesis: cell plasticity, linking embryonal stem cells to adult stem cell reservoirs and metastatic cancer cells? *Exp Hematol.* 2005; 33:738–746.

20. Christofori G, Semb H. The role of the cell-adhesion molecule E-cadherin as a tumor-suppressor gene. *Trends Biochem Sci.* 1999; 24:73–76.

21. Kang Y, Massague J. Epithelial–mesenchymal transitions: twist in development and metastasis. *Cell.* 2004; 118:277–279.

22. Christiansen JJ, Rajasekaran AK. Reassessing epithelial to mesenchymal transition as a prerequisite for carcinoma invasion and metastasis. *Cancer Res.* 2006; 66:8319–8326.

23. Tarin D, Thompson, EW, Newgreen DF. The fallacy of epithelial mesenchymal transition in neoplasia. *Cancer Res.* 2005; 65:5996–6001.

24. Thompson EW, Newgreen DF, Tarin D. Carcinoma invasion and metastasis: a role for epithelial–mesenchymal transition? *Cancer Res.* 2005; 65:5991–5995.

25. Peinado H, Portillo F, Cano A. Transcriptional regulation of cadherins during development and carcinogenesis. *Int J Dev Biol.* 2004; 48:365–375.

26. Huber MA, Kraut N, Beug H. Molecular requirements for epithelial–mesenchymal transition during tumor progression. *Curr Opin Cell Biol.* 2005; 17:548–558.

27. Thiery JP, Sleeman JP. Complex networks orchestrate epithelial–mesenchymal transitions. *Nat Rev Mol Cell Biol.* 2006; 7:131–142.

28. Brabletz T, Hlubek F, Spaderna S, et al. Invasion and metastasis in colorectal cancer: epithelial–mesenchymal transition, mesenchymal–epithelial transition, stem cells and beta-catenin. *Cells Tissues Organs.* 2005; 79:56–65.

29. Bilder D. Epithelial polarity and proliferation control: links from the *Drosophila* neoplastic tumor suppressors. *Genes Dev.* 2004; 18:1909–1925.

30. Besson A, Assoian RK, Roberts JM. Regulation of the cytoskeleton: an oncogenic function for CDK inhibitors? *Nat Rev Cancer.* 2004; 4:948–955.

31. Stoker M, Perryman M. An epithelial scatter factor released by embryo fibroblasts. *J Cell Sci.* 1985; 77:209–223.

32. Birchmeier C, Birchmeier W, Gherardi E, Vande Woude GF, Met, metastasis, motility and more. *Nat Rev Mol Cell Biol.* 2003; 4:915–925.

33. Boccaccio C, Comoglio PM. Invasive growth: a MET-driven genetic programme for cancer and stem cells. *Nat Rev Cancer.* 2006; 6:637–645.

34. De Craene B, van Roy F, Berx G. Unraveling signalling cascades for the Snail family of transcription factors. *Cell Signal.* 2005; 17:535–547.

35. Barrallo-Gimeno A, Nieto MA. The *Snail* genes as inducers of cell movement and survival: implications in development and cancer. *Development.* 2005; 132:3151–3161.

36. Bissell MJ, Labarge MA. Context, tissue plasticity, and cancer: Are tumor stem cells also regulated by the microenvironment? *Cancer Cell.* 2005; 7:17–23.

37. Willert K, Jones KA. Wnt signaling: Is the party in the nucleus? *Genes Dev.* 2006; 20:1394–1404.

38. Zavadil J, Cermak L, Soto-Nieves N, Bottinger EP. Integration of TGF-beta/Smad and Jagged1/Notch signalling in epithelial-to-mesenchymal transition. *EMBO J.* 2004; 23:1155–1165.

39. Gumbiner BM. Cell adhesion: the molecular basis of tissue architecture and morphogenesis. *Cell.* 1996; 84:345–357.

40. Frixen UH, Behrens J, Sachs M, et al. E-Cadherin-mediated cell–cell adhesion prevents invasiveness of human carcinoma cells. *J Cell Biol.* 1991; 113:173–185.

41. Navarro P, Gomez M, Pizarro A, Gamallo C, Quintanilla M, Cano A. A role for the E-cadherin cell–cell adhesion molecule during tumor progression of mouse epidermal carcinogenesis. *J Cell Biol.* 1991; 115:517–533.

42. Takeichi M. Cadherins in cancer: implications for invasion and metastasis. *Curr Opin Cell Biol.* 1993; 5:806–811.

43. Vleminckx K, Vakaet L Jr, Mareel M, Fiers W, van Roy F. Genetic manipulation of E-cadherin expression by epithelial tumor cells reveals an invasion suppressor role. *Cell.* 1991; 66:107–119.

44. Behrens J, Birchmeier W. Cell–cell adhesion in invasion and metastasis of carcinomas. *Cancer Treat Res.* 1994; 71:251–266.

45. Peinado H, Cano A. Epigenetic regulation of the E-cadherin cell–cell adhesion molecule. In: Esteller M, ed. *DNA Methylation, Epigenetics and Metastasis.* Utrecht, Germany: Springer-Verlag; 2005: 157–190.

46. Strathdee G. Epigenetic versus genetic alterations in the inactivation of E-cadherin. *Semin Cancer Biol.* 2002; 2:373–379.

47. Batlle E, Sancho E, Franci C, et al. The transcription factor snail is a repressor of E-cadherin gene expression in epithelial tumor cells. *Nat Cell Biol.* 2000; 2:84–89.

48. Cano A, Perez-Moreno MA, Rodrigo I, et al. The transcription factor snail controls epithelial–mesenchymal transitions by repressing E-cadherin expression. *Nat Cell Biol.* 2000; 2:76–83.

49. Bolos V, Peinado H, Perez-Moreno MA, Fraga MF, Esteller M, Cano A. The transcription factor Slug represses E-cadherin expression and induces epithelial to mesenchymal transitions: a comparison with Snail and E47 repressors. *J Cell Sci.* 2003; 116:499–511.

50. Hajra KM, Chen DY, Fearon ER. The Slug zinc-finger protein represses E-cadherin in breast cancer. *Cancer Res.* 2002; 62:1613–1618.

51. Carver EA, Jiang R, Lan Y, Oram KF, Gridley T. The mouse snail gene encodes a key regulator of the epithelial–mesenchymal transition. *Mol Cell Biol.* 2001; 21:8184–8188.

52. Comijn J, Berx G, Vermassen P, et al. The two-handed E box binding zinc finger protein SIP1 downregulates E-cadherin and induces invasion. *Mol Cell.* 2001; 7:1267–1278.

53. Eger A, Aigner K, Sonderegger S, et al. DeltaEF1 is a transcriptional repressor of E-cadherin and regulates epithelial plasticity in breast cancer cells. *Oncogene*. 2005; 24:2375–2385.

54. Perez-Moreno MA, Locascio A, Rodrigo I, Dhondt G, Portillo F, Nieto MA, Cano A. A new role for E12/E47 in the repression of E-cadherin expression and epithelial–mesenchymal transitions. *J Biol Chem*. 2001; 276:27424–27431.

55. Kondo M, Cubillo E, Tobiume K, et al. A role for Id in the regulation of TGF-beta-induced epithelial–mesenchymal transdifferentiation. *Cell Death Differ*. 2004; 11:1092–1101.

56. Castanon I, Baylies MK. A Twist in fate: evolutionary comparison of Twist structure and function. *Gene*. 2002; 287:11–22.

57. Yang J, Mani SA, Donaher JL, et al. Twist, a master regulator of morphogenesis, plays an essential role in tumor metastasis. *Cell*. 2004; 117:927–939.

58. Al-Hajj M, Wicha MS, Benito-Hernandez A, Morrison SJ, Clarke MF. Prospective identification of tumorigenic breast cancer cells. *Proc Natl Acad Sci USA*. 2003; 100:3983–3988.

59. Bonnet D, Dick JE. Human acute myeloid leukemia is organized as a hierarchy that originates from a primitive hematopoietic cell. *Nat Med*. 1997; 3:730–737.

60. Huntly BJ, Gilliland DG. Leukaemia stem cells and the evolution of cancer-stem-cell research. *Nat Rev Cancer*. 2005; 5:311–321.

61. Vescovi AL, Galli R, Reynolds BA. Brain tumor stem cells. *Nat Rev Cancer*. 2006; 6:425–436.

62. Bjerkvig R, Tysnes BB, Aboody KS, Najbauer J, Terzis AJ. Opinion: The origin of the cancer stem cell: current controversies and new insights. *Nat Rev Cancer*. 2005; 5:899–904.

63. Reya T, Morrison SJ, Clarke MF, Weissman IL. Stem cells, cancer, and cancer stem cells. *Nature*. 2001; 414:105–111.

64. Ramaswamy S, Ross KN, Lander ES, Golub TR. A molecular signature of metastasis in primary solid tumors. *Nat Genet*. 2003; 33:49–54.

65. van't Veer LJ, Dai H, van de Vijver MJ, et al. Gene expression profiling predicts clinical outcome of breast cancer. *Nature*. 2002; 415:530–536.

66. Weigelt B, Peterse JL, van't Veer LJ. Breast cancer metastasis: markers and models. *Nat Rev Cancer*. 2005; 5:591–602.

67. Kang Y, Siegel PM, Shu W, et al. A multigenic program mediating breast cancer metastasis to bone. *Cancer Cell*. 2003; 3:537–549.

68. Minn AJ, Gupta GP, Siegel PM, et al. Genes that mediate breast cancer metastasis to lung. *Nature*. 2005; 436:518–524.

69. Brabletz, T, Jung A, Spaderna S, Hlubek F, Kirchner T. Opinion: migrating cancer stem cells—an integrated concept of malignant tumor progression. *Nat Rev Cancer*. 2005; 5:744–749.

70. Brabletz T, Herrmann K, Jung A, Faller G, Kirchner T. Expression of nuclear beta-catenin and c-myc is correlated with tumor size but not with proliferative activity of colorectal adenomas. *Am J Pathol*. 2000; 156:865–870.

71. Streuli CH, Gilmore AP. Adhesion-mediated signaling in the regulation of mammary epithelial cell survival. *J Mammary Gland Biol Neoplasia*. 1999; 4:183–191.

72. Mehlen P, Puisieux A. Metastasis: a question of life or death. *Nat Rev Cancer*. 2006; 6:449–458.

73. Vega S, Morales AV, Ocana OH, Valdes F, Fabregat I, Nieto MA. Snail blocks the cell cycle and confers resistance to cell death. *Genes Dev*. 2004; 18:1131–1143.

74. Inoue A, Seidel MG, Wu W, et al. Slug, a highly conserved zinc finger transcriptional repressor, protects hematopoietic progenitor cells from radiation-induced apoptosis in vivo. *Cancer Cell*. 2002; 2:279–288.

75. Kajita M, McClinic KN, Wade PA. Aberrant expression of the transcription factors snail and slug alters the response to genotoxic stress. *Mol Cell Biol*. 2004; 24:7559–7566.

76. Pérez-Losada J, Sánchez-Martin M, Pérez-Caro M, Pérez-Mancera PA, Sánchez-García I. The radioresistance biological function of the SCF/kit signaling pathway is mediated by the zinc-finger transcription factor Slug. *Oncogene*. 2003; 22:4205–4211.

77. Moody SE, Perez D, Pan TC, et al. The transcriptional repressor Snail promotes mammary tumor recurrence. *Cancer Cell*. 2005; 8:197–209.

78. Olmeda D, Jordá M, Peinado H, Fabra A, Cano A. Snail silencing effectively suppresses tumor growth and invasiveness. *Oncogene*. 2007; 26(13): 1862–1874.

79. De Craene B, Berx G. Snail in the frame of malignant tumor recurrence. *Breast Cancer Res*. 2006; 8:105–108.

80. Jung A, Schrauder M, Oswald U, et al. The invasion front of human colorectal adenocarcinomas shows co-localization of nuclear beta-catenin, cyclin D1, and p16INK4A and is a region of low proliferation. *Am J Pathol*. 2001; 159:1613–1617.

81. Kim PJ, Plescia J, Clevers H, Fearon ER, Altieri DC. Survivin and molecular pathogenesis of colorectal cancer. *Lancet*. 2003; 362:205–209.

82. Zhang T, Otevrel T, Gao Z, Ehrlich SM, Fields JZ, Boman BM. Evidence that APC regulates survivin expression: a possible mechanism contributing to the stem cell origin of colon cancer. *Cancer Res*. 2001; 61:8664–8667.

83. Conacci-Sorrell M, Simcha I, Ben-Yedidia T, Blechman J, Savagner P, Ben-Ze'ev A. Autoregulation of E-cadherin expression by cadherin–cadherin interactions: the roles of beta-catenin signaling, Slug, and MAPK. *J Cell Biol*. 2003; 163, 847–857.

84. Ozanne BW, Spence HJ, McGarry LC, Hennigan RF. Invasion is a genetic program regulated by transcription factors. *Curr Opin Genet Dev*. 2006; 16:65–70.

85. Peinado H, Marin F, Cubillo E, et al. Snail and E47 repressors of E-cadherin induce distinct invasive and angiogenic properties in vivo. *J Cell Sci*. 2004; 117:2827–2839.

86. Moreno-Bueno G, Cubillo E, Sarrio E, et al. Genetic profiling of epithelial cells expressing E-cadherin repressors reveals a distinct role for Snail, Slug and E47 factors in epithelial–mesenchymal transition. *Cancer Res*. 2006; 66(19): 9543–9556.

87. De Craene B, Gilbert B, Stove C, Bruyneel E, van Roy F, Berx G. The transcription factor snail induces tumor cell invasion through modulation of the epithelial cell differentiation program. *Cancer Res*. 2005; 65:6237–6244.

88. Pérez-Losada, J, Sánchez-Martin M, Rodriguez-Garcia A, et al. Zinc-finger transcription factor Slug contributes to the function of the stem cell factor c-kit signaling pathway. *Blood*. 2002; 100:1274–1286.

89. Bermejo-Rodriguez C, Perez-Caro M, Perez-Mancera PA, Sanchez-Beato M, Piris MA, Sánchez-García I. Mouse cDNA microarray analysis uncovers Slug targets in mouse embryonic fibroblasts. *Genomics*. 2006; 87:113–118.

90. Franci C, Takkunen M, Dave N, et al. Expression of Snail protein in tumor–stroma interface. *Oncogene*. 2006; 25:5134–5144.

91. Tavassoli FA, Devilee P, eds. Pathology and Genetics: Tumors of the Breast and Female Genital Organs. *WHO Classification of Tumors*. Lyon, France: IARC Press; 2003.

92. Ishida H, Ohsawa T, Nakada H, et al. Carcinosarcoma of the rectosigmoid colon: report of a case. *Surg Today*. 2003; 33:545–549.

93. Hu TH, Chien CS, Hsieh MJ, Lin CC. Carcinosarcoma of esophagus. report of one case and review of the literature. *Changgeng Yi Xue Za Zhi*. 1992; 15:161–166.

94. Fujii H, Yoshida M, Gong ZX, et al. Frequent genetic heterogeneity in the clonal evolution of gynecological carcinosarcoma and its influence on phenotypic diversity. *Cancer Res*. 2000; 60:114–120.

95. Gorai I, Doi C, Minaguchi H. Establishment and characterization of carcinosarcoma cell line of the human uterus. *Cancer*. 1993; 71:775–786.

96. de Brito PA, Silverberg SG, Orenstein JM. Carcinosarcoma (malignant mixed mullerian (mesodermal) tumor) of the female genital tract: immunohistochemical and ultrastructural analysis of 28 cases. *Hum Pathol*. 1993; 24:32–142.

97. Dacic S, Finkelstein SD, Sasatomi E, Swalsky PA, Yousem SA. Molecular pathogenesis of pulmonary carcinosarcoma as determined by microdissection-based allelotyping. *Am J Surg Pathol*. 2002; 26:510–516.

98. Lien HC, Lin CW, Mao TL, Kuo SH, Hsiao CH, Huang CS. p53 overexpression and mutation in metaplastic carcinoma of the breast: genetic evidence for a monoclonal origin of both the carcinomatous and the heterogeneous sarcomatous components. *J Pathol*. 2004; 204:131–139.

99. Matsumoto T, Fujii H, Arakawa A, et al. Loss of heterozygosity analysis shows monoclonal evolution with frequent genetic progression and divergence in esophageal carcinosarcoma. *Hum Pathol*. 2004; 35:322–327.

100. Gorai I, Yanagibashi T, Taki A, et al. Uterine carcinosarcoma is derived from a single stem cell: an in vitro study. *Int J Cancer*. 1997; 72:821–827.

101. Moinfar F, Man YG, Arnould L, Bratthauer GL, Ratschek M, Tavassoli FA. Concurrent and independent genetic alterations in the stromal and epithelial cells of mammary carcinoma: implications for tumorigenesis. *Cancer Res*. 2000; 60:2562–2566.

102. Wernert N, Locherbach C, Wellmann A, Behrens P, Hugel A. Presence of genetic alterations in microdissected stroma of human colon and breast cancers. *J Mol Med*. 2000; 78:B30.

103. Peinado P, Quintanilla M, Cano A. Transforming growth factor beta 1 induces Snail transcription factor in epithelial cell lines: mechanisms for epithelial mesenchymal transitions. *J Biol Chem*. 2003; 278:21113–21123.

104. Domagala W, Lasota J, Dukowicz A, et al. Vimentin expression appears to be associated with poor prognosis in node-negative ductal NOS breast carcinomas. *Am J Pathol*. 1990; 137:1299–1304.

105. Korsching E, Packeisen J, Liedtke C, et al. The origin of vimentin expression in invasive breast cancer: epithelial–mesenchymal transition, myoepithelial histogenesis or histogenesis from progenitor cells with bilinear differentiation potential? *J Pathol*. 2005; 206:451–457.

106. Hajra KM, Fearon ER. Cadherin and catenin alterations in human cancer. *Genes Chromosomes Cancer*. 2002; 34:255–268.

107. Hazan RB, Qiao R, Keren R, Badano I, Suyama K. Cadherin switch in tumor progression. *Ann N Y Acad Sci.* 2004; 1014:155–163.

108. Tang A, Eller MS, Hara M, Yaar M, Hirohashi S, Gilchrest BA. E-Cadherin is the major mediator of human melanocyte adhesion to keratinocytes in vitro. *J Cell Sci.* 1994; 107(Pt 4): 983–992.

109. Hsu MY, Meier FE, Nesbit M, E-Cadherin expression in melanoma cells restores keratinocyte-mediated growth control and down-regulates expression of invasion-related adhesion receptors. *Am J Pathol.* 2000; 156:1515–1525.

110. Li G, Satyamoorthy K, Herlyn M. N-Cadherin-mediated intercellular interactions promote survival and migration of melanoma cells. *Cancer Res.* 2001; 61:3819–3825.

111. Jaggi M, Nazemi T, Abrahams NA, et al. N-Cadherin switching occurs in high Gleason grade prostate cancer. *Prostate.* 2006; 66:193–199.

112. Tomita K, van Bokhoven A, van Leenders GJ, et al. Cadherin switching in human prostate cancer progression. *Cancer Res.* 2000; 60:3650–3654.

113. Hazan RB, Phillips GR, Qiao RF, Norton L, Aaronson SA. Exogenous expression of N-cadherin in breast cancer cells induces cell migration, invasion, and metastasis. *J Cell Biol.* 2000; 148:779–790.

114. Pishvaian MJ, Feltes CM, Thompson P, Bussemakers MJ, Schalken JA, Byers SW. Cadherin-11 is expressed in invasive breast cancer cell lines. *Cancer Res.* 1999; 59:947–952.

115. Maeda G, Chiba T, Okazaki M, et al. Expression of SIP1 in oral squamous cell carcinomas: implications for E-cadherin expression and tumor progression. *Int J Oncol.* 2005; 27:1535–1541.

116. Rosivatz E, Becker I, Specht K, et al. Differential expression of the epithelial–mesenchymal transition regulators Snail, SIP1, and Twist in gastric cancer. *Am J Pathol.* 2002; 161:1881–1891.

117. Sugimachi K, Tanaka S, Kameyama T, et al. Transcriptional repressor *snail* and progression of human hepatocellular carcinoma. *Clin Cancer Res.* 2003; 9:2657–2664.

118. Uchikado Y, Natsugoe S, Okumura H, et al. Slug expression in the E-cadherin preserved tumors is related to prognosis in patients with esophageal squamous cell carcinoma. *Clin Cancer Res.* 2005; 11:1174–1180.

119. Blanco MJ, Moreno-Bueno G, Sarrio D, et al. Correlation of Snail expression with histological grade and lymph node status in breast carcinomas. *Oncogene.* 2002; 21:3241–3246.

120. Cheng CW, Wu PE, Yu JC, et al. Mechanisms of inactivation of E-cadherin in breast carcinoma: modification of the two-hit hypothesis of tumor suppressor gene. *Oncogene.* 2001; 20:3814–3823.

121. Elloul S, Elstrand MB, Nesland JM, et al. Snail, Slug, and Smad-interacting protein 1 as novel parameters of disease aggressiveness in metastatic ovarian and breast carcinoma. *Cancer.* 2005; 103:1631–1643.

122. Fujita N, Jaye DL, Kajita M, Geigerman C, Moreno CS, Wade PA. MTA3, a Mi-2/NuRD complex subunit, regulates an invasive growth pathway in breast cancer. *Cell.* 2003; 113:207–219.

123. Martin TA, Goyal A, Watkins G, Jiang WG. Expression of the transcription factors Snail, Slug, and Twist and their clinical significance in human breast cancer. *Ann Surg Oncol.* 2005; 12:488–496.

124. Toyama T, Zhang Z, Iwase H, et al. Low expression of the *snail* gene is a good prognostic factor in node-negative invasive ductal carcinomas. *Jpn J Clin Oncol.* 2006; 36:357–363.

125. Palmer HG, Larriba MJ, Garcia JM, et al. The transcription factor Snail represses vitamin D receptor expression and responsiveness in human colon cancer. *Nat Med.* 2004; 10:917–919.

126. Peña C, Garcia JM, Silva J, et al. E-cadherin and vitamin D receptor regulation by Snail and ZEB1 in colon cancer: clinicopathological correlations. *Hum Mol Genet.* 2005; 14:3361–3370.

127. Larriba MJ, Muñoz A. Snail vs vitamin D receptor expression in colon cancer: therapeutics implications. *Br J Cancer.* 2005; 92:985–989.

128. Rosivatz E, Becker KF, Kremmer E, et al. Expression and nuclear localization of Snail, an E-cadherin repressor, in adenocarcinomas of the upper gastrointestinal tract. *Virchows Arch.* 2006; 448:277–287.

129. Parker BS, Argani P, Cook BP, et al. Alterations in vascular gene expression in invasive breast carcinoma. *Cancer Res.* 2004; 64:7857–7866.

130. Gupta PB, Kuperwasser C, Brunet JP, et al. The melanocyte differentiation program predisposes to metastasis after neoplastic transformation. *Nat Genet.* 2005; 37:1047–1054.

131. Shih JY, Tsai MF, Chang TH, et al. Transcription repressor *slug* promotes carcinoma invasion and predicts outcome of patients with lung adenocarcinoma. *Clin Cancer Res.* 2005; 11:8070–8078.

132. Shioiri M, Shida T, Koda K, et al. Slug expression is an independent prognostic parameter for poor survival in colorectal carcinoma patients. *Br J Cancer.* 2006; 94:816–822.

133. Rosivatz E, Becker I, Bamba M, et al. Neoexpression of N-cadherin in E-cadherin positive colon cancers. *Int J Cancer.* 2004; 111:711–719.

134. Domagala W, Wozniak L, Lasota J, Weber K, Osborn M. Vimentin is preferentially expressed in high-grade ductal and medullary, but not in lobular breast carcinomas. *Am J Pathol.* 1990; 137:1059–1064.

135. Sarrió D, Moreno-Bueno G, Hardisson D, et al. Epigenetic and genetic alterations of *APC* and *CDH1* genes in lobular breast cancer: relationships with abnormal E-cadherin and catenin expression and microsatellite instability. *Int J Cancer.* 2003; 106:208–215.

136. Mironchik Y, Winnard PT Jr Vesuna F, et al. Twist overexpression induces in vivo angiogenesis and correlates with chromosomal instability in breast cancer. *Cancer Res.* 2005; 65:10801–10809.

137. Kwok WK, Ling MT, Lee TW, et al. Up-regulation of Twist in prostate cancer and its implication as a therapeutic target. *Cancer Res.* 2005; 65:5153–5162.

138. Friedl P, Hegerfeldt Y, Tusch M. Collective cell migration in morphogenesis and cancer. *Int J Dev Biol.* 2004; 48:441–449.

139. Perou CM, Sorlie T, Eisen MB, et al. Molecular portraits of human breast tumors. *Nature.* 2000; 406:747–752.

140. Sorlie T, Perou CM, Tibshirani R, et al. Gene expression patterns of breast carcinomas distinguish tumor subclasses with clinical implications. *Proc Natl Acad Sci U S A.* 2001; 98:10869–10874.

141. Sorlie T, Tibshirani R, Parker J, et al. Repeated observation of breast tumor subtypes in independent gene expression data sets. *Proc Natl Acad Sci U S A*. 2003; 100:8418–8423.

142. Jones C, MacKay A, Grigoriadis A, et al. Expression profiling of purified normal human luminal and myoepithelial breast cells: identification of novel prognostic markers for breast cancer. *Cancer Res*. 2004; 64:3037–3045.

143. Livasy CA, Karaca G, Nanda R, et al. Phenotypic evaluation of the basal-like subtype of invasive breast carcinoma. *Mod Pathol*. 2006; 19:264–271.

144. Rodríguez-Pinilla SM, Sarrió D, Honrado E, et al. Prognostic significance of basal-like phenotype and fascin expression in node-negative invasive breast carcinomas. *Clin Cancer Res*. 2006; 12:1533–1539.

145. Fulford LG, Easton DF, Reis-Filho JS, et al. Specific morphological features predictive for the basal phenotype in grade 3 invasive ductal carcinoma of breast. *Histopathology*. 2006; 49:22–34.

146. Leibl S, Gogg-Kammerer M, Sommersacher A, Denk H, Moinfar F. Metaplastic breast carcinomas: Are they of myoepithelial differentiation?—Immunohistochemical profile of the sarcomatoid subtype using novel myoepithelial markers. *Am J Surg Pathol*. 2005; 29:347–353.

147. Reis-Filho JS, Milanezi F, Steele D, et al. Metaplastic breast carcinomas are basal-like tumors. *Histopathology*. 2006; 49:10–21.

148. Asselin-Labat ML, Shackleton M, Stingl J, et al. Steroid hormone receptor status of mouse mammary stem cells. *J Natl Cancer Inst*. 2006; 98:1011–1014.

149. Tischkowitz MD, Foulkes WD. The basal phenotype of BRCA1-related breast cancer: past, present and future. *Cell Cycle*. 2006; 5:963–967.

150. Smalley M, Ashworth A. Stem cells and breast cancer: a field in transit. *Nat Rev Cancer*. 2003; 3:832–844.

151. Woodward WA, Chen MS, Behbod F, Rosen JM. On mammary stem cells. *J Cell Sci*. 2005; 118:3585–3594.

152. Sarrió D, Rodriguez-Pinilla SM, Hardisson D, et al. Epithelial-mesenchymal transition in breast cancer relates to the basal-like phenotype. *Cancer Res*. 2008; 68:989–997.

10 Epigenetics in Cancer Stem Cell Development

KENNETH NEPHEW, CURT BALCH, TIM H.-M. HUANG,
ZHANG SHU, MICHAEL CHAN, and PEARLLY YAN

10.1 INTRODUCTION

It is now well known that cancer pathogenesis is strongly associated with abnormal epigenetic (i.e., DNA sequence-external) alterations, represented by both losses and gains of DNA methylation and deviant patterns of histone modifications.[1,2] These chromatin modifications allow for the stable inheritance of various aberrant cellular properties without involving changes in DNA sequence (e.g., deletions, mutations) or the amount of DNA (e.g., gene amplification) in cancer cells. An emerging hypothesis of tumor propagation is based on the activity of a subpopulation of *cancer stem cells* (CSCs) capable of an atypical type of differentiation that results in tumor formation, invasion, and metastasis to distant sites.[3] Such abnormal progenitors are believed to strongly resemble normal tissue stem cells, possessing defining characteristics such as asymmetric cell division, self-renewal, and the maintenance of an undifferentiated phenotype.[3-5] As normal differentiation is regulated by epigenetic modifications to chromatin,[6,7] it is almost certain that epigenetics also contributes to the aberrant differentiation present in tumors, including the maintenance of multipotency and self-renewal of their progenitor cells. Indeed, one recent review suggested a causal role for repressive epigenetic modifications in the establishment and propagation of CSCs.[8] Consequently, knowledge of the epigenetic events associated with tumor stem cell behavior will result in a more complete understanding of carcinogenesis in general, in addition to providing insight into new therapeutic targets and clinical risk assessment. In addition, based on the cancer stem cell hypothesis (outlined in more detail in Chapter 1), it is believed that the inadequacy of standard chemotherapies for many solid tumors is due largely to their failure to target CSCs,[9] resulting in inevitable disease relapse. Thus, therapeutic reversal of epigenetic modifications that define the undifferentiated state of those malignant progenitor cells (including a drug-resistant phenotype) could allow for *epigenetic resensitization* of recalcitrant tumors to conventional agents.

10.2 CHARACTERIZATION OF CANDIDATE CANCER STEM CELLS

To date, isolation of CSCs has been based largely on their probable phenotypic similarity to normal stem cells: demonstrating such defining characteristics as anchorage-independent formation of spheres, dye exclusion (due to overexpression of efflux transporters), expression of cell surface differentiation markers, and clonogenicity.[3] The most significant characteristic, however, is amplified tumorigenic potential, as defined by the capacity of significantly small numbers of cells to form tumors in animals (see Chapter 1), and the ability to recapitulate the phenotype of the original tumor from which they were derived.[10] Additionally, CSCs probably express self-renewal-associated genes found in normal stem cells, including *NANOG, SOX2, Oct-4, SCF-1*, and *Bmi-1*, encoding a transcriptional repressor previously established as necessary for self-renewal in hematopoietic stem cells (HSCs).[11] Suggestive of a role in tumor "stemness," Bmi-1 has now been demonstrated as overexpressed in a number of malignancies, as has another HSC transcriptional repressor, the Enhancer of Zeste homolog-2 (EZH2).[12] Genes that are commonly *down-regulated* in normal (and probably, cancer) stem cells are typically those associated with a commitment to differentiation, and include those of the *CDX* family, involved in intestinal cell determination[13]; the *Myo* genes, which contribute to muscle differentiation[14]; and *early B cell factors*, which play a role in B-cell lineage commitment.[15] Gene expression microarrays have recently been used to globally determine specific markers/pathways up- and down-regulated in normal mammary stem cells,[16] an approach that will undoubtedly yield useful information in global analyses of cancer stem cells.

10.3 POSSIBLE ORIGINS OF CANCER STEM CELLS

As a consequence of likely CSC genotypes, several hypotheses have been put forth regarding the origin of tumorigenic progenitors. One supposition is that these arise from normal tissue stem cells, which are long-lived, often express membrane efflux transporters, and are highly efficient at DNA repair.[3] For example, it is now strongly believed that acute myelogenous leukemia (AML) stem cells are derived from normal hematopoietic stem cells.[17] Another scenario sets forth that CSCs arise from *transit-amplifying cells* (unipotent progenitors).[18] Support for this hypothesis is found in chronic myeloid leukemia, shown to derive from cells committed to the granulocyte–macrophage lineage.[19] It is also possible that some cancers originate from mature cells that subsequently reacquire stem cell properties (i.e., de-differentiate).[18] In a third scenario, in some gastric cancers, there is now evidence that infiltrating bone marrow–derived cells are transformed into CSCs within the epithelial lining, following infection with the bacterium *Helicobacter pylori*.[20] This hypothesis could be extended to other epithelial malignancies associated with chronic inflammation, including esophageal, colorectal, and ovarian cancers.[21] CSCs could also arise by gain of whole or partial chromosomes by cell–cell fusion or horizontal gene transfer.[18] Indeed, hematopoietic stem cells can fuse to a variety of differentiated cells, and one hypothesis for tumorigenesis is based on fusion of tissue stem and differentiated cells, resulting in possible genomic instability.[18] In fact, CD44, a cell surface marker of putative breast, prostate, and colon cancer stem cells, has previously been demonstrated to be a fusogen in macrophages.[22] Horizontal gene transfer by phagocytosis of fragmented DNA from apoptotic cells could also result in the conferral of stemlike properties to recipient cells, and one tumor cell type, invasive glioma cells, has indeed been demonstrated to be highly phagocytic.[23]

10.4 EPIGENETICS IN NORMAL DEVELOPMENT

One biochemical phenomenon intricately linked to differentiation is *epigenetic* modification of chromatin, referring to heritable changes external to DNA nucleotide sequence.[24] These modifications, including methylation of deoxycytosine and ubiquitination, methylation, phosphorylation, and acetylation of histone lysines, are responsible for altered gene expression patterns that allow for specific cell or organotypic phenotypes.[24] One well-known epigenetic phenomenon is random inactivation of one X chromosome in female somatic cells, a process known as *dosage compensation*, for which complete failure is lethal.[25] Further, epigenetic "programs" are probably the primary impetus for both normal development and differentiation, as all somatic cells (with the exception of lymphocytes) possess identical genomic DNA, yet possess widely divergent patterns of gene expression.[7] In fact, the very term "epigenetics" was used originally to describe the process by which genotype gives rise to phenotype.[7,26] In mice, it is well established that during preimplantation embryonic development, a large degree of "epigenetic reprogramming" occurs, including active demethylation of the paternal genome and passive, DNA replication–dependent demethylation of the maternal genome.[27] Following implantation, remethylation occurs, in distinct patterns that facilitate commitment to specific tissue lineages.[28] A lack of epigenetic reprogramming is strongly believed to be responsible for the low success rate of somatic cell cloning (i.e., transplantation of an adult cell nucleus into an enucleated unfertilized egg cell), as it appears that differentiation-related epigenetic marks must be "erased" to restore totipotency.[29] Indeed, it was found that methylation patterns from cloned genomes more closely resembled the adult cells from which they derived than embryonic genomes.[28] Epigenetic reprogramming might also allow for *trans-differentiation*, a clinical ambition for the use of adult tissue stem cells to serve as precursors for the generation of an unrelated tissue cell type (e.g., adipocyte differentiation into neurons).[30] The use of epigenetic inhibitors could conceivably allow for such therapies in the future. Finally, a role for DNA methylation in gene imprinting, defined as a difference in expression between the paternal and maternal alleles of distinct genes, has long been established.[31]

As epigenetics is intricately linked to normal differentiation, disruption of normal epigenetic processes has been strongly associated with a variety of developmental disorders. Specifically, dysregulated genomic imprinting has been correlated with numerous neurodevelopmental diseases, including Prader–Willi and Angelman syndromes, and the growth disorder Beckwith–Wiedemann syndrome.[32] Further, genetic or functional knockout of several chromatin-modifying enzymes, including histone and DNA methyltransferases, is lethal or results in greatly altered development.[33,34] Mutations in a DNA methyltransferase, gene *DNMT3b*, result in a condition known as *immunodeficiency, centromeric region instability, and facial anomalies* (ICF) *syndrome*; one manifestation of ICF syndrome is impaired immune system development.[35] Another epigenetics-based disorder, Rett syndrome, resulting from an X-linked mutation in a gene encoding a methylcytosine-binding protein (MeCP2), is also characterized by developmental defects, particularly in the nervous system.[32] In adult mammals, a *histone* methyltransferase, EZH2, is essential for B-cell development[36] and the avoidance of hematopoietic stem cell senescence.[37] In fact, epigenetics is now believed to contribute to the majority of noninfectious diseases, as evidenced by discordance of various disorders in identical twins (i.e., epigenetic drift) and the delayed (adult) onset of inherited genetic diseases such as familial amyotrophic lateral sclerosis (ALS) and Huntington's chorea.[32,38] Other conditions, including Alzheimer's disease, Down's syndrome, coronary artery disease, and a developmental syndrome, neural tube defect, have been associated with deficiencies in the metabolism of folate, a dietary precursor for S-adenosylmethionine, the DNA methyltransferase cofactor that donates the methyl group for methylation of 5-deoxycytosine.[39]

10.5 EPIGENETIC REGULATION OF THE CANCER STEM CELL PHENOTYPE

In association with its intricate role in differentiation, it is also now well established that epigenetics contributes significantly to tumorigenesis. Although genetic contributions have been well documented in neoplasia, they are probably inadequate alone to explain the origin of most tumors.[8] As it has been proposed that five or six biallelic mutations are necessary for a fully malignant phenotype,[40] some have questioned the sufficiency of a hypothesized somatic cell mutation rate (10^{-7} per gene per cell division[41]) to achieve that level of variance in a single cell in a human lifetime.[42] Consequently, a possible *mutator phenotype* has been invoked to be responsible for amplified genetic alteration in tumors.[43] Although many tumors do exhibit increased genomic instability, this finding is not universal,[44] and others have noted the possible deleterious effects of a high mutation rate on tumorigenesis (e.g., apoptosis of highly mutated tumor cells).[45] Other potential shortcomings of pure genetic models of cancer are that the classical progression-related genes are not always found mutated, and that no reproducible mutations have been identified outside the well-known *genetic gatekeepers*.[8]

To further support a role for epigenetics in cancer development, it was demonstrated that a transplanted mouse melanoma nucleus was capable of development into an entire mouse, suggesting that a large degree of reprogramming of tumor-associated gene expression is possible by epigenetic modification.[46] Moreover, a recent study of glioblastoma found extensive phenotypic and molecular similarities between normal neural stem cells and tumor stem cells.[47] That report further demonstrates that CSCs harbor a number of genetic aberrations found within their parental tumors, suggesting that stemness properties could be largely unrelated to primary DNA sequence, strongly supporting a role for nongenetic (i.e., epigenetic) events in the multipotency of malignant progenitors. Indeed, it is well established that tumor cells, arising within a specific organ, possess patterns of epigenetic modifications quite different from the normal cells within that organ. These alterations include global DNA hypomethylation, with localized hypermethylation, and aberrant patterns of histone modifications, such as atypical lysine acetylation and methylation.[2] To identify possible "epigenetic signatures" of advanced-stage ovarian cancer, our group has performed comprehensive analyses of chromatin modifications in cell lines and tumors using microarrays for the global examination of DNA methylation and histone modifications (the sum of such modifications now referred to as the *epigenome*).[48,49] Many cancer-specific chromatin modifications are repressive of gene expression and act to silence tumor suppressor genes (TSGs) such as *RASSF1A*, $p16^{INK4}$, and *hMLH1*[50,51]; indeed, many such TSGs are now known to contribute to differentiation programs.[52] Additionally, a number of epigenetic gatekeeper genes have recently been identified; these are similar to the genetic gatekeepers described previously to be intimately involved in regulating cell proliferation (see above).[53] These epigenetic gatekeepers include genes encoding SFRP, a negative regulator of the Wnt cell fate determination cascade, the differentiation-conferring transcription factors GATA-4 and -5, and the cell cycle regulator $p16^{INK4}$.[54,55] It is also well established that epigenetic alterations typically occur early in carcinogenesis[56] (probably even prior to genetic changes),[8] and cancer-associated DNA methylation alterations are often present in normal tissue adjacent to tumors, a phenomenon known as a *field effect*.[57]

To account for this emerging body of evidence, Feinberg et al. recently put forth an epigenetic progenitor model of human cancer, in which early epigenetic events result in expansion of a progenitor cell pool, followed by hyperplasia and an "initiating" genetic or epigenetic mutation.[8] Further tumor progression is the result of heterogeneity due primarily to epigenetics, which can beget permanent genetic mutations. A variation of this model is shown in Fig. 10.1, illustrating that initiating *epimutations* could be responsible for the generation of tumor progenitors from normal stem cells, transit-amplifying cells, or bone marrow–derived infiltrating cells, followed by expansion and abnormal differentiation into a full-fledged tumor. Although comprehensive identification of these epimutations has not yet been reported, the recent development of several high-throughput microarray analyses could be utilized to establish genome-wide "signatures" of epigenetic changes that occur in CSCs. Indeed, a similar study has now been performed to compare

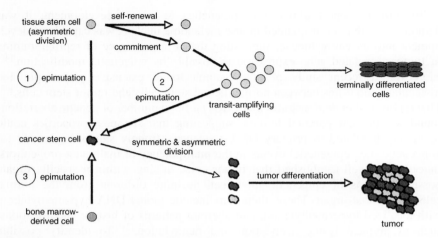

FIGURE 10.1 Cancer stem cells might originate by epigenetic alteration of normal tissue stem cells (epimutation 1), transit-amplifying cells (epimutation 2), or bone marrow–derived cells (epimutation 3).

methylation signatures of embryonic stem cells, fully differentiated cells, and cancer cell lines, demonstrating considerable differences in epigenetic profiles.[58]

Polycomb Repressor Complexes in Cancer Stem Cells   In addition to the aforementioned cell fate signaling cascades, another group of genes, discovered originally in *Drosophila* development, are those of the *Polycomb group* (PcG).[59] In that insect, PcG proteins were found intricately involved in segmentation by the repression of a group of developmental regulators known as *homeobox* (encoding proteins possessing a conserved DNA-binding *homeodomain*).[60] The action of PcG proteins (gene repression) was found in *Drosophila* to be directly opposed by another gene family, the *trithorax* group (trxG) (i.e., trxG proteins act as *activators* of homeobox family genes).[60] It is now known that both PcG and trxG proteins, as developmental regulators, exert their actions by epigenetic mechanisms. TrxG proteins exist in large complexes that modify chromatin epigenetically into an "open" configuration permissive of gene expression.[61] One trxG gene, *Brahma*, encodes a chromatin-remodeling protein that disrupts histone–DNA interactions,[62] a process necessary for active gene expression,[63] while another family member, *TRX*, encodes a histone methyltransferase capable of catalyzing the trimethylation of histone 3 lysine 4 (H3K4) to form a known transcriptionally permissive chromatin "mark" (H3K4me3). In direct opposition to trxG complexes, PcG proteins exert their effects (also in multisubunit complexes) by modifying chromatin into a transcriptionally repressive state.[64] Two such Polycomb repressive complexes, PRC1 and PRC2, have now been well characterized. PRC2 is believed to be responsible for the establishment of stem cell self-renewal and consists of a histone methyltransferase, known in mammals as Enhancer of Zeste homolog 2 (EZH2), that modifies histone H3 lysine 27 (H3K27), also by trimethylation, to generate the histone mark H3K27me3.[12] In contrast to H3K4me3, H3K27me3 is a transcriptionally *repressive* chromatin mark[65,66] that we recently demonstrated to be capable of being genetically manipulated to study the "histone code" in cancer cells.[67] Recent findings have also shown that a number of genes frequently found DNA-methylated in adult tumors possess a bivalent pattern of both transcriptionally activating (H3K4me3) and repressive (H3K27me3) chromatin marks in embryonic stem cells, suggesting that this specific epigenotype holds those genes in a transcription-ready state for further repression or activation during differentiation.[68–71] However, it has also been hypothesized that during tumorigenesis, further epigenetic repression (by the addition of other silencing chromatin marks, including DNA methylation) of those differentiation-associated genes may occur in CSCs committed to the malignant phenotype.[71–73] Although a direct association of DNA methylation with the PRC2 complex has been reported,[74] our group[67] and others[71,75] have failed to observe an immediate linkage of these two repressive chromatin marks (H3K27me3 and DNA methylation). However, an alternative hypothesis is that the H3K27me3 mark may *precede* DNA methylation,[71,72] and indeed, a recent report demonstrated that genes possessing that specific histone modification (H3K27me3) were 12-fold more likely to acquire DNA methylation in adult tumors.[76]

In addition to EZH2, two other core PRC2 PcG proteins are *Suppressor of Zeste-12 (Suz12)* and *embryonic ectoderm development (EED)*,[64] and another PRC2 subunit is a histone-binding protein, RBP4.[77] Two more recently discovered Polycomb repressor complexes, PRC3 and PRC4, differ slightly from PRC2 in that they possess various isoforms of EED.[78,79] Although EZH2 appears to be the PRC2 enzyme directly responsible for H3K27 methylation,[77] both EED and Suz12 are also essential for that activity.[80,81] In support of a role for Polycomb complex activity in CSCs, EZH2 was found highly overexpressed in advanced breast and prostate cancers.[12] Specifically, PRC2 may contribute to the maintenance of undifferentiated tumor progenitors, as demonstrated by EZH2 association with poorly differentiated malignancies.[82] Although PRC2 plays a role in the establishment of stem cell self-renewal, PRC1 is responsible for the maintenance of that process. In humans, PRC1 consists of the core PcG proteins B lymphoma Mo-MLV insertion region-1 (Bmi-1), human polyhomeotic (HPH), human Polycomb (HPC), and a histone ubiquitin ligase known as Ring2 (also known as RTF and RING1B).[64,83] Similar to EZH2, Bmi-1 has also been reported as overexpressed in a number of malignancies, including neuroblastoma and nasopharyngeal and breast cancers.[12] Recently, a microarray approach was used to identify an 11-gene expression "death-from-cancer signature" associated with shortened survival in 11 different malignancies; those 11 genes appear to constitute a Bmi-1-regulated pathway, as the signature did not occur in a Bmi-1$^{-/-}$ genetic background.[84]

In regulating stem cell self-renewal, there is now emerging evidence that PRC1 is actually linked to the action of PRC2, probably through binding of its component HPC proteins to H3K27me3.[66] Additionally, it appears that recruitment of the PRC1 member Bmi-1 to specific heterochromatic sites is contingent on the action of the PRC2 components EED and EZH2, in addition to DNA methyltransferase-1, an enzyme also reported to associate with PRC2.[74,85] To further investigate mechanisms of both PRC1- and PRC2-mediated repression in mammals, genome-wide microarray analyses were performed to identify target promoters bound by PRC1 (Phc1, Ring2, and Bmi-1) and PRC2 (EED, EZH2, and Suz12) components in mouse and human normal embryonic stem cells.[86–88] As might be predicted, a large number of stem cell–and differentiation-specific genes were bound (500 to 1000 Polycomb target genes) by these specific PcG proteins, confirming that both PRC1 and PRC2 target key pathways central to cell fate decisions and development.[86,87] Those mammalian findings concurred well with previous global studies in *Drosophila*, which demonstrated a change in PcG protein distribution during development.[89] Thus, it appears that in mammals, as well as in *Drosophila*, PcG repressors, via their ability to initiate and maintain specific chromatin modifications (e.g., histone methylation), epigenetically regulate stem cell establishment and self-renewal.[86–88] As described above, the aberrant overexpression of both EZH2 (component of PRC2) and Bmi-1 (component of PRC1) has been linked to several malignancies,[12] and additionally, Bmi-1 has been associated specifically with the proliferation of leukemic stem cells.[90] Consequently, it is highly likely that PRC1 and PRC2 activity, involved in the molecular regulation or normal stem cells, may become dysregulated in CSCs. Thus, genome-wide approaches for discovering PRC1 and PRC2 targets in CSCs may result in the identification of pathways responsible for their defining characteristic of malignant stemness.

10.6 CONTRIBUTIONS OF EPIGENETICS TO DRUG RESISTANCE IN CANCER STEM CELLS

Several mechanisms have been hypothesized as a basis for drug resistance in solid tumors, including drug inactivation, quiescence, and increased DNA repair in chemoresistant tumor cells.[91] The classical drug-resistance model invokes drug selection of clones possessing mutations (i.e., increased DNA repair, quiescence, etc.) that confer a survival advantage from various drug insults (e.g., platinum damage of DNA); each cell within the tumor would have an equal probability of gaining such mutations.[10] However, the above-mentioned phenotypes also describe tissue stem cells. These normal progenitor cells are generally long-lived (thus quiescent or slowly proliferating) and therefore require survival-conferring characteristics such as enhanced DNA repair and the expression of membrane transporters that allow increased resistance to environmental toxins.[3] As a consequence of the cancer stem cell theory, it is believed that malignancies that initially undergo complete remission but subsequently relapse to a completely refractory state (e.g., ovarian cancer[92] and anti-hormone-refractory breast[93] and prostate[94] cancers) are more likely to possess tumor stem cells than are cancers that do not respond well to primary therapy (e.g., pancreatic[95] cancer).[9] In that model, shown in Fig. 10.2, chemotherapeutics preferentially target transit-amplifying cells, causing tumor regression, but fail to eradicate drug-resistant CSCs. Following "complete" remission of detectable disease, CSCs proliferate and probably confer the drug-resistant phenotype to their progeny, resulting in a tumor that is now fully refractory to further treatment (Fig. 10.2), as occurs in patients suffering from several advanced-stage malignancies.[92–94] Consequently, therapies are greatly needed that specifically target the small percentage of tumor progenitors in addition to the nontumorigenic progeny that comprise the bulk of the tumor mass. As epigenetic modifications are the primary driving force for differentiation in normal somatic cells,[7,79] it is highly likely that these also play a role in maintaining differentiation states in tumors (see below), necessitating a comprehensive study of such modifications (and agents capable of their reversal) in malignant progenitors. As the Polycomb repressor EZH2 is widely implicated in aggressive malignancies,[12] it is likely that this enzyme plays a significant role in CSC self-renewal and multipotency. In this regard, it was recently demonstrated that in vivo delivery of a small inhibitory RNA against E2H2 could abolish human prostate xenograft bone metastases in athymic mice.[96] Moreover, as histone deacetylases (HDACs) and DNA methyltransferases (DNMTs) may associate with the PRC2 complex,[74,97] it is feasible that inhibition of these enzymes could disrupt PRC2 activity and, consequently, stem cell establishment. In support of such a hypothesis, it was demonstrated that treatment with HDAC inhibitors could deplete EZH2 and other PRC2 component proteins from acute leukemia cells.[98] We[99] and others[100,101] have also shown that HDAC and DNMT inhibitors, alone and in combination with conventional chemotherapies, are strongly antiproliferative to drug-resistant human cancer cells, suggesting that these therapies may hold the potential to directly target tumor stem cells.

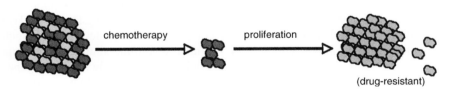

chemotherapy   proliferation

(drug-resistant)

FIGURE 10.2 Based on several defining characteristics, cancer stem cells could probably resist conventional proliferative cell-targeting chemotherapies, survive, and repopulate a now drug-resistant tumor (red cells represent stem cells, yellow cells represent transit amplifying cells, and blue cells represent differentiated tumor cells). In the new tumor, the chemoresistant phenotype is passed on to stem cell progeny (gray). (*See insert for color representation of figure.*)

10.7 GENOME-WIDE INTERROGATION OF EPIGENETIC MODIFICATIONS IN CANCER STEM CELLS

We and others have successfully categorized global epigenomic alterations in cancer cells and tumors, using a variety of high-throughput, genome-wide epigenetic analyses.[49,102,103] One such method, pioneered by our group, is differential methylation hybridization (DMH), is a microarray-based technique now capable of analyzing simultaneously over 240,000 CpG-rich fragments (promoter-associated). In general, "CpG islands" (CpGi), promoter-associated CG-rich regions, are normally unmethylated.[104] However, in tumors, numerous CpGi are found aberrantly methylated, and consequently, these sequences represent potential biomarkers and/or therapeutic targets.[2] We previously employed DMH to show that promoter-associated hypermethylation, within specific CpGi, is widespread in ovarian cancer genomes; we also identified specific methylated loci correlated with poor prognosis in advanced ovarian cancer patients.[49,103] Those studies demonstrate that distinct hypermethylated CpGi may be important prognostic markers for this disease. In fact, as a number of technologies have recently been developed to detect large numbers of differentially methylated CpGi,[102] these methods could be exploited to better understand the role of DNA methylation in CSC biology.

In addition to DMH, another high-throughput epigenetic analysis, *chromatin immunoprecipitation-to-microarray hybridization* (ChIP-on-chip), is now widely utilized by many groups, including ours. In that technique, antibodies against specific chromatin-modifying enzymes (e.g., EZH2) or histone modifications (e.g., trimethylated histone 3 lysine 27) are used to coprecipitate DNA associated with those chromatin components. The precipitated DNA is then isolated, PCR amplified, fluorescently labeled, and hybridized to "tiled" high-density microarrays.[105] Hybridized "spots" thus represent individual promoters that possess the specific chromatin modification under study.[105,106] Our research group used ChIP-on-chip to establish estrogen receptor-alpha (ER) target genes in breast cancer cells[106,107] and, using bioinformatics approaches, determined specific ordered transcription factor–binding sites associated with those ER targets.[108]

This global approach was also used for genome-wide mapping of Polycomb repressor targets in normal embryonic stem cells and resulted in the comprehensive identification of genes repressed by both PRC2 (SUZ12, EED, and EZH2) and PRC1 (Bmi-1, PHC1, and RING2) component PcG proteins.[86–88] Such an approach would almost certainly be useful to determine similar epigenomic changes that occur in CSCs. By revealing specific epigenetically repressed genes in CSCs, combined with powerful bioinformatics analyses, it would be possible to build a comprehensive model(s) of altered pathways in tumor progenitors. Such a model may provide a systematic understanding of pathways associated with CSC tumorigenic potential, self-renewal properties, and the maintenance of an undifferentiated phenotype, and may also provide insight into various targets for intervention.

10.8 EPIGENETIC THERAPIES AGAINST POORLY DIFFERENTIATED CANCER CELLS

In the last 15 years, several agents capable of reversing epigenetic repression have demonstrated effectiveness in clinical trials for various solid and hematologic cancers.[50,109,110] These include HDAC inhibitors, a group of drugs that were originally discovered as compounds capable of differentiating erythroleukemia cells.[111] Although the precise mechanism(s) of these agents remains under study, it is likely that they affect numerous signaling pathways unrelated to histone acetylation.[112] Another class of epigenetic therapies includes inhibitors of DNA methylation, representing another repressive epigenetic modification[1,2,56]; DNA methylation inhibitors have similarly been well demonstrated as cancer-differentiating agents.[113,114] We and others have also affirmed that methylation inhibitors are capable of reversing platinum resistance in chemoresistant cancer cells.[99,101] It is also well established that *combinations* of DNA methylation and HDAC inhibitors are more potent for gene reexpression than is either alone[48,110,115]; indeed, several such combinations are now in cancer clinical trials.[110] It is possible that these epigenetic therapies could target CSCs in several ways. First, as the PRC2 complex has been reported to associate with HDACs[97] and DNMTs[74] and one DNMT is also linked to PRC1 recruitment to heterochromatin,[85] both DNMTs and HDACs may play roles in suppressing differentiation-related genes.[74,97] Consequently, inhibition of those epigenetic repressors could conceivably allow for CSC "escape" to committed pathways, as shown in Fig. 10.3 (upper path). Second, as tumor stem cells maintain the ability to differentiate, agents that induce differentiation (e.g., HDAC and DNA

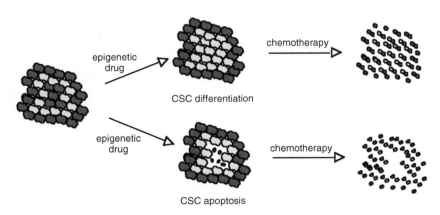

CSC differentiation

CSC apoptosis

FIGURE 10.3 Based on the disruption of DNA methylation and/or histone deacetylation necessary for cancer stem cell (CSC)–initiated tumorigenesis, epigenetic therapies could induce CSC differentiation (top path) or apoptosis (lower path). Conventional chemotherapy could then eliminate the remaining non-stem cell population. (*See insert for color representation of figure.*)

methylation inhibitors) may have a therapeutic role to play against the cancer stem cell compartment. Moreover, although conventional chemotherapies, which preferentially target cells with a high mitotic index, may be ineffective against the slower-proliferating or quiescent stem cells (Fig. 10.2), it has now been reported that HDAC inhibitors can target both proliferating and nonproliferating malignant cells.[116] When combined with conventional therapies such as platinum agents, complete tumor regression could conceivably occur. Finally, disruption of Polycomb action could abrogate self-renewal, resulting in apoptosis (Fig. 10.3, lower path), as both DNA methylation and HDAC inhibitors have demonstrated an ability to induce programmed cell death in cancer cells.[50,109,112] Thus, combinations of epigenetic drugs and conventional chemotherapies may be capable of direct targeting of CSCs, although this has yet to be demonstrated.

Although methylation and HDAC inhibitors have been shown previously to elicit differentiation in various cancers, some studies have reported contradictory outcomes (i.e., de-differentiation or stem cell expansion). In one study, human embryoid bodies (formed by in vitro differentiation of embryonic stem cells) treated with the demethylating agent 5-azacytidine altered their morphology and expressed genetic markers of undifferentiated cells.[117] Similarly, others reported expansion of hematopoietic stem cells following treatment with methylation and histone deacetylase inhibitors,[118] while another study noted that severe hypomethylation in DNMT$^{-/-}$ embryonic stem (ES) cells results in complete inhibition of ES cell differentiation.[119] However, it appears unlikely that clinically used DNMT inhibitors can actually achieve the same degree of hypomethylation as knockout or severe knockdown of DNMT genes.[120] Moreover, other studies have reported opposite effects (stem/progenitor cell differentiation) of epigenetic drugs in stem cell populations, including commitment of ES cells into cardiomyocytes[121,122] and the induction of myogenesis in mesenchymal progenitors.[123] In general, these agents have been well established as differentiating agents of various malignancies, both in vitro and in vivo, and indeed, many HDAC inhibitors were originally characterized solely on that basis (see ref. 111 and the discussion above). Additionally, it was suggested recently that differentiating agents might be most effective as therapies targeting the cancer stem cell compartment, based on a study of glioblastoma tumor stem cells demonstrating a likely *epigenetic* (not genetic) basis for the altered stem cell phenotype.[47] It is likely that future epigenetic analyses of isolated cancer stem cells, including manipulation with DNA methylation and histone deacetylase inhibitors, will shed light on this important controversy.

10.9 CONCLUSIONS AND FUTURE PERSPECTIVES

Epigenetic alterations, including DNA methylation accompanied by local changes in histone modification and chromatin structure, have been firmly demonstrated to play a role in tumor initiation, development, and progression. As epigenetics is the primary catalyst for establishing and maintaining normal differentiation states in somatic cells, it is likely that these are significantly altered in cancer stem cells.

Future analyses may include identifying epigenomic profiles of cancer stem cells and establishing differentiation/self-renewal pathways that are disrupted in those cells during tumorigenesis. Such information will allow further identification of possible targets to alter aberrant self-renewal and the poor differentiation capacity of tumor progenitors. Explicitly targeting such abnormal stem cells with epigenetic therapies, possibly resulting in the induction of apoptosis and/or differentiation to a chemosensitive phenotype, is also feasible. In summary, the establishment of cancer stem cell epigenotypes may lead to the identification of cancer pathways and potential therapeutic targets of the most primitive cells within a tumor, cells with the capacity to propagate the entire malignant phenotype. Such findings may allow elucidation of strategies for the targeted eradication of those cells by specifically directed therapies.

Acknowledgments

The authors gratefully acknowledge grant support from the U.S. National Institutes of Health, National Cancer Institute grants CA 085289 (to K.N.) and CA 113001 (to T.H.), Ovar'coming Together, Indianapolis, IN (to C.B.), and the Walther Cancer Institute, Indianapolis, IN (to K.N.).

REFERENCES

1. Feinberg AP, Tycko B. The history of cancer epigenetics. *Nat Rev Cancer*. 2004; 4:143–153.
2. Jones PA, Baylin SB. The fundamental role of epigenetic events in cancer. *Nat Rev Genet*. 2002; 3:415–428.
3. Wicha MS, Liu S, Dontu G. Cancer stem cells: an old idea—a paradigm shift. *Cancer Res*. 2006; 66:1883–1890; commentary, 1890.
4. Polyak K, Hahn WC. Roots and stems: stem cells in cancer. *Nat Med*. 2006; 12:296–300.
5. Reya T, Morrison SJ, Clarke MF, Weissman IL. Stem cells, cancer, and cancer stem cells. *Nature*. 2001; 414:105–111.
6. Jaenisch R, Bird A. Epigenetic regulation of gene expression: how the genome integrates intrinsic and environmental signals. *Nat Genet*. 2003; 33 Suppl:245–254.
7. Morange M. The relations between genetics and epigenetics: a historical point of view. *Ann N Y Acad Sci*. 2002; 981:50–60.
8. Feinberg AP, Ohlsson R, Henikoff S. The epigenetic progenitor origin of human cancer. *Nat Rev Genet*. 2006; 7:21–33.
9. Dean M, Fojo T, Bates S. Tumor stem cells and drug resistance. *Nat Rev Cancer*. 2005; 5:275–284.
10. Clarke MF, Dick JE, Dirks PB, et al. Cancer stem cells: perspectives on current status and future directions. AACR Workshop on Cancer Stem Cells. *Cancer Res*. 2006; 66:9339–9344.
11. Park IK, Qian D, Kiel M, et al. Bmi-1 is required for maintenance of adult self-renewing haematopoietic stem cells. *Nature*. 2003; 423:302–305.

12. Sparmann A, van Lohuizen M. Polycomb silencers control cell fate, development and cancer. *Nat Rev Cancer*. 2006; 6:846–856.

13. Lorentz O, Duluc I, Arcangelis AD, Simon- Assmann P, Kedinger M, Freund JN. Key role of the *Cdx2* homeobox gene in extracellular matrix-mediated intestinal cell differentiation. *J Cell Biol*. 1997; 139:1553–1565.

14. Berkes CA, Tapscott SJ. MyoD and the transcriptional control of myogenesis. *Semin Cell Dev Biol*. 2005; 16:585–595.

15. Maier H, Hagman J. Roles of EBF and Pax-5 in B lineage commitment and development. *Semin Immunol*. 2002; 14:415–422.

16. Behbod F, Xian W, Shaw CA, Hilsenbeck SG, Tsimelzon A, Rosen JM. Transcriptional profiling of mammary gland side population cells. *Stem Cells*. 2006; 24:1065–1074.

17. Hope KJ, Jin L, Dick JE. Acute myeloid leukemia originates from a hierarchy of leukemic stem cell classes that differ in self-renewal capacity. *Nat Immunol*. 2004; 5:738–743.

18. Bjerkvig R, Tysnes BB, Aboody KS, Najbauer J, Terzis AJ. Opinion: The origin of the cancer stem cell—current controversies and new insights. *Nat Rev Cancer*. 2005; 5:899–904.

19. Jamieson CH, Ailles LE, Dylla SJ, et al. Granulocyte–macrophage progenitors as candidate leukemic stem cells in blast-crisis CML. *N Engl J Med*. 2004; 351:657–667.

20. Houghton J, Stoicov C, Nomura S, et al. Gastric cancer originating from bone marrow–derived cells. *Science*. 2004; 306:1568–1571.

21. Marx J. Cancer research. Inflammation and cancer: the link grows stronger. *Science*. 2004; 306:966–968.

22. Vignery A. Osteoclasts and giant cells: macrophage–macrophage fusion mechanism. *Int J Exp Pathol*. 2000; 81:291–304.

23. Bjerknes R, Bjerkvig R, Laerum OD. Phagocytic capacity of normal and malignant rat glial cells in culture. *J Natl Cancer Inst*. 1987; 78:279–288.

24. Jablonka E, Lamb MJ. The changing concept of epigenetics. *Ann N Y Acad Sci*. 2002; 981:82–96.

25. Li E. Chromatin modification and epigenetic reprogramming in mammalian development. *Nat Rev Genet*. 2002; 3:662–673.

26. Waddington C. The epigenotype. *Endeavor*. 1942; 1:18–20.

27. Surani MA. Reprogramming of genome function through epigenetic inheritance. *Nature*. 2001; 414:122–128.

28. Reik W, Dean W, Walter J. Epigenetic reprogramming in mammalian development. *Science*. 2001; 293:1089–1093.

29. Dean W, Santos F, Reik W. Epigenetic reprogramming in early mammalian development and following somatic nuclear transfer. *Semin Cell Dev Biol*. 2003; 14:93–100.

30. Liu Y, Rao MS. Transdifferentiation: fact or artifact. *J Cell Biochem*. 2003; 88:29–40.

31. Rainier S, Feinberg AP. Genomic imprinting, DNA methylation, and cancer. *J Natl Cancer Inst*. 1994; 86:753–759.

32. Feinberg AP. Phenotypic plasticity and the epigenetics of human disease. *Nature*. 2007; 447:433–440.

33. Li E, Bestor TH, Jaenisch R. Targeted mutation of the DNA methyltransferase gene results in embryonic lethality. *Cell*. 1992; 69:915–926.

34. Okano M, Bell DW, Haber DA, Li E. DNA methyltransferases Dnmt3a and Dnmt3b are essential for de novo methylation and mammalian development. *Cell.* 1999; 99:247–257.

35. Ehrlich M. The ICF syndrome, a DNA methyltransferase 3B deficiency and immunodeficiency disease. *Clin Immunol.* 2003; 109:17–28.

36. Su IH, Basavaraj A, Krutchinsky AN, et al. Ezh2 controls B cell development through histone H3 methylation and Igh rearrangement. *Nat Immunol.* 2003; 4:124–131.

37. Kamminga LM, Bystrykh LV, de Boer A, et al. The Polycomb group gene Ezh2 prevents hematopoietic stem cell exhaustion. *Blood.* 2006; 107:2170–2179.

38. Bjornsson HT, Fallin MD, Feinberg AP. An integrated epigenetic and genetic approach to common human disease. *Trends Genet.* 2004; 20:350–358.

39. Loenen WA. S-Adenosylmethionine: jack of all trades and master of everything? *Biochem Soc Trans.* 2006; 34:330–333.

40. Cho KR, Vogelstein B. Genetic alterations in the adenoma–carcinoma sequence. *Cancer.* 1992; 70:1727–1731.

41. Hethcote HW, Knudson AG Jr. Model for the incidence of embryonal cancers: application to retinoblastoma. *Proc Natl Acad Sci U S A.* 1978; 75:2453–2457.

42. Loeb LA. Mutator phenotype may be required for multistage carcinogenesis. *Cancer Res.* 1991; 51:3075–3079.

43. Jackson AL, Loeb LA. The mutation rate and cancer. *Genetics.* 1998; 148:1483–1490.

44. Pihan GA, Purohit A, Wallace J, et al. Centrosome defects and genetic instability in malignant tumors. *Cancer Res.* 1998; 58:3974–3985.

45. Tomlinson I, Bodmer W. Selection, the mutation rate and cancer: ensuring that the tail does not wag the dog. *Nat Med.* 1999; 5:11–12.

46. Hochedlinger K, Blelloch R, Brennan C, et al. Reprogramming of a melanoma genome by nuclear transplantation. *Genes Dev.* 2004; 18:1875–1885.

47. Lee J, Kotliarova S, Kotliarov Y, et al. Tumor stem cells derived from glioblastomas cultured in bFGF and EGF more closely mirror the phenotype and genotype of primary tumors than do serum-cultured cell lines. *Cancer Cell.* 2006; 9:391–403.

48. Shi H, Wei SH, Leu YW, et al. Triple analysis of the cancer epigenome: an integrated microarray system for assessing gene expression, DNA methylation, and histone acetylation. *Cancer Res.* 2003; 63:2164–2171.

49. Wei SH, Chen CM, Strathdee G, et al. Methylation microarray analysis of late-stage ovarian carcinomas distinguishes progression-free survival in patients and identifies candidate epigenetic markers. *Clin Cancer Res.* 2002; 8:2246–2252.

50. Balch C, Montgomery JS, Paik HI, et al. New anti-cancer strategies: epigenetic therapies and biomarkers. *Front Biosci.* 2005; 10:1897–1931.

51. Jones PA, Laird PW. Cancer epigenetics comes of age. *Nat Genet.* 1999; 21:163–167.

52. Cowell JK. Tumor suppressor genes. *Ann Oncol.* 1992; 3:693–698.

53. Kinzler KW, Vogelstein B. Lessons from hereditary colorectal cancer. *Cell.* 1996; 87:159–170.

54. Baylin SB, Ohm JE. Epigenetic gene silencing in cancer: a mechanism for early oncogenic pathway addiction? *Nat Rev Cancer.* 2006; 6:107–116.

55. Jones PA, Baylin SB. The epigenomics of cancer. *Cell.* 2007; 128:683–692.

56. Das PM, Singal R. DNA methylation and cancer. *J Clin Oncol*. 2004; 22:4632–4642.

57. Giovannucci E, Ogino S. DNA methylation, field effects, and colorectal cancer. *J Natl Cancer Inst*. 2005; 97:1317–1319.

58. Bibikova M, Chudin E, Wu B, et al. Human embryonic stem cells have a unique epigenetic signature. *Genome Res*. 2006; 16:1075–1083.

59. Hanson RD, Hess JL, Yu BD, et al. Mammalian trithorax and Polycomb-group homologues are antagonistic regulators of homeotic development. *Proc Natl Acad Sci U S A*. 1999; 96:14372–14377.

60. Kennison JA. The Polycomb and trithorax group proteins of Drosophila: trans-regulators of homeotic gene function. *Annu Rev Genet*. 1995; 29:289–303.

61. Cernilogar FM, Orlando V. Epigenome programming by Polycomb and trithorax proteins. *Biochem Cell Biol*. 2005; 83:322–331.

62. Collins RT, Treisman JE. Osa-containing Brahma chromatin remodeling complexes are required for the repression of wingless target genes. *Genes Dev*. 2000; 14:3140–3152.

63. Mellor J. The dynamics of chromatin remodeling at promoters. *Mol Cell*. 2005; 19:147–157.

64. Otte AP, Kwaks TH. Gene repression by Polycomb group protein complexes: a distinct complex for every occasion? *Curr Opin Genet Dev*. 2003; 13:448–454.

65. Dillon N. Heterochromatin structure and function. *Biol Cell*. 2004; 96:631–637.

66. Kuzmichev A, Nishioka K, Erdjument-Bromage H, Tempst P, Reinberg D. Histone methyltransferase activity associated with a human multiprotein complex containing the Enhancer of Zeste protein. *Genes Dev*. 2002; 16:2893–2905.

67. Abbosh PH, Montgomery JS, Starkey JA, et al. Dominant-negative histone H3 lysine 27 mutant derepresses silenced tumor suppressor genes and reverses the drug-resistant phenotype in cancer cells. *Cancer Res*. 2006; 66:5582–5591.

68. Bernstein BE, Mikkelsen TS, Xie X, et al. A bivalent chromatin structure marks key developmental genes in embryonic stem cells. *Cell*. 2006; 125:315–326.

69. Gan Q, Yoshida T, McDonald OG, Owens GK. Concise review: epigenetic mechanisms contribute to pluripotency and cell lineage determination of embryonic stem cells. *Stem Cells*. 2007; 25:2–9.

70. Mikkelsen TS, Ku M, Jaffe DB, et al. Genome-wide maps of chromatin state in pluripotent and lineage-committed cells. *Nature*. 2007; 448:553–560.

71. Ohm JE, McGarvey KM, Yu X, et al. A stem cell–like chromatin pattern may predispose tumor suppressor genes to DNA hypermethylation and heritable silencing. *Nat Genet*. 2007; 39:237–242.

72. Balch C, Nephew KP, Huang TH, Bapat SA. Epigenetic "bivalently marked" process of cancer stem cell-driven tumorigenesis. *Bioessays*. 2007; 29:842–845.

73. Ting AH, McGarvey KM, Baylin SB. The cancer epigenome: components and functional correlates. *Genes Dev*. 2006; 20:3215–3231.

74. Vire E, Brenner C, Deplus R, et al. The Polycomb group protein EZH2 directly controls DNA methylation. *Nature*. 2006; 439:871–874.

75. McGarvey KM, Greene E, Fahrner JA, Jenuwein T, Baylin SB. DNA methylation and complete transcriptional silencing of cancer genes persist after depletion of EZH2. *Cancer Res*. 2007; 67:5097–5102.

76. Widschwendter M, Fiegl H, Egle D, et al. Epigenetic stem cell signature in cancer. *Nat Genet*. 2007; 39:157–158.

77. Cao R, Zhang Y. SUZ12 is required for both the histone methyltransferase activity and the silencing function of the EED-EZH2 complex. *Mol Cell*. 2004; 15:57–67.

78. Kuzmichev A, Jenuwein T, Tempst P, Reinberg D. Different EZH2-containing complexes target methylation of histone H1 or nucleosomal histone H3. *Mol Cell*. 2004; 14:183–193.

79. Kuzmichev A, Margueron R, Vaquero A, et al. Composition and histone substrates of Polycomb repressive group complexes change during cellular differentiation. *Proc Natl Acad Sci U S A*. 2005; 102:1859–1864.

80. Montgomery ND, Yee D, Chen A, et al. The murine Polycomb group protein Eed is required for global histone H3 lysine-27 methylation. *Curr Biol*. 2005; 15:942–947.

81. Pasini D, Bracken AP, Jensen MR, Lazzerini Denchi E, Helin K. Suz12 is essential for mouse development and for EZH2 histone methyltransferase activity. *EMBO J*. 2004; 23:4061–4071.

82. Raaphorst FM, Meijer CJ, Fieret E, et al. Poorly differentiated breast carcinoma is associated with increased expression of the human Polycomb group EZH2 gene. *Neoplasia*. 2003; 5:481–488.

83. Li Z, Cao R, Wang M, Myers MP, Zhang Y, Xu RM. Structure of a Bmi-1-Ring1B Polycomb group ubiquitin ligase complex. *J Biol Chem*. 2006; 281:20643–20649.

84. Glinsky GV, Berezovska O, Glinskii AB. Microarray analysis identifies a death-from-cancer signature predicting therapy failure in patients with multiple types of cancer. *J Clin Invest*. 2005; 115:1503–1521.

85. Hernandez-Munoz I, Taghavi P, Kuijl C, Neefjes J, van Lohuizen M. Association of BMI1 with Polycomb bodies is dynamic and requires PRC2/EZH2 and the maintenance DNA methyltransferase DNMT1. *Mol Cell Biol*. 2005; 25:11047–11058.

86. Boyer LA, Plath K, Zeitlinger J, et al. Polycomb complexes repress developmental regulators in murine embryonic stem cells. *Nature*. 2006; 441(7091): 349–353.

87. Bracken AP, Dietrich N, Pasini D, Hansen KH, Helin K. Genome-wide mapping of Polycomb target genes unravels their roles in cell fate transitions. *Genes Dev*. 2006; 20:1123–1136.

88. Lee TI, Jenner RG, Boyer LA, et al. Control of developmental regulators by Polycomb in human embryonic stem cells. *Cell*. 2006; 125:301–313.

89. Negre N, Hennetin J, Sun LV, et al. Chromosomal distribution of PcG proteins during Drosophila development. *PLoS Biol*. 2006; 4:e170.

90. Lessard J, Sauvageau G. Bmi-1 determines the proliferative capacity of normal and leukaemic stem cells. *Nature*. 2003; 423:255–260.

91. Agarwal R, Kaye SB. Ovarian cancer: strategies for overcoming resistance to chemotherapy. *Nat Rev Cancer*. 2003; 3:502–516.

92. Bhoola S, Hoskins WJ. Diagnosis and management of epithelial ovarian cancer. *Obstet Gynecol*. 2006; 107:1399–1410.

93. Roy V, Perez EA. New therapies in the treatment of breast cancer. *Semin Oncol*. 2006; 33:S3–S8.

94. Kasamon KM, Dawson NA. Update on hormone-refractory prostate cancer. *Curr Opin Urol*. 2004; 14:185–193.

95. Li D, Xie K, Wolff R, Abbruzzese JL. Pancreatic cancer. *Lancet.* 2004; 363:1049–1057.

96. Takeshita F, Minakuchi Y, Nagahara S, et al. Efficient delivery of small interfering RNA to bone-metastatic tumors by using atelocollagen invivo. *Proc Natl Acad Sci U S A.* 2005; 102:12177–12182.

97. van der Vlag J, Otte AP. Transcriptional repression mediated by the human Polycomb-group protein EED involves histone deacetylation. *Nat Genet.* 1999; 23:474–478.

98. Fiskus W, Pranpat M, Balasis M, et al. Histone deacetylase inhibitors deplete Enhancer of Zeste 2 and associated Polycomb repressive complex 2 proteins in human acute leukemia cells. *Mol Cancer Ther.* 2006; 5:3096–3104.

99. Balch C, Yan P, Craft T, et al. Antimitogenic and chemosensitizing effects of the methylation inhibitor zebularine in ovarian cancer. *Mol Cancer Ther.* 2005; 4:1505–1514.

100. Kim MS, Blake M, Baek JH, Kohlhagen G, Pommier Y, Carrier F. Inhibition of histone deacetylase increases cytotoxicity to anticancer drugs targeting DNA. *Cancer Res.* 2003; 63:7291–7300.

101. Plumb JA, Strathdee G, Sludden J, Kaye SB, Brown R. Reversal of drug resistance in human tumor xenografts by 2′-deoxy-5-azacytidine-induced demethylation of the hMLH1 gene promoter. *Cancer Res.* 2000; 60:6039–6044.

102. Callinan PA, Feinberg AP. The emerging science of epigenomics. *Hum Mol Genet.* 2006; 15 Spec No 1: R95–R101.

103. Wei SH, Balch C, Paik HH, et al. Prognostic DNA methylation biomarkers in ovarian cancer. *Clin Cancer Res.* 2006; 12:2788–2794.

104. Gardiner-Garden M, Frommer M. CpG islands in vertebrate genomes. *J Mol Biol.* 1987; 196:261–282.

105. Horak CE, Snyder M. ChIP-chip: a genomic approach for identifying transcription factor binding sites. *Methods Enzymol.* 2002; 350:469–483.

106. Cheng AS, Jin VX, Fan M, et al. Combinatorial analysis of transcription factor partners reveals recruitment of c-MYC to estrogen receptor-alpha responsive promoters. *Mol Cell.* 2006; 21:393–404.

107. Leu YW, Yan PS, Fan M, et al. Loss of estrogen receptor signaling triggers epigenetic silencing of downstream targets in breast cancer. *Cancer Res.* 2004; 64:8184–8192.

108. Li L, Cheng AS, Jin VX, et al. A mixture model based discriminate analysis for identifying ordered transcription factor binding site pairs in gene promoters directly regulated by estrogen receptor-alpha. *Bioinformatics.* 2006.

109. Lyko F, Brown R. DNA methyltransferase inhibitors and the development of epigenetic cancer therapies. *J Natl Cancer Inst.* 2005; 97:1498–1506.

110. Yoo CB, Jones PA. Epigenetic therapy of cancer: past, present and future. *Nat Rev Drug Discov.* 2006; 5:37–50.

111. Reboulleau CP, Shapiro HS. Chemical inducers of differentiation cause conformational changes in the chromatin and deoxyribonucleic acid of murine erythroleukemia cells. *Biochemistry.* 1983; 22:4512–4517.

112. Minucci S, Pelicci PG. Histone deacetylase inhibitors and the promise of epigenetic (and more) treatments for cancer. *Nat Rev Cancer.* 2006; 6:38–51.

113. de Vos D. Epigenetic drugs: a longstanding story. *Semin Oncol.* 2005; 32:437–442.

114. Momparler RL. Epigenetic therapy of cancer with 5-aza-2′-deoxycytidine (decitabine). *Semin Oncol.* 2005; 32:443–451.

115. Jones PA. Overview of cancer epigenetics. *Semin Hematol.* 2005; 42:S3–S8.

116. Burgess A, Ruefli A, Beamish H, et al. Histone deacetylase inhibitors specifically kill nonproliferating tumor cells. *Oncogene.* 2004; 23:6693–6701.

117. Tsuji-Takayama K, Inoue T, Ijiri Y, et al. Demethylating agent, 5-azacytidine, reverses differentiation of embryonic stem cells. *Biochem Biophys Res Commun.* 2004; 323:86–90.

118. Milhem M, Mahmud N, Lavelle D, et al. Modification of hematopoietic stem cell fate by 5-aza-2′-deoxycytidine and trichostatin A. *Blood.* 2004; 103:4102–4110.

119. Jackson M, Krassowska A, Gilbert N, et al. Severe global DNA hypomethylation blocks differentiation and induces histone hyperacetylation in embryonic stem cells. *Mol Cell Biol.* 2004; 24:8862–8871.

120. Yang AS, Estecio MR, Garcia-Manero G, Kantarjian HM, Issa JP. Comment on "Chromosomal instability and tumors promoted by DNA hypomethylation" and "Induction of tumors in nice by genomic hypomethylation." *Science.* 2003; 302:1153; author reply, 1153.

121. Yoon BS, Yoo SJ, Lee JE, You S, Lee HT, Yoon HS. Enhanced differentiation of human embryonic stem cells into cardiomyocytes by combining hanging drop culture and 5-azacytidine treatment. *Differentiation.* 2006; 74:149–159.

122. Xu C, Police S, Rao N, Carpenter MK. Characterization and enrichment of cardiomyocytes derived from human embryonic stem cells. *Circ Res.* 2002; 91:501–508.

123. Kuwana M, Okazaki Y, Kodama H, et al. Human circulating CD14+ monocytes as a source of progenitors that exhibit mesenchymal cell differentiation. *J Leukoc Biol.* 2003; 74:833–845.

11   Cancer Stem Cells and New Therapeutic Approaches

MICHAEL DEAN

11.1 CANCER STEM CELLS

The recent identification of a small population of cancer stem cells (CSCs) in solid tumors such as breast and brain has changed the way that many scientists view cancer.[1,2] These cells represent only 1% of the tumor mass and are the only cells capable of transplanting the tumor into nude mice. Additional studies have presented data that long-established cell lines, even HeLa cells, contain a minor population of cells with some of the same properties as stem cells.[3,4] Many researchers now suspect that all cancers are composed of a mixture of stem cells and proliferative cells with a limited life span. It has also been known for decades that there exists a proportion of cells in a tumor capable of surviving radiation treatment and cytotoxic drug exposure.[5] Stem cells must also survive many genetic insults during a human life and express drug transporters and DNA repair systems. Stem cells are necessarily refractory to programmed cell death, and can be quiescent for long periods of time, properties that would allow a cancer cell to resist standard therapeutic approaches.[6–8] Hence, it is likely that the surviving cells in tumors may share certain attributes with stem cells.

The implications of the concept of CSCs are far reaching. The regression of many cancers following chemotherapy could result from the survival of cancer stem cells. This is paralleled in the body with the regrowth of hair due to the survival of hair follicles and the recovery of blood cells due to the survival of hematopoetic stem cells. Can these results be extrapolated to most or all solid tumors? Are there therapeutic approaches targeting these cancer stem cells with application to a wide array of cancers? These are critical questions remaining to be addressed in the cancer stem cell field.

Cancer Stem Cells: Identification and Targets, Edited by Sharmila Bapat
Copyright © 2009 John Wiley & Sons, Inc.

11.2 ACTIVATION OF STEM CELLS AND CANCER

Most stem cells in the body remain in a dormant state. These cells are surrounded by other, differentiated cells within the tissue in a structure known as the *niche*. The cells of the niche regulate the stem cells via cell–cell contacts, interactions with the extracellular matrix, and secretion of inhibitory factors. Disruption of the niche through infection, inflammation, tissue damage, or chemical assault can activate the division of the stem cells. The activated stem cell gives rise to additional stem cells as well as cells committed to differentiate. These new cells repair the damaged area of tissues and the stem cell returns to its quiescent state. Virtually all of the agents described to confer a risk for cancer also result in tissue alteration (and therefore activation of stem cells), including radiation, wounding, chemical damage, infectious agents, and inflammation.

Therefore, cancer can be thought of as a disease resulting from the abnormal growth of stem cells, resulting from chronic activation of stem cells (caused by disruption of the niche), and leading to the long-term proliferation of the stem cells (Fig. 11.1). Chronically dividing stem cells are a target for additional mutagenic agents, resulting in genetic damage to the cell (mutation of tumor suppressor genes and activation of oncogenes), generating first a premalignant stem cell, and subsequently, a malignant stem cell (Fig. 11.1). The disruption of a niche and stem cell activation could occur by hormonal stimulation, tissue damage caused by inflammation, radiation, chemicals, infections, or inactivation of certain tumor suppressor genes (Table 11.1). The abnormally dividing stem cell could be subject

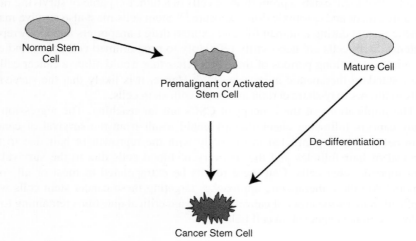

FIGURE 11.1   Formation of cancer stem cells. Cancer requires the generation of an activated or premalignant stem cell. Mutations in tumor suppressor genes and oncogenes can convert an activated stem cell into a premalignant cancer stem cell. Further mutations are required for tumor progression and metastasis. An alternative pathway has a mature tissue cell acquiring the ability of self-renewal.

TABLE 11.1   Stem Cell Activation and Common Cancers[a]

Class of Agent	Action	Target Cell	Cancer Type
None	Naturally activated stem cells	Retinoblast	Retinoblastoma
		Kidney stem cell	Wilm's tumor
		Testes	Testicular germ cell
Inflammation	Activates cell growth	Colon	Colon cancer
		Pancreas	Pancreatic cancer
Hormones	Stimulate tissue growth	Breast	Breast cancer
		Ovary	Ovarian cancer
		Prostate	Prostate cancer
Viral infection	Tissue damage	Liver	Hepatocellular carcinoma
		Cervix	Cervical cancer
Bacterial infection	Inflammation	Stomach	Gastric cancer
Tumor suppressor	Stem cell proliferation	Colon	Colon cancer
Inactivation chemical irritants	Inflammation	Skin	Basal cell carcinoma
		Lung	Lung cancer

[a]Endogenous and environmental agents postulated to play a role in the activation of stem cells are shown along with their mode of action, and one or two examples of a target cell type and resulting cancer.

to additional genetic events, leading to autonomous growth, the loss of cell cycle regulation, and resistance to apoptosis—all well-understood properties of cancer cells.[9] However, it is also possible that a more mature, committed cell could acquire the property of self-renewal (de-differentiation) and therefore become a malignant stem cell (Fig. 11.1).

11.2.1 Initiation and Promotion Revised

One of the most important discoveries in cancer research has been the elucidation of the multistep nature of cancer and the development of experimental systems that allowed the nature of carcinogenic agents to be explored. In 1942, Peyton Rous demonstrated that mouse skin could be exposed to mutagens and that tumors would not result unless there was subsequent wounding of the site. This later event could be carried out long after the original exposure. Subsequently, the terms initiator and promoter were coined. *Initiators* are agents that mice had to be exposed to initially in order to develop a tumor. Although many exposures to a tumor initiator could be shown to cause cancer, the typical experiment involved a single exposure to the initiator and multiple subsequent exposures with a tumor promoter (Fig. 11.2). The

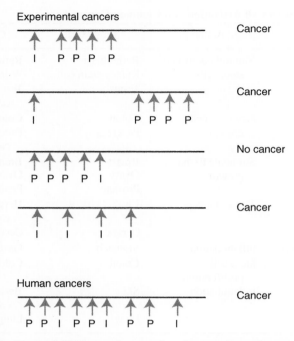

FIGURE 11.2 Tumor initiation and promotion. In experimental systems the administration of an initiator (mutagen) when followed by multiple administrations of a tumor promoter results in tumors. If the promoter is given first, no tumor results, although many administrations of the initiator can cause cancer. In the human system, chronic inflammation and/or exposure to tissue-disrupting agents, combined with mutagen exposure, causes cancer. (*See insert for color representation of figure.*)

promoter is thus required to be given multiple times and within a fairly constrained window.

The mechanism of action of tumor initiators was the first to be elucidated. These were shown to be any agents damaging the DNA and causing mutations. Tumor *promoters* were more elusive and included agents such as phorbol esters and mineral oils. In addition, physiological processes such as wounding were also shown to cause tumor promotion. Eventually, some tumor promoters were shown to activate cells via the protein kinase C pathway.

Tumor promoters can then be thought of as agents that disrupt the tissue and activate stem cells, whereas tumor initiators damage DNA and mutate specific genes. The only difference between the experimental mouse models of tumorigenesis and the human situation is the fact that human cancers involve chronic exposure to tissue-damaging agents and mutagens.[10] Most of the environmental agents that lead to stem cell activation would be classified as promoters, whereas non-DNA reactive agents that enhance the tumorigenicity of initiators would be classified as mutagens. Some environmental agents, such as tobacco smoke and ultraviolet irradiation, contribute to both initiation and promotion, as they stimulate tissue proliferation and damage DNA.

In light of the identification of cancer stem cells, the initiation and promotion scheme can be updated. Cancer can be envisioned to be caused by a combination of agents disrupting tissue architecture and activating stem cells, and agents that damage and mutate DNA. This paradigm explains the role of known agents implicated as causing cancer and focuses the problem on a subset of abnormal cells that could be targeted specifically, resulting in more effective therapies.

11.2.2 Stem Cell Activation and Specific Cancers

Although a model of a small population of self-renewing cells as the key to all cancers is an attractive idea, can the model be extended to the wide variety of tumor types and specific agents implicated in causing these tumors? Broadly, three distinct origins of cancers have been described: embryonic, conditional growth, and renewal[11] (Fig. 11.3). Embryonic cancers derive from rapidly dividing embryonic tissue and therefore contain a population of actively dividing stem cells. The prototype embryonic cancer is retinoblastoma,[12] but Wilms' tumor, Ewing's sarcoma, and childhood bone and brain cancers are also known to be embryonic in origin. Retinoblastoma arises in embryonic cells in the developing eye known as retinoblasts. These cells are highly proliferative and represent naturally activated stem cells. Mutations or loss of the *RB1* gene transforms these embryonic stem cells into CSCs. These cells would be expected to have lost the response to growth regulatory signals shutting down the stem cell once eye development is completed. Other childhood cancers could involve multipotent stem cells in other tissues suffering genetic damage during development. These cancers require the fewest number of genetic events, because the target cell is a fully activated stem cell.

The identification of the *Patched* (*PTCH*) gene as a tumor suppressor gene directly connected early development to tumorigenesis.[13–15] The PTCH protein is a membrane protein with 12 membrane spanning segments and is the receptor for the Hedgehog (HH) family of signaling molecules. PTCH and the HHs play a central role in the cell fate and patterning of early embryonic cells. Alteration in the growth regulation of stem cells was invoked to explain the role of the *PTCH* gene in causing basal cell carcinomas in patients with nevoid basal cell carcinoma syndrome.[16]

Stem cells can also be activated during the normal process of expansion of certain tissues due to the action of hormones, particularly during puberty (conditional growth tissues).[11] Examples are the breast and prostate, which undergo dramatic expansion and growth during puberty under the control of estrogen, testosterone, and other hormones.[17] Activated stem cells in the breast represent the target cells for breast cancer. Inactivation of specific tumor suppressor genes such as *TP53* would transform the breast tissue stem cells into unregulated cells, resulting initially in premalignant lesions. These uncontrolled stem cells would be the targets for additional events, leading to the progression of the premalignant lesion into a fully malignant tumor.

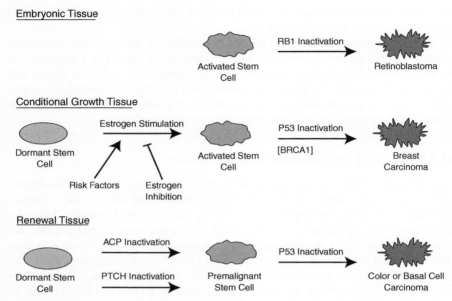

FIGURE 11.3   Tumor types and stem cell activation. Embryonic tumors arise from normal embryonic stem cells. These rapidly dividing cells require few alterations to become malignant. Conditional growth tissues, such as the breast, ovary, and prostate, are activated by hormones during puberty and during reproductive life. Activated stem cells in these tissues that suffer tumor suppressor gene mutations can give rise to premalignant lesions that can progress to malignancy. Hormonal risk factors such as age at menarche and age at first pregnancy affect the window of susceptibility of these cells, and agents that inhibit estrogen production can suppress cancer risk. Through inhibition of DNA repair, *BRCA1* and *BRCA2* mutations can accelerate the progression into a carcinoma of an activated or premalignant stem cell. Renewal tissues such as the colon and skin have dormant or slowly replicating stem cells. Certain tumor suppressor gene mutations can give rise to premalignant stem cells with uncontrolled replication. Subsequent mutations give rise to tumor progression.

Consistent with this model, the major risk factors for breast cancer risk involve hormonal and reproductive variables (Fig. 11.1).[18] Women with an early onset of puberty have a higher rate of breast and ovarian cancer than those with later menarche. Pregnancies, especially those starting at a relatively younger age, decrease cancer risk. These factors influence either the number or activation of breast stem cells. Several drugs have been developed that are able to decrease cancer risk and/or cancer recurrence. These include agents reducing the production of estrogen or blocking its action on cells. Similarly, removal of the ovaries reduces cancer risk in those with an extensive family history of breast and ovarian cancer.[19] Mutations in the *BRCA1* and *BRCA2* genes dramatically increase the risk of breast cancer. However, unlike many other tumor suppressor genes, *BRCA1* or *BRCA2* mutations are not commonly found in sporadic breast tumors. The BRCA1 and BRCA2 proteins play a role in the DNA repair process. These mutations can be thought of as increasing the probability of genetic events associated with tumor progression. Since these genes are not in the main pathway leading to breast cancer, they are not frequently mutated in sporadic tumors, but when mutated do increase a person's risk of disease.

In tissues such as the colon or skin, which undergo frequent self-renewal and turnover of cells, the tissue stem cells also keep dividing at a slow but constant rate. In the absence of disruption of these tissues, the risk of cancer is low. However, activation of stem cells in such tissues with a high cell turnover can occur either by inflammation or by tissue damage caused by radiation, mutagens, and irritants (Fig. 11.2). Chronic tissue damage would cause an increased rate of division of the renewing tissue stem cells and a corresponding increase in the number of target cells available for transformation. Colon cancer is considered a classical example that highlights the influence of inflammation on cancer. Inflammatory diseases such as Crohn's disease and inflammatory bowel disease result in dramatically increased risk for colon cancer.[20] Patients with mutations in the *APC* gene have familial adenomatous polyposis coli (FAP) syndrome and suffer from large numbers of colon polyps.[21,22] Polyps are premalignant lesions and FAP subjects have an elevated risk for colon cancer.[23] Individuals with FAP, as well as mice with mutations in *APC*, are associated with an increased cellular proliferation compartment in the colon.[24-26] The effect of *APC* mutations has been proposed to be to increase the proliferation of colon stem cells.[11] Consistent with this model, the *APC* gene is mutated in the vast majority of sporadic colon tumors.[27] Additional downstream events in colon cancer are very well characterized and include mutations in *p*53, *RAS*, and other genes.[27]

11.3 MAJOR CANCERS AND RISK FACTORS

Environmental risk factors have been identified for most of the most common cancers. These risk factors can be classified by their potential role in either activating stem cells or mutating target genes.

11.3.1 Liver Cancer

Liver cancer is one of the most common cancers in the world and is the third most common cause of cancer death. Liver cancer is particularly prevalent in eastern and southeastern Asia and in central and eastern Africa.[28] Hepatitis B and C viruses are the principal risk factor for hepatocellular carcinoma.[29,30] HBV or HCV do not contain genes known to inactivate tumor suppressor genes directly. However, patients persistently infected with these viruses typically suffer from liver inflammation, extensive cirrhosis, and abnormal liver function. The liver responds to damage by regenerating and repairing the damaged tissue. Therefore, persistent infection leads to chronic inflammation and continual stimulation of hepatic stem cells. HBV or HBC infection would be classified as a stem cell–activating agent. Other risk factors for liver cancer include aflatoxin B1, one of the most potent cancer-causing agents known. Aflatoxin is produced by the mold *Aspergillus flavus*, found in such foods as peanuts, rice, soybeans, corn, and wheat. Aflatoxin is a mutagen known to generate mutations in the *p53* tumor suppressor gene.

11.3.2 Lung Cancer

Lung cancer is considered to be a deadly cancer, with over 3 million deaths worldwide each year. The highest risk factor for lung cancer is recognized to be tobacco smoke. Tobacco smoke contains thousands of chemicals and several known carcinogens. The two primary carcinogens in tobacco smoke are nitrosamines and polycyclic aromatic hydrocarbons, and these agents are known to cause mutagen specific mutations.[31] Many agents in smoke damage the lung tissue and cause inflammation. Smoking causes the most common respiratory disease, chronic obstructive pulmonary disease (COPD). Recently, nicotine has been shown to promote the survival of lung cells, allowing the accumulation of genetic changes necessary for cancer formation.[32] Therefore, tobacco smoke contains both agents promoting cell death and tissue damage compounds capable of mutating DNA.

Another prevalent lung cancer carcinogen is asbestos. Asbestos is a metabolically inert compound remaining essentially permanently in the lungs of persons exposed to large amounts of the fibers. Asbestos exposure causes asbestosis, a fibrotic disease of the lung and is a risk factor for mesothelioma, a cancer of the mesothelial layer of the lung.[33] Asbestos also works synergistically with tobacco in increasing the risk of other lung cancers.[34] The chronic fibrosis and irritation caused by asbestos could be expected to stimulate stem cells in the lung continuously leading to an increased risk of mutations and the formation of premalignant lesions.

11.3.3 Gastric Cancer

Gastric cancer is the second highest cause of cancer death in the world and is highly prevalent in Japan, Great Britain, South America, and Iceland. Stomach cancers are highly associated with infection of the bacterium *Helicobacter pylori* and metal dust exposure.[35, 36] *H. pylori* infection leads to inflammation of the cells of the stomach lining and damage to the tissue. Dietary carcinogens are another risk factor for gastric cancer, and the combination of *H. pylori* infection and carcinogen ingestion is thought to account for the majority of this disease.

11.3.4 Pancreatic Cancer

Pancreatic cancer is a common disease in most of the world's populations, and the fatality rate of this malignancy is high, due to the failure to detect it early in most cases. The most common environmental factor associated with pancreatic cancer is smoking, and it is estimated that 25% of all pancreatic tumors are caused by smoking.[37] In animals, pancreatic tumors can be induced with tobacco-specific N-nitrosamines or the administration of other N-nitroso compounds.[38] Chronic pancreatitis and diabetes mellitus are also associated with pancreatic cancer,[39] and persons with mutations in the *BRCA2* gene have a 10-fold higher risk of the disease.[40] As in the other cancers described so far, the major risk factors are inflammatory conditions and mutagenic agents.

11.3.5 Cervical Cancer

Cervical cancer has been largely eliminated from developed countries by the routine screening of women and the administration of PAP smears to detect early lesions. However, cervical cancer remains one of the most common cancer killers of women in the world. Some environmental agents associated with cancer may contribute both to stem cell activation and to downstream mutations and tumor suppressor gene inactivation. Human papilloma viruses, particularly HPV 16 and 18, are highly associated with cervical cancer.[41] These viruses contain proteins (E6 and E7) that interact with and inhibit the P53 and RB1 tumor suppressor proteins.[42,43] Most women with infection by pathogenic HPV strains do not develop cervical cancer; therefore, additional factors are also likely to be involved. HPV infection leads to an increased proliferation of cervical stem cells. Although HPV infection produces a minimal immune response[44] and infected women do not have a visibly inflamed cervix, the immune system does recognize HPV infected cells, and immune-compromised patients are at increased risk for cervical cancer.[45] This chronic proliferation provides an increased pool of cells that are susceptible to viral integration and disease progression.

11.4 TREATMENT IMPLICATIONS

If it is true that most if not all solid tumors are composed of a minor population of self-renewing (stem) cells and a large fraction of non-renewing cells, tumors regressing following radiation and chemotherapy treatments are not the result of a rare cell evolving from within the tumor, but are the regrowth of the cancer stem cells. Of course, tumor stem cells could accumulate genetic changes, rendering them even more drug resistant, radiation resistant, or aneuploid. Because cures are achieved for many types of cancer, the cancer stem cells must be eliminated by some therapeutic strategies. The committed cells in a tumor are likely to play a role in supporting or stimulating the stem cells, forming a *tumor niche*. Rapid regression of a tumor could lead to disruption of the tumor niche and elimination of the cancer stem cells. Immune surveillance is clearly important in many cancers,[46] and reducing the mass of the tumor may allow the immune system to recognize the remaining cells efficiently.

Targeted therapies that suppress or kill tumor stem cells directly may synergize with established therapies to provide increased efficacy. Angiogenesis is likely to be critical to provide blood supply to the tumor stem cells, and strategies to inhibit the development of blood vessels are likely to be effective.[47]

1. *Drug transporters as chemotherapeutic targets.* One of the protective mechanisms of stem cells against toxins is the expression of one or more ATP-binding cassette (ABC) efflux transporters. These pumps protect stem cells from xenobiotic toxins.[48] The *ABCG2* and *ABCB1/MDR1* genes are expressed in the majority of stem cells and in most tumor stem cells.[6-8] These transporters can efflux fluorescent dyes such as rhodamine and Hoecht 33342, and this property allows stem cells to be separated from non-stem cells on a cell sorter.[49] The combined use of chemotherapy drugs and ATP-binding cassette (ABC) transporter inhibitors could be used to specifically target cancer stem cells.[50] There are highly specific inhibitors of ABCB1 in clinical use and ABCG2 inhibitors in development.[51] Transporter inhibition therapies are likely to have toxic effects on a patient's normal stem cells, and both ABCG2 and ABCB1 play a role in the blood–brain barrier. Therefore, this approach would have to be adjusted carefully to avoid excessive toxicity.

2. *Signaling molecules as chemotherapeutic targets.* Another approach to inhibiting cancer stem cells is to target the proteins essential for the growth and maintenance of stem cells. Because of the fundamental research in *Drosophila*, mice, *Caenorhabditis elegans*, zebrafish, and other developmental systems, a tremendous amount is known about the growth regulatory pathways functioning in embryonic cells.[52] One pathway, controlled by the Hedgehog (HH) and WNT signaling molecules, contains several genes functioning as either tumor suppressor genes or oncogenes.[16] For example, *Patched* (*PTCH*) is the receptor for HH molecules, and *PTCH* is mutated in patients with nevoid basal cell carcinoma syndrome.[14,53] The *PTCH* gene is also mutated in virtually all sporadic basal cell carcinomas and in some medulloblastomas, rhabdomyomas, and rhabdomyosarcomas.[54,55] The mammalian *HH* genes (*IHH, SHH, DHH*) are overexpressed in a large number of cancers, including small cell lung, pancreas, gastric, breast, and prostate.[56-59] *HH* ligand overexpression and *PTCH* mutation both have the effect of constitutive expression of *smoothened* (*SMO*), a G protein–coupled receptor family protein, a key signaling protein in the pathway. Constitutive *HH* expression could be an important component in stem cell activation in many cancers and therefore represents an attractive target for cancer therapy.

Cyclopamine is a compound discovered in the corn lily (*Veratrum californicum*), a plant teratogenic to sheep.[60] Cyclopamine binds to and inhibits the SMO protein and suppresses the growth of cells and tumors with activated HH signaling.[61] Human prostate tumor cell lines grown as xenografts in mice were eliminated following 21 days of treatment with cyclopamine,[59] and ultraviolet-induced basal cell carcinomas were suppressed in mice given low levels of cyclopamine in their drinking water.[62] Recently, it has been demonstrated that vitamin D_3 is a critical signaling molecule between PTCH and SMO. PTCH normally secretes vitamin D_3, and this molecule inhibits SMO on that cell as well as in adjacent cells.[63] HHs inhibit this secretion and cause a release from repression. Cyclopamine competes for the binding of vitamin D_3, on SMO and so appears to act in a similar manner. It is likely that vitamin D_3 and/or other steroidal analog could have a similar effect and be a candidate anticancer compound.

Other pathways critical to embryonic development and potentially important in cancer have also been described and include the WNT and NOTCH pathways. A number of experimental inhibitors of these pathways have been developed. There pathways are also the subject of drug development for a number of conditions, one example being the drug MK0752, which is in clinical trials for the treatment of acute T-cell lymphoblastic leukemia, myelogenous leukemia, chronic lymphocytic leukemia, and myelodysplastic syndrome. Gamma-secretase is required for maturation of the NOTCH protein, and secretase inhibitors have been developed for a number of pathological conditions. In a recent study, one such γ-secretase inhibitor was effective in the inhibition of stemlike cells in embryonal brain tumors.[64]

11.5 FUTURE PERSPECTIVES

The identification of cancer stem cells in solid tumors has important implications for basic cancer research. Most analyses of tumors, such as gene expression, microarrays, proteomic, and many phenotypic assays, have been performed on whole tumors and have not revealed data on the small fraction of tumor stem cells. In addition, screens for cancer cytotoxic drugs have involved cell cultures treated over short time periods.[65] Drugs specifically targeting cancer stem cells may display modest activity in short-term proliferation assays and be rejected for further follow-up study in animals or humans. Several important questions remain from the current data. Do the side population cells isolated from cell lines[3,4] bear a relationship to cancer stem cells? In principle, in any permanent cell line there must be a self-renewing cell population. If the characterization of the SP cells in cell lines could be applied to cancer stem cells, this could advance understanding rapidly. One property of cancer cells, like stem cells, is their ability to grow in soft agar cultures.[66] It has been found that only a fraction of cells in a tumor cell culture can form a colony in soft agar. Are the cells forming soft agar colonies cancer stem cells? This would be a logical conclusion from the information at hand. It is known that the clonogenicity varies substantially between different tumor cell lines. If clonogenicity it related to self-renewing cells in the culture, assays based on colony formation may be useful for screening for stemcell targeting therapies. Such assays would be more time consuming and have a lower throughput, but might in the end prove more informative.

Metastasis is the most troublesome property of tumor cells. The vast majority of cancer fatalities are due to the effect of the spread of the initial tumor to other sites.

Although they can be quite invasive, basal cell carcinomas virtually never metastasize and are rarely fatal. Stem cells, especially certain cells of the neural crest, possess the ability to migrate through the developing embryo. The neural crest gives rise to the precursors of melanoma, neuroblastoma, and small cell lung cancers, all highly metastatic tumors. Is the metastatic ability of tumor cells related to an innate property of the cancer stem cells to migrate? If true, further characterization of germ cell migration could lead to new insights into metastasis. Anti–cancer stem cell therapies might find their best application in the restriction of metastasis. If metastasis could be prevented, even if the primary tumor remains intact, the patient might still experience a substantial increase in survival time.

11.6 CONCLUSIONS

The identification of cancer stem cells in certain solid tumors has created notable excitement in the field and generated new research possibilities. If these results can be extended to most or all cancer cell types, a considerable advance in understanding will be achieved. Separating the cancer process into a stem cell activation phase and a tumor progression phase allows an understanding of how the myriad cancer-causing agents can have an effect on specific tissues. Research efforts directed at understanding the growth requirements of tumor stem cells as well as identifying tumor stem cell antigens could lead to new targeted approaches.

The isolation and characterization of cancer stem cells from other tissues will be a great aid in cancer diagnostics, cancer prevention, and therapeutics. Normal stem cell–based approaches are being developed intensively as an aid in replacing damaged cells and tissues in the body. The insight from the growth and characterization of normal stem cells will aid in the understanding of cancer stem cells and in new therapeutic approaches.

REFERENCES

1. Al-Hajj M, Wicha MS, Benito-Hernandez A, Morrison SJ, Clarke MF. Prospective identification of tumorigenic breast cancer cells. *Proc Natl Acad Sci U S A*. 2003; 100:3983–3988.

2. Hemmati HD, Nakano I, Lazareff JA, et al. Cancerous stem cells can arise from pediatric brain tumors. *Proc Natl Acad Sci U S A*. 2003; 100:15178–15183.

3. Hirschmann-Jax C, Foster AE, Wulf GG, et al. A distinct "side population" of cells with high drug efflux capacity in human tumor cells. *Proc Natl Acad Sci U S A*. 2004; 101:14228–14233.

4. Kondo T, Setoguchi T, Taga T. Persistence of a small subpopulation of cancer stemlike cells in the C6 glioma cell line. *Proc Natl Acad Sci U S A*. 2004; 101:781–786.

5. Thomlinson RH, Gray LH. The histological structure of some human lung cancers and the possible implications for radiotherapy. *Br J Cancer*. 1955; 9:539–549.

6. Zhou S, Schuetz JD, Bunting KD, et al. The ABC transporter Bcrp1/ABCG2 is expressed in a wide variety of stem cells and is a molecular determinant of the side-population phenotype. *Nat Med.* 2001; 7:1028–1034.

7. Kim M, Turnquist H, Jackson J, et al. The multidrug resistance transporter ABCG2 (breast cancer resistance protein 1) effluxes Hoechst 33342 and is overexpressed in hematopoietic stem cells. *Clin Cancer Res.* 2002; 8:22–28.

8. Scharenberg CW, Harkey MA, Torok-Storb B. The ABCG2 transporter is an efficient Hoechst 33342 efflux pump and is preferentially expressed by immature human hematopoietic progenitors. *Blood.* 2002; 99:507–512.

9. Beachy PA, Karhadkar SS, Berman DM. Tissue repair and stem cell renewal in carcinogenesis. *Nature.* 2004; 432:324–331.

10. Tennant R. What is a tumor promoter? *Environ Health Perspect.* 1999; 107:A390–A391.

11. Knudson AG. Antioncogenes and human cancer. *Proc Natl Acad Sci U S A.* 1993; 90:10914–10921.

12. Knudson AG. Mutation and cancer: statistical study of retinoblastoma. *Proc Natl Acad Sci U S A.* 1971; 68:820–823.

13. Hahn H, Christiansen J, Wicking C, et al. A mammalian *patched* homolog is expressed in target tissues of Sonic Hedgehog and maps to a region associated with developmental anomalies. *J Biol Chem.* 1996; 271:12125–12128.

14. Hahn H, Wicking C, Zaphiropoulous PG, et al. Mutations of the human homolog of *Drosophila* patched in the nevoid basal cell carcinoma syndrome. *Cell.* 1996; 85:841–851.

15. Johnson RL, Rothman AL, Xie J, et al. Human homolog of *patched*, a candidate gene for the basal cell nevus syndrome. *Science.* 1996; 272:1668–1671.

16. Dean M. Towards a unified model of tumor suppression: lessons learned from the human *patched* gene. *Biochim Biophys Acta.* 1997; 1332:M43–M52.

17. Trichopoulos D, Lagiou P, Adami HO. Towards an integrated model for breast cancer etiology: the crucial role of the number of mammary tissue–specific stem cells. *Breast Cancer Res.* 2005; 7:13–17.

18. Gail MH, Brinton LA, Byar DP, et al. Projecting individualized probabilities of developing breast cancer for white females who are being examined annually. *J Natl Cancer Inst.* 1989; 81:1879–1886.

19. Kauff ND, Satagopan JM, Robson ME, et al. Risk-reducing salpingo-oophorectomy in women with a BRCA1 or BRCA2 mutation. *N Engl J Med.* 2002; 346:1609–1615.

20. Eaden J. Colorectal carcinoma and inflammatory bowel disease. *Aliment Pharmacol Ther.* 2004; 20:24–30.

21. Groden J, Thliveris A, Samowitz W, et al. Identification and characterization of the familial adenomatous polyposis *coli* gene. *Cell.* 1991; 66:589–600.

22. Nishisho I, Nakamura Y, Miyoshi Y, et al. Mutations of chromosome *5q21* genes in FAP and colorectal cancer patients. *Science.* 1991; 253:665–669.

23. Gardner EJ. Follow-up study of a family group exhibiting dominant inheritance for a syndrome including intestinal polyps, osteomas, fibromas and epidermal cysts. *Am J Hum Genet.* 1962; 14:376–390.

24. Deschner EE, Lipkin M. Proliferative patterns in colonic mucosa in familial polyposis. *Cancer* 1975; 35(2): 413–418.

25. Moser AR, Dove WF, Roth KA, Gordon JI. The Min (multiple intestinal neoplasia) mutation: its effect on gut epithelial cell differentiation and interaction with a modifier system. *J Cell Biol.* 1992; 116:1517–1526.

26. Su LK, Kinzler KW, Vogelstein B, et al. Multiple intestinal neoplasia caused by a mutation in the murine homolog of the *APC* gene. *Science.* 1992; 256:668–670.

27. Kinzler KW, Vogelstein B. Lessons from hereditary colorectal cancer. *Cell.* 1996; 87:159–170.

28. Parkin DM, Bray F, Ferlay J, Pisani P. Global cancer statistics, 2002. *CA Cancer J Clin.* 2005; 55:74–108.

29. Beasley RP, Hwang LY, Lin CC, Chien CS. Hepatocellular carcinoma and hepatitis B virus: a prospective study of 22,707 men in Taiwan. *Lancet* 1981; 2:1129–1133.

30. Yu MC, Tong MJ, Coursaget P, Ross RK, Govindarajan S, Henderson BE. Prevalence of hepatitis B and C viral markers in black and white patients with hepatocellular carcinoma in the United States. *J Natl Cancer Inst.* 1990; 82:1038–1041.

31. Millward MJ, Cantwell BM, Munro NC, Robinson A, Corris PA, Harris AL. Oral verapamil with chemotherapy for advanced non-small cell lung cancer: a randomised study. *Br J Cancer.* 1993; 67:1031–1035.

32. West KA, Brognard J, Clark AS, et al. Rapid Akt activation by nicotine and a tobacco carcinogen modulates the phenotype of normal human airway epithelial cells. *J Clin Invest.* 2003; 111:81–90.

33. Hughes J, Weil H. Lung biology in health and disease. In Samet J, ed. *Epidemiology of Lung Cancer.* New York: Marcel Dekker; 1994; 74:185–205.

34. Knudson A. Asbestos and mesothelioma: genetic lessons from a tragedy. *Proc Natl Acad Sci U S A.* 1995; 92:10819–10820.

35. Wu-Williams AH, Yu MC, Mack TM. Life-style, workplace, and stomach cancer by subsite in young men of Los Angeles County. *Cancer Res.* 1990; 50:2569–2576.

36. Uemura N, Okamoto S, Yamamoto S, et al. *Helicobacter pylori* infection and the development of gastric cancer. *N Engl J Med.* 2001; 345:784–789.

37. Coughlin SS, Calle EE, Patel AV, Thun MJ. Predictors of pancreatic cancer mortality among a large cohort of United States adults. *Cancer Causes Control.* 2000; 11:915–923.

38. Rivenson A, Hoffmann D, Prokopczyk B, Amin S, Hecht SS. Induction of lung and exocrine pancreas tumors in F344 rats by tobacco-specific and Areca-derived N-nitrosamines. *Cancer Res.* 1988; 48:6912–6917.

39. Michaud DS. Epidemiology of pancreatic cancer. *Minerva Chir.* 2004; 59:99–111.

40. Goggins M, Schutte M, Lu J, et al. Germline *BRCA2* gene mutations in patients with apparently sporadic pancreatic carcinomas. *Cancer Res.* 1996; 56:5360–5364.

41. Zur Hausen H, Meinhof W, Scheiber W, Bornkamm GW. Attempts to detect virus-specific DNA in human tumors: I. Nucleic acid hybridizations with complementary RNA of human wart virus. *Int J Cancer.* 1974; 13:650–656.

42. Werness BA, Levine AJ, Howley PM. Association of human papillomavirus types 16 and 18 E6 proteins with p53. *Science.* 1990; 248:76–79.

43. Tommasino M, Crawford L. Human papillomavirus E6 and E7: proteins which deregulate the cell cycle. *Bioessays.* 1995; 17:509–518.

44. Tindle RW. Immune evasion in human papillomavirus-associated cervical cancer. *Nat Rev Cancer*. 2002; 2:59–65.

45. Sasadeusz J, Kelly H, Szer J, Schwarer AP, Mitchell H, Grigg A. Abnormal cervical cytology in bone marrow transplant recipients. *Bone Marrow Transplant*. 2001; 28:393–397.

46. Nakachi K, Hayashi T, Imai K, Kusunoki Y. Perspectives on cancer immunoepidemiology. *Cancer Sci*. 2004; 95:921–929.

47. Folkman J. Role of angiogenesis in tumor growth and metastasis. *Semin Oncol*. 2002; 29:15–18.

48. Gottesman MM, Fojo T, Bates SE. Multidrug resistance in cancer: role of ATP-dependent transporters. *Nat Rev Cancer*. 2002; 2:48–58.

49. Goodell MA, Brose K, Paradis G, Conner AS, Mulligan RC. Isolation and functional properties of murine hematopoietic stem cells that are replicating in vivo. *J Exp Med*. 1996; 183:1797–1806.

50. Dean M, Fojo T, Bates S. Tumor stem cells and drug resistance. *Nat Rev Cancer*. 2005; 5:275–284.

51. Henrich CJ, Bokesch HR, Dean M, et al. A high-throughput cell-based assay for inhibitors of ABCG2 activity. *J Biomol Screen*. 2006; 11:176–183.

52. Nusslein-Volhard C, Wieschaus E. Mutations affecting segment number and polarity in *Drosophila*. *Nature*. 1980; 287:795–801.

53. Chidambaram A, Goldstein AM, Gailani MR, et al. Mutations in the human homolog of the *Drosophila patched* gene in Caucasian and African American nevoid basal cell carcinoma syndrome patients. *Cancer Res*. 1996; 56:4599–4601.

54. Gailani MR, Stahle-Backdahl M, Leffell DJ, et al. The role of the human homologue of *Drosophila patched* in sporadic basal cell carcinomas. *Nat Genet*. 1996; 14:78–81.

55. Bale AE, Yu KP. The Hedgehog pathway and basal cell carcinomas. *Hum Mol Genet*. 2001; 10:757–762.

56. Berman DM, Karhadkar SS, Maitra A, et al. Widespread requirement for Hedgehog ligand stimulation in growth of digestive tract tumors. *Nature*. 2003; 425:846–851.

57. Thayer SP, di Magliano MP, Heiser PW, et al. Hedgehog is an early and late mediator of pancreatic cancer tumorigenesis. *Nature*. 2003; 425:851–856.

58. Watkins DN, Berman DM, Burkholder SG, Wang B, Beachy PA, Baylin SB. Hedgehog signalling within airway epithelial progenitors and in small-cell lung cancer. *Nature*. 2003; 422:313–317.

59. Karhadkar SS, Bova GS, Abdallah N, et al. Hedgehog signalling in prostate regeneration, neoplasia and metastasis. *Nature*. 2004; 431:707–712.

60. James LF, Panter KE, Gaffield W, Molyneux RJ. Biomedical applications of poisonous plant research. *J Agric Food Chem*. 2004; 52:3211–3230.

61. Chen JK, Taipale J, Cooper MK, Beachy PA. Inhibition of Hedgehog signaling by direct binding of cyclopamine to *smoothened*. *Genes Dev*. 2002; 16:2743–2748.

62. Athar M, Li C, Tang X, et al. Inhibition of smoothened signaling prevents ultraviolet B-induced basal cell carcinomas through regulation of Fas expression and apoptosis. *Cancer Res*. 2004; 64:7545–7552.

63. Bijlsma MF, Spek CA, Zivkovic D, van de Water S, Rezaee F, Peppelenbosch MP. Repression of *smoothened* by *patched* -dependent (pro-)vitamin D3 secretion. *PLoS Biol.* 2006; 4.

64. Fan X, Matsui W, Khaki L, et al. Notch pathway inhibition depletes stem-like cells and blocks engraftment in embryonal brain tumors. *Cancer Res.* 2006; 66:7445–7452.

65. Alley MC, Scudiero DA, Monks A, et al. Feasibility of drug screening with panels of human tumor cell lines using a microculture tetrazolium assay. *Cancer Res.* 1988; 48:589–601.

66. Hamburger AW, Salmon SE. Primary bioassay of human tumor stem cells. *Science.* 1977; 197:461–463.

12 Immunobiology of Cancer Stem Cells

SHUBHADA V. CHIPLUNKAR

12.1 CANCER STEM CELLS

The characteristics that define stem cells are self-renewal and multipotentiality. A *cancer stem cell* is a cell within a tumor that possesses the capacity to self-renew and to cause the heterogeneous lineages of cancer cells that comprise the tumor. Such cells regenerate a replica of the original tumor when transplanted and show differential chemo- and radiosensitivity.[1] The concept of cancer stem cells was first documented in 1994 in human acute lymphocytic leukemias.[2] A series of reports demonstrated the presence of self-renewing cancer stem cells in solid tumors such as brain, breast, multiple myeloma, and ovarian cancer.[3–5]

Cancer stem cells resemble normal stem cells as they exist in a quiescent state but are relatively resistant to drugs that target dividing cells. Relapse of tumors following a chemotherapeutic regimen indicates a possibility that cancer stem cells survive within the host and cannot easily be eliminated. These cells have the capacity to metastasize and multiply. The antigenic differences between cancer stem cells and normal stem cells are not well defined; in fact, it is also not understood whether antigenic differences exist between cancer stem cells and other cancer cells. Cancer stem cells represent 1 % of tumor cells and it is therefore unlikely that the antigens expressed on cancer stem cells may be the same as those expressed on tumor cells.[6]

This brings us to the important question of what makes tumors antigenic and how the break of tolerance (if any) is achieved. Mutation may be responsible for the immunogenicity of tumor-specific antigens, while the aberrant expression of nonmutated self-antigens makes them candidate tumor antigens. Immune response to self-antigens is generated because of overexpression of the self-antigens, which overcomes the threshold of antigen concentration at which an immune response is generated.[7]

Cancer Stem Cells: Identification and Targets, Edited by Sharmila Bapat
Copyright © 2009 John Wiley & Sons, Inc.

Use of SEREX technology in screening of cDNA expression libraries with patient sera has identified overexpressed antigen on tumor cells.[8] There are two major groups of self-tumor antigens, which include conventional antigens, such as proteins, encoded by genes with exon–intron organization and translated in a primary open reading frame. The second group includes unconventional cryptic peptide antigens: cryptic antigens encoded in (1) introns of genes, (2) exon–intron junctional regions, (3) alternative reading frames, or (4) subdominant open reading frames located in the 5′ untranslated region or 3′ UTR of the primary open reading frame, chromosome rearrangement, and aberrant processing.[9] Since the unconventional antigen peptides are not expressed by normal stem cells or normal tissue cells but are able to generate an immune response, these may belong to cancer stem cells and would be ideal targets for future immunotherapy.

Following the success of molecular profiling in identifying prognostic signatures for many cancers, researchers have initiated profiling of cancer stem cells as well. From the perspective of systems biology, a global assessment of cancer stem cells and their microenvironments (niche) at the level of complete transcriptome, proteome and epigenome using high-throughput technologies is now possible. Emerging proteomic technologies employing mass spectrometry and protein chip platforms would permit identification of better cell surface markers of cancer stem cells. The antigenic profiling of cancer stem cells would help in targeting cancer stem cells for therapeutic benefits.

12.2 CANCER STEM CELLS AND LYMPHOCYTES

A new model for genesis of cancer proposed by Grandics in 2006 explored the similarities between lymphocytes and cancer cells.[10] The development of cancer requires infection(s), during which antigenic determinants from pathogens mimicking self-antigens are co-presented to the immune system, leading to break of tolerance. Thus, the close relationship that exists among infection, autoimmunity, and cancer has been identified in several instances.[11] Hepatocellular carcinoma develops on a background of hepatitis B and hepatitis C virus infection.[12] Epstein–Barr virus and human T-lymphotropic virus type 1 (HTLV-1) infections are associated with autoimmune phenomena, including abnormal lymphoproliferation, Hodgkin's lymphoma, Sjogren's syndrome, and arthritis.[13] Cytomegalovirus infection is linked to autoimmunity and testicular cancer.[14,15] The association of *Helicobacter pylori* with gastric cancer is well established. Additionally, it can lead to autoimmune neutropenia and MALT-lymphoma.[16]

It is thought that microbes can play a direct role in induction of autoimmunity, for instance by molecular mimicry or bystander activation of autoreactive T-cells.[17] Viruses play a possible role as an environmental factor in autoimmune diseases. These viruses include mumps,[18] rubella,[19] enteroviruses such as Coxsackie B4[20] and cytomegalovirus,[21] which are related to induction of type 1 diabetes. Pro-inflammatory cytokines are induced by virus infection and may be responsible for bystander activation of self-reactive lymphocytes.[22]

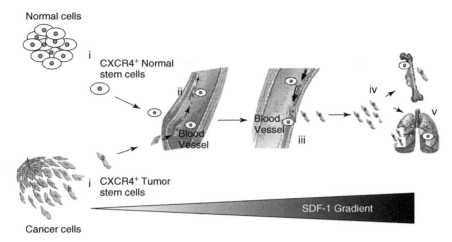

FIGURE 12.1 Migration of normal and cancer stem cells is a multistep process which involves: (i) Egress of CXCR4 expressing stem cells from the tissue; (ii) adhesion of stem cells to the endothelium; (iii) crossing the endothelial barrier; (iv) homing to the target tissues chemoattracted by the SDF-1 released; and (v) expansion of stem cells at the site that provides a favorable niche. (*See insert for color representation of figure.*)

The presence of autoreactive T-cells has been described in healthy persons, as some degree of autoreactive T-cells are essential in immune defense mechanisms. It has been reported that T-cells which recognize variants of self-antigens are of lower avidity than those recognizing non-self-antigens.[23,24] An effective response against pathogenic microorganisms requires the detection of activated T- and B-cells that are reactive to self-antigens.[25] Any defects in the programmed cell death of these lymphocytes may result in autoimmunity[26,27] and cancer.[28,29]

Markers for apoptosis, such as the Fas receptor (FasR), are expressed on a wide variety of cell types, whereas Fas ligand (FasL) is expressed mainly on T-cells[27] and "immune-privileged" tissues such as brain, testes, and eyes. Mutation in FasR or FasL may result in autoimmune disorders.[30,31] It has been argued from the expression of such apoptosis related molecules on the surface of cancer cells and the presence of lymphocytes infiltrating the tumors whether neoplastic cells are formed from cytotoxic T-lymphocytes by a premature termination of the apoptosis mechanism.[10]

Some similarities exist between neoplastic cells and cytotoxic T-lymphocytes; both express FasL and are capable of inducing the apoptotic death of activated T-cells, as well as other cancer cells that carry a functional FasR.[32,33] It was therefore proposed that an aberration in the process of apoptosis could lead to the formation of cancer stem cells from autoreactive T-cells.[10]

Several examples of lymphocyte markers expressed on cytotoxic T-lymphocytes, leukemia, and solid tumors establish the link between lymphocytes and cancer. Cytotoxic T-lymphocyte-associated antigen 4 (CTLA-4), a regulator of the effector function of T-cells, is expressed on various leukemias and solid tumors.[34] Melanoma and glioma cell lines express the common acute lymphoblastic leukemia antigen.[35–37] The leukocyte common antigen CD45 is expressed on seminoma,[38] rhabdomyosarcoma,[39] and neuroendocrine carcinomas.[40] The myeloid antigen Leu 7 expressed on NK-cells is also expressed on a variety of solid tumors[41] and Hodgkin's lymphomas.[42]

The resulting cancer stem cell preserves some function of effector T-cells, such as homing to inflammatory sites and secretion of inflammatory cytokines, which leads to disruption of the local immune system and apoptosis. Owing to the defective constitutive production of inflammatory cytokines and other growth factors, a stroma is built at the site of inflammation similar to the temporary stroma built during the process of wound healing. The cancer stem cells grow inside this stroma, forming a tumor that has its own vascular supply and offers protection from cellular immune responses.[10]

12.3 TRAFFICKING OF NORMAL STEM CELLS AND METASTASIS OF CANCER STEM CELLS

Chemokines are key factors involved in regulation of the immune response, through the activation and control of leukocyte traffic, lymphopoiesis, and immune surveillance. Chemokines are small pro-inflammatory chemoattractant cytokines that bind to specific G protein–coupled seven-span transmembrane receptors present on target cells. The human chemokine family currently includes more than 50 chemokines and 20 chemokine receptor families. Chemokine receptors are defined by their ability to induce directional migration of cells toward a gradient of a chemotactic cytokine. Chemokine receptors are present on different cell types and are broadly divided into two main subfamilies, CC and CXC, based on the position of two conserved cysteine residues and the intervening amino acid present in the extracellular domain.[43]

The interactions between the chemokine and its respective receptors help coordinate the trafficking and organization of cells within various tissue compartments. Stromal cell–derived factor (SDF-1 now designated CXCL12) is a homeostatic chemokine that signals through CXCR4. The role of the SDF-1-CXCR4 axis was studied initially in hematopoiesis, development, and organization of the immune system.[44,45] This axis regulates trafficking and homing of CXCR4 hematopoietic stem and progenitor cells, pre-B-lymphocytes, and T-lymphocytes.[44–47]

Evidence indicates that besides the hematopoietic stem cells, CXCR4 is expressed on nonhematopoietic tissue/organ-committed progeneitor stem cells.[48] Perturbation of the SDF-1/CXCR4 axis by mobilizing agents is essential for the egress of hematopoietic stem/progenitor cells from the bone marrow into peripheral blood.[49,50] Evidence accumulates that CXCR4, which is expressed on normal stem cells of different organs and tissues, is also expressed on several tumors that are derived from these cells. This implies that the SDF-1/CXCR4 axis plays an important role in the metastasis of cancer stem cells to organs that express high SDF-1 (e.g., lymph nodes, liver, lungs, bones) Classical properties of normal stem cells are strikingly reminiscent of the experimental and clinical behavior of metastatic cancer cells. The common features are unlimited capacity of self-renewal and the requirement for a specific microenvironment that can help cancer stem cells to grow.

Another important function of the CXCR4/CXCL12 axis is related to tissue repair and regeneration.[51-53] Hypoxia-inducible factor-1 (HIF-1), a key player of tissue hypoxia, is known to induce CXCL12 expression in ischemic areas in direct proportion to reduced oxygen tension in vivo. [51] HIF-1 is also known to enhance the expression and function of CXCR4 on normal and malignant cells.[51-53] In patients with renal cell carcinoma that harbor VHL gene mutation, HIF-1 accumulates and induces expression of CXCR4. This provides a possible mechanism to explain how CXCR4 is induced during tumor cell evolution, which explains the egress of neoplastic cells from areas of low oxygen tension.[54] The SDF-1 promoter contains two HIF-1-binding sites, and HIF-1 is known to bind to this sequence on SDF-1.[51]

Mobilization of cancer stem cells and metastasis is a multistep process and is very similar to that in normal stem cells.[49,55] The cancer stem cells egress from the primary tumor mass and enter peripheral blood and lodge in lymph nodes, liver, lung, or bone marrow. The stem cells sense the chemo-attracting gradient of CXCL12 within the tumor microenvironment, which directs their homing (normal stem cells) metastasis (cancer stem cells) (Fig. 12.1). The mobilization and egress of metastasizing cancer cells are not well understood. However, mobilization of normal hematopoietic stem cells (HSPCs) from their tissue niches provides important clues to our understanding of the egress of cancer stem cells. CXCL12 regulates tethering or adhesion of hematopoietic stem cells to the endothelium,[56,57] the expression of basement degrading enzymes matrix metalloproteinases (MMPs), and other processes that are essential to HSPC homing and engraftment.

After the stem cells reach circulation, they respond to CXCL12, which directs them to a new stem cell niche (normal stem cell) or to the organ where the cells metastasize (cancer stem cell).[48] CXCL12 is known to increase cell motility as changes in rearrangements of cytoskeletal proteins exposed to CXCL12 have been reported. SDF-1-mediated cell motility and chemotaxis are affected by inhibitors of p13-AKT, MAPK p42/44, and phosphatases.[58-60]

SDF-1 modulates the adhesion of cells to the endothelium in the organ where they home or metastasize. SDF-1 modulates activation of various adhesion molecules (i.e., LFA-1, VLA-4, VLA-5) on the surface of target and $\alpha 11b/\beta 3$ cells.[61] SDF-1 induces adhesion and transendothelial migration of human CD34+ hematopoietic cells. Increased migration of tumor cells to endothelium has been reported in the presence of SDF-1. After exposure to SDF-1 tumor cells secrete more MMPs, nitric oxide, and VEGF.[57,62] These factors may help stem cells to cross the endothelial barrier and are important in trafficking of normal stem cells and metastasis of cancer stem cells. In the presence of CXCL12 signal, the stem cells attach firmly to endothelium via CXCR4, followed by penetration of the microvessel wall. These stem cells would lodge in tissues or organs that would provide a protective niche for the cells. The microenvironment has to protect the stem cells from apoptosis. In addition, the stromal cells within the metastatic site regulate tumor progression. Adhesion to stromal cells supports the growth of neoplastic cells through high-level expression of CXCL12 and integrins that can activate CXCR4.[63]

The SDF-1/CXCR4 axis has emerged as an important regultor of trafficking of normal and malignant stem cells. It has been demonstrated that breast cancer cells treated with a CXCR4 inhibitor showed significantly inhibited metastatic ability. In breast cancer, CXCR4 expression correlates with the cancer stem cells content and aggressiveness of the cancer cell lines.[64] In pancreatic cancers, only the CD133+ and CXCR4+ cancer stem cells were able to metastasize.[65] The SDF/CXCR4 axis is therefore a prime target for the development of new drugs that would have the ability to control the metastatic behavior of tumor stem cells expressing CXCR4.

To conclude, the future therapies for cancer would not rely only on elimination of cancer cells, as this would affect the immune system negatively. The focus should now be directed at cancer stem cells, which would eliminate cancer at the precancerous stage. The cancer stem cell–directed approach would be the new cancer vaccine strategy. Monoclonal gammopathy of undetermined significance (MGUS) represents a precursor legion of multiple myeloma (MM).[66] Although the bulk tumor in MM consists of plasma cells that express CD138, recent studies have suggested that the clonogenic growth may be enriched in a fraction missing this marker.[67] It was demonstrated that the expression of an embryonal stem cell marker, SOX2 specifically marks the clonogenic CD138 compartment in MGUS patients. These patients mount cell-mediated and humoral responses to SOX2.[67] This report establishes that harnessing immunity to antigens expressed on tumor progenitor (stem) cells may be critical for prevention and therapy of human cancer.

REFERENCES

1. Lobo NA, Shimono Y, Qian D, Clarke MF. The biology of cancer stem cells. *Annu Rev Cell Dev Biol.* 2007; 23:675–699.

2. Lapidot T, Sirard C, Vormoor J, et al. A cell initiating human acute myeloid leukemia after transplantation into SCID mice. *Nature.* 1994; 367:645–648.

3. Hemmati HD, Nakano I, Lazareff JA, et al. Cancerous stem cells can arise from pediatric brain tumors. *Proc Natl Acad Sci U S A.* 2003; 100:15178–15183.

4. Al-Hajj M, Clarke MF. Self renewal and solid tumor stem cells. *Oncogene.* 2004; 23:7274–7282.

5. Bapat SA, Mali AM, Koppikar CB, Kurrey NK. Stem and progenitor-like cells contribute to the aggressive behavior of human epithelial ovarian cancer. *Cancer Res.* 2005; 65:3025–3029.

6. Lou H, Dean M. Targeted therapy for cancer stem cells: the *patched* pathway and ABC transporters. *Oncogene.* 2007; 26:1357–1360.

7. Zinkernagel RM. Immunity against solid tumors? *Int J Cancer.* 2001; 93:1–5.

8. Sahin U, Tureci O, Schmitt H, et al. Human neoplasms elicit multiple specific immune responses in the autologous host. *Proc Natl Acad Sci USA.* 1995; 92:11810–11813.

9. Yang XF. Immunology of stem cells and cancer stem cells. *Cell Mol Immunol.* 2007; 4:161–171.

10. Grandics P. The cancer stem cell: evidence for its origin as an injured autoreactive T cell. *Mol Cancer.* 2006; 5:6.

11. Groux H, Cottrez F. The complex role of interleukin-10 in autoimmunity. *J Autoimmun.* 2003; 20:281–285.

12. Pellegris G, Ravagnani F, Notti P, Fissi S, Lombardo C. B and C hepatitis viruses, HLA-DQI and -DR3 alleles and autoimmunity in patients with hepatocellular carcinoma. *J Hepatol.* 2002; 36:521–526.

13. Venables P. Epstein–Barr virus infection and autoimmunity in rheumatoid arthritis. *Ann Rheum Dis.* 1988; 47:265–269.

14. Andreev VC, Zlatkov NB. Systemic lupus erythematosus and neoplasia of the lymphoreticular system. *Br J Dermatol.* 1968; 80:503–508.

15. Mueller N, Hinkula J, Wahren B. Elevated antibody titers against cytomegalovirus among patients with testicular cancer. *Int J Cancer.* 1988; 41:399–403.

16. Gupta V, Eden AJ, Mills MJ. *Helicobacter pylori* and autoimmune neutropenia. *Clin Lab Haematol.* 2002; 24:183–185.

17. Pedersen AE. The potential for induction of autoimmune disease by a randomly mutated self antigen. *Med Hypothesis.* 2007; 68:1240–1246.

18. Harris HF. A case of diabetes mellitus quickly following mumps. *Boston Med Surg J.* 1898; 140:465–469.

19. Ginsberg-Fellner F, Witt ME, Yagihashi S, et al. The interrelationship of congenital rubella. *Diabetologia.* 1984; 27:87–89.

20. Gamble DR, Taylor KW, Cumming H. Coxsackie viruses and diabetes mellitus. *Br Med J.* 1973; 4:260–262.

21. Hiemstra HS, Schloot NC, Veelen PA. Cytomegalovirus in autoimmunity: T cell cross reactivity to viral antigen and autoantigens to glutamic acid decarboxylase. *Proc Natl Acad Sci U S A.* 2001; 98:398–399.

22. Shi F, Ljunggren HG, Sarvetnick N. Innate immunity and autoimmunity: from self-protection to self-destruction. *Trends Immunol.* 2001; 22:97–101.

23. Pircher H, Rohrer UH, Moskophidis D, Zinkernagel RM, Hengartner H. Lower receptor avidity required for thymic clonal deletion than for effector T-cell function. *Nature.* 1991; 351:482–485.

24. Sandberg JK, Franksson L, Sundback J, et al. T cell tolerance based on avidity thresholds rather than complete deletion allows maintenance of maximal repertoire diversity. *J Immunol.* 2000; 165:25–33.

25. Strasser A, Whittingham S, Vaux DL, et al. Enforced BCL2 expression in B-lymphoid cells prolongs antibody responses and elicits autoimmune disease. *Proc Natl Acad Sci U S A.* 1991; 88:8661–8665.

26. Watanabe-Fukunaga R, Brannan CI, Copeland NG, Jenkins NA, Nagata S. Lymphoproliferation disorder in mice explained by defects in Fas antigen that mediates apoptosis. *Nature.* 1992; 356:314–317.

27. Grodzicky T, Elkon KB. Apoptosis: a case where too much or too little can lead to autoimmunity. *Mount Sinai J Med.* 2002; 69:208–219.

28. Sachs L. The control of hematopoiesis and leukaemia: from basic biology to the clinic. *Proc Natl Acad Sci U S A.* 1996, 93:4742–4749.

29. Strasser A, Harris AW, Bath ML, Cory S. Novel primitive lymphoid tumours induced in transgenic mice by cooperation between myc and bcl-2. *Nature*. 1990; 348:331–333.

30. Drappa J, Vaishnaw AK, Sullivan KE, Chu JL, Slkon KB. Fas gene mutations in the Canale–Smith syndrome, an inherited lymphoproliferative disorder associated with autoimmunity. *N Engl J Med*. 1996; 335:1643–1649.

31. Fisher GH, Rosenberg FJ, Straus SE, et al. Dominant interfering *Fas* gene mutations impair apoptosis in a human autoimmune lymphoproliferative syndrome. *Cell*. 1995; 81:935–946.

32. Strand S, Hofmann WJ, Hug H, et al. Lymphocyte apoptosis induced by CD95 (APO-I/Fas) ligand-expressing tumor cells-a mechanism of immune evasion? *Nat Med*. 1996; 2:1361–1366.

33. Chappell DB, Restifo NP. T cell-tumor cell: a fatal interaction? *Cancer Immunol Immunother*. 1998; 47:65–71.

34. Paydas S, Tanriverdi K, Yavuz S, Disel U, Baslamisli F, Burgut R. PRAME mRNA levels in cases with acute leukaemia: clinical importance and future prospects. *Am J Hematol*. 2005; 79:257–261.

35. Contardi E, Palmisano GL, Tazzari PL, et al. CTLA-4 is constitutively expressed on tumor cells and can trigger apoptosis upon ligand interaction. *Int J Cancer*. 2005; 117:538–550.

36. Carrel S, De Tribolet N, Gross N. Expression of HLA-DR and common acute lymphoblastic leukaemia antigens on glioma cells. *Eur J Immunol*. 1982; 12:354–357.

37. Carrel S, Schmidt-Kessen A, Mach JP, Heumann D, Girardet C. Expression of common acute lymphoblastic leukaemia antigen (CALLA) on human malignant melanoma cell lines. *J Immunol*. 1983; 130:2456–2460.

38. Warnke RA, Rouse RV. Limitations encounter in the application of tissue section immunodiagnosis to the study of lymphomas and related disorders. *Hum Pathol*. 1985; 16:326–331.

39. McDonnell JM, Beschorner WE, Kuhajda FP, Dement SH. Common leukocyte antigen staining sarcoma. *Cancer*. 1987; 59:1438–1441.

40. Nandedkar MA, Palazzo J, Abbondanzo SL, Lasota J, Miettinen M. CD45 (leukocyte common antigen) immunoreactivity in metastatic undifferentiated and neuroendocrine carcinoma: a potential diagnostic pitfall. *Mod Pathol*. 1998; 11:1204–1210.

41. Lipinski M, Braham K, Caillaud J-M, Tursz T. HNK-1 antibody detects an antigen expressed on neuroectodermal cells. *J Exp Med*. 1983; 158:1775–1780.

42. Papadimitriou CS, Bai MK, Kotsianti AJ, Costopoulos JS, Hytiroglou P. Phenotype of Hodgkin and Sternberg–Reed cells and expression of CD57 (LEU7) antigen. *Leuk Lymphoma*. 1995; 20:125–130.

43. Baggiolini M. Chemokines and leukocyte traffic. *Nature*. 1998; 392:565–568.

44. Aiuti A, Webb IJ, Bleul C, Springer T, Gutierrez-Ramos JC. The chemokine SDF-1 is a chemoattractant for human CD34$^+$ hematopoietic progenitor cells and provides a new mechanism to explain the mobilization of CD34$^+$ progenitors to peripheral blood. *J Exp Med*. 1997; 185:111–120.

45. Ma Q, Jones D, Borghesani PR, et al. Impaired B-lymphopoiesis, myelopoiesis and derailed cerebellar neuron migration in CXCR4- and SDF-1-deficient mice. *Proc Natl Acad Sci U S A*. 1998; 95:9448–9453.

46. Nagasawa T, Hirota S, Tachibana K, et al. Defects of B-cell lymphopoiesis and bone-marrow myelopoiesis in mice lacking the CXC chemokine PBSF/SDF-1. *Nature*. 1996; 382:635–638.

47. Kim CH, Broxmeyer HE. Chemokines: signal lamps for trafficking of T and B cells for development and effector function. *J Leukoc Biol*. 1999; 65:6–15.

48. Kucia M, Reca R, Miekus K, et al. Trafficking of normal stem cells and metastasis of cancer stem cells involve similar mechanisms: pivotal role of the SDF1–CXCR4 axis. *Stem Cells*. 2005; 23:879–894.

49. Petit I, Szyper-Kravitz M, Nagler A, et al. G-CSF induces stem cell mobilization by decreasing bone marrow SDF-1 and up-regulating CXCR4. *Nat Immunol*. 2002; 3:687–694.

50. Liles WC, Broxmeyer HE, Rodger E, et al. Mobilization of hematopoietic progenitor cells in healthy volunteers by AMD3100, a CXCR4 antagonist. *Blood*. 2003; 102:2728–2730.

51. Ceradini DJ, Kulkarni AR, Callaghan MJ, et al. Progenitor cell trafficking is regulated by hypoxic gradients through HIF-1 induction of SDF-1. *Nat Med*. 2004; 10:858–864.

52. Schioppa T, Uranchimeg B, Saccani A, et al. Regulation of the chemokine receptor CXCR4 by hypoxia. *J Exp Med*. 2003; 198:1391–1402.

53. Zagzag D, Krishnamachary B, Yee H, et al. Stromal cell–derived factor-1 α and CXCR4 expression in hemangioblastoma and clear cell–renal cell carcinoma: von Hippel–Lindau loss-of-function induces expression of a ligand and its receptor. *Cancer Res*. 2005; 65:6178–6188.

54. Burger JA, Kipps TJ. CXCR4: a key receptor in the crosstalk between tumor cells and their microenvironment. *Blood*. 2006; 107:1761–1767.

55. Lapidot T, Petit I. Current understanding of stem cell mobilization: the roles of chemokines, proteolytic enzymes, adhesion molecules, cytokines, and stromal cells. *Exp Hematol*. 2002; 30:973–981.

56. Grabovsky V, Feigelson S, Chen C, et al. Subsecond induction of alpha4 integrin clustering by immobilized chemokines stimulates leukocyte tethering and rolling on endothelial vascular cell adhesion molecule 1 under flow conditions. *J Exp Med*. 2000; 192:495–506.

57. Kijowski J, Baj-Krzyworzeka M, Majka M, et al. The SDF-1–CXCR4 axis stimulates VEGF secretion and activates integrins but does not affect proliferation and survival in lymphohematopoietic cells. *Stem Cells*. 2001; 19:453–466.

58. Ganju RK, Brubaker SA, Meyer J, et al. The α-chemokine stromal cell-derived factor–1α binds to the transmembrane G-protein-coupled CXCR-4 receptor and activates multiple signal transduction pathways. *J Biol Chem*. 1998; 273:23169–23175.

59. Fernandis AZ, Cherla RP, Ganju RK. Differential regulation of CXCR4- mediated T-cell chemotaxis and mitogen-activated protein kinase activation by the membrane tyrosine phosphatase, CD45. *J Biol Chem*. 2003; 278:9536–9543.

60. Sotsios Y, Whittaker GC, Westwick J, Ward SG. The CXC chemokine stromal cell–derived factor activates a Gi-coupled phosphoinositide 3-kinase in T lymphocytes. *J Immunol*. 1999; 163:5954–5963.

61. Peled A, Kollet O, Ponomaryov T, et al. The chemokine SDF-1 activates the integrins LFA-1, VLA-4, and VLA-5 on immature human CD34(+) cells: role in transendothelial/stromal migration and engraftment of NOD/SCID mice. *Blood*. 2000; 95:3289–3296.

62. Janowska-Wieczorek A, Marquez LA, Dobrowsky A, Ratajczak MZ, Cabuhat ML. Differential MMP and TIMP production by human marrow and peripheral blood CD34(+) cells in response to chemokines. *Exp Hematol.* 2000; 28:1274–1285.

63. Orimo A, Gupta PB, Sgroi DC, et al. Stromal fibroblasts present in invasive human breast carcinomas promote tumor growth and angiogenesis through elevated SDF-1/CXCL12 secretion. *Cell.* 2005; 121:335–348.

64. Croker AK, Allan AL. Cancer stem cells: implications for the progression and treatment of metastatic disease. *J Cell Mol Med.* 2008; 12:374–390.

65. Hermann PC, Huber SL, Herrler T, et al. Distinct populations of cancer stem cells determine tumor growth and metastatic activity in human pancreatic cancer. *Cell Stem Cell.* 2007; 1:313–323.

66. Kyle RA, Therneau TM, Rajkumar SV, et al. Prevalence of monoclonal gammopathy of undetermined significance. *N Engl J Med.* 2006; 354:1362–1369.

67. Spisek R, Kukreja A, Chen L, et al. Frequent and specific immunity to the embryonal stem cell–associated antigen SOX2 in patients with monoclonal gammopathy. *J Exp Med.* 2007; 204:831–840.

INDEX

ABCB1 (MDR1), 41, 75, 79, 226
ABCG2 (BCRP), 41, 78, 89, 226
Aberrant differentiation, 12, 14, 30,197
Aberrant hematopoietic tissue, 30
Activation of oncogene(s), 8, 218
Adherens junctions, 175, 185
Adult stem cells, 4, 7, 9, 58, 73, 77, 104, 105
Aldehyde dehydrogenase (ALDH1), 35, 81
AML-initiating cells, 35
Androgen ablation therapy, 111
Androgen receptor (AR), 113, 115, 168, 226
Angiogenesis, 4, 5, 8, 15, 28, 29, 30, 33, 34, 37,
    38, 41, 44, 113, 119, 145, 149, 177, 200,
    207, 219, 235, 237
Astrocytes, 64, 65, 66, 67
Asymmetrical cell division, 5, 9, 13, 73, 197
Autoreactive T-cell, 235

BCR-ABL, 33, 36, 39, 43
Benign prostate hyperplasia (BPH), 112, 113, 184
Bipotent progenitors, 102, 100
Blastocyst, 2, 4, 13, 32, 33, 36, 37, 39, 40
Block of differentiation (maturation arrest), 30,
    31, 33
Bone marrow-derived stem cells, 3, 8, 9
Brain tumor stem cells, 66, 67, 99, 100
Breast-cancer initiating cells, 61
Breast cancer stem cells, 57, 59, 61, 62, 63, 66, 79

Cadherin switching, 181, 182
Cancer/testis antigens, 2
CD133, 10, 11, 36, 66, 67, 68, 69, 79, 91, 107,
    111, 117, 122, 123, 124, 177, 201, 209
CD34 expression, 10, 11, 35, 36, 37, 39, 40, 78,
    79, 81, 107, 237
Cell-cell interactions, 78
Cell cycle, 8, 14, 38, 40, 73, 75, 78, 106, 118, 121,
    140, 141, 144, 145, 177, 201, 219
Cell differentiation, 3, 4, 38, 43, 45, 115, 139,
    140, 144, 152, 208
Cell extrinsic, 14, 16

Cell intrinsic (autocrine) factors, 14, 30, 40
Cellular proliferation, 1, 140, 144, 145, 223
Cellular resistance, 10
Chemokine receptors, 236
Chemotherapeutic targets, 226
Chromatin immunoprecipitaion-to-microarray
    hybridization (ChIP-on-chip), 206
Chromatin-regulating genes, 41
Chromatin remodelling, 16, 43, 202
Chronic activation, 15
Clone amplification, 29, 30
Colony-forming units (CFUs), 27, 34, 37
Colorectal tumors, 176
Copy number of genes, 14
Covalent histone tail modifications, 40
CpG island methylation, 37, 40, 206
Cryptic peptide antigens, 234
Cytogenetic abnormalities, 10, 32
Cytokines, 10, 33, 39, 124, 140, 141, 144, 235,
    236
Cytotoxic T-lymphocytes, 235

De-differentiation, 120, 171, 208, 209, 219
Developmentally regulated genes, 16
Differential methylation hybridization (DMH),
    206
Differentiated somatic cells, 8
DNA damage, 12
Drosophila germline and somatic stem cells, 13
Drug inactivation, 204
Drug resistance, 12, 75, 81, 89, 92, 204
Dye efflux, 78, 107, 117

Embryo, 1, 2, 3, 4, 6, 96, 104, 136, 140, 144, 147,
    155, 170, 172, 173, 175
Embryogenesis, 4, 13, 17, 101, 145
Embryonic stem cells, 15, 89, 99, 104, 120, 126,
    152, 202, 203, 206, 208, 221
Epethelial-mesenchymal transition (EMT), 4, 97,
    167, 168, 170, 171–187
Epigenetic reprogamming, 199

Printed and bound by CPI Group (UK) Ltd, Croydon, CR0 4YY

16/04/2025

14658518-0002